An Undergraduate
Introduction to
Financial Mathematics

Second Edition

An Undergraduate Introduction to Financial Mathematics

Second Edition

J Robert Buchanan

Millersville University, USA

NEW JERSEY · LONDON · SINGAPORE · BEIJING · SHANGHAI · HONG KONG · TAIPEI · CHENNAI

Published by

World Scientific Publishing Co. Pte. Ltd.

5 Toh Tuck Link, Singapore 596224

USA office: 27 Warren Street, Suite 401-402, Hackensack, NJ 07601

UK office: 57 Shelton Street, Covent Garden, London WC2H 9HE

Library of Congress Cataloging-in-Publication Data
Buchanan, J. Robert
    An undergraduate introduction to financial mathematics, 2nd ed. / J. Robert Buchanan.
     p. cm.
    Includes bibliographical references and index.
    ISBN-13 978-981-2835-35-2
    ISBN-10 981-2835-35-0
    1. Business mathematics. I. Title.
    HF5691 .B875 2008
    330.01'513

                        2009285578

British Library Cataloguing-in-Publication Data
A catalogue record for this book is available from the British Library.

First published 2008
Reprinted 2011

Printed in Singapore by World Scientific Printers.

Dedication

For my wife, Monika.

Preface

This second edition of *An Undergraduate Introduction to Financial Mathematics* extends significantly the material found in the first edition. Owners of the first edition will find the second contains corrections and clarifications of the contents of the first as well as additional examples, exercises, and two entirely new chapters. As carefully as I proof-read the manuscript of the first edition, an embarrassingly large number of typos and garbled sentences managed to pass through my filter. Fortunately several of the readers of the first edition took the time to compile and send to me a list of errors and other suggestions. The improvements in the second edition are a result of the set of corrections and comments made by the readers. Two individuals stand out for the volume of suggestions and help they gave, Prof. M.M. Chawla and Prof. Josef Dick.

To the ten chapters of the first edition have been added two more. The first addition is on the topic of "Forwards and Futures" and constitutes the new Chapter 6. This topic allows the reader to exercise their newly obtained knowledge of Brownian motion, stochastic processes, and arbitrage at an earlier stage of the book than in the first edition. Previously these various threads were woven together in the chapters on options and solving the Black-Scholes equation. The earlier application of these topics may help the reader to gain greater mastery and to feel more comfortable using these tools. Chapter 6 also includes a discussion of the practice of "Marking to Market" for futures. This is provided as a preview of the process of hedging for portfolios of securities and options which appears in a later chapter. The second addition is on the topic of "American Options". In the first edition of the text, American options were mentioned and briefly described mainly to give the reader a sense of the broad array of financial instruments found in the world of investment and risk management. In the second edition, properties of American options are more fully explored and an elementary algorithm for pricing a type of American option is explained. This material forms the new Chapter 12 of the second edition. Chapters 6–10 of the first edition are now Chapters 7–11 of the second.

The chapters returning in the second edition from the first edition should not disappoint the reader as they have been corrected, expanded, and polished. New examples, exercises, and higher quality graphics appear in the returning chapters.

Since the appearance of the first edition, I have taught a course for undergraduates using the first edition as the textbook. I appreciate the comments of the students I faced in the classroom and those of the students at other institutions who emailed me. Student Catherine Albright from the fall semester 2007 read the first edition with a careful eye and brought to my attention numerous typographical errors.

It remains the author's hope that this text is an accurate, accessible introduction for undergraduates to the mathematics of options and derivatives. The prerequisite mathematical background (multivariable calculus) has been kept the same as in the first edition. If a reader has corrections or suggestions to share with me, or to check the latest list of errata, please consult the links found at the web site:

http://banach.millersville.edu/~bob/book/

J. Robert Buchanan
Wyomissing, PA, USA
August 12, 2008

Preface to the First Edition

This book is intended for an audience with an undergraduate level of exposure to calculus through elementary multivariable calculus. The book assumes no background on the part of the reader in probability or statistics. One of my objectives in writing this book was to create a readable, reasonably self-contained introduction to financial mathematics for people wanting to learn some of the basics of option pricing and hedging. My desire to write such a book grew out of the need to find an accessible book for undergraduate mathematics majors on the topic of financial mathematics. I have taught such a course now three times and this book grew out of my lecture notes and reading for the course. New titles in financial mathematics appear constantly, so in the time it took me to compose this book there may have appeared several superior works on the subject. Knowing the amount of work required to produce this book, I stand in awe of authors such as those.

This book consists of ten chapters which are intended to be read in order, though the well-prepared reader may be able to skip the first several with no loss of understanding in what comes later. The first chapter is on interest and its role in finance. Both discretely compounded and continuously compounded interest are treated there. The book begins with the theory of interest because this topic is unlikely to scare off any reader no matter how long it has been since they have done any formal mathematics.

The second and third chapters provide an introduction to the concepts of probability and statistics which will be used throughout the remainder of the book. Chapter Two deals with discrete random variables and emphasizes the use of the binomial random variable. Chapter Three introduces continuous random variables and emphasizes the similarities and differences between discrete and continuous random variables. The normal random variable and the closely related lognormal random variable are introduced and explored in the latter chapter.

In the fourth chapter the concept of arbitrage is introduced. For readers already well versed in calculus, probability, and statistics, this is the first material which may be unfamiliar to them. The assumption that financial calculations are carried out in an "arbitrage free" setting pervades the remainder of the book. The lack of arbitrage opportunities in financial transactions ensures that it is not possible to make a risk free profit. This chapter includes a discussion of the result from linear algebra and operations research known as the Duality Theorem of Linear Programming.

The fifth chapter introduces the reader to the concepts of random walks and Brownian motion. The random walk underlies the mathematical model of the value of securities such as stocks and other financial instruments whose values are derived from securities. The choice of material to present and the method of presentation is difficult in this chapter due to the complexities and subtleties of stochastic processes. I have attempted to introduce stochastic processes in an intuitive manner and by connecting elementary stochastic models of some processes to their corresponding deterministic counterparts. Itô's Lemma is introduced and an elementary proof of this result is given based on the multivariable form of Taylor's Theorem. Readers whose interest is piqued by material in Chapter Five should consult the bibliography for references to more comprehensive and detailed discussions of stochastic calculus.

Chapter Six introduces the topic of options. Both European and American style options are discussed though the emphasis is on European options. Properties of options such as the Put/Call Parity Formula are presented and justified. In this chapter we also derive the partial differential equation and boundary conditions used to price European call and put options. This derivation makes use of the earlier material on arbitrage, stochastic processes and the Put/Call Parity Formula.

The seventh chapter develops the solution to the Black-Scholes PDE. There are several different methods commonly used to derive the solution to the PDE and students benefit from different aspects of each derivation. The method I choose to solve the PDE involves the use of the Fourier Transform. Thus this chapter begins with a brief discussion of the Fourier and Inverse Fourier Transforms and their properties. Most three- or four-semester elementary calculus courses include at least an optional section on the Fourier Transform, thus students will have the calculus background necessary to follow this discussion. It also provides exposure to the Fourier

Transform for students who will be later taking a course in PDEs and more importantly exposure for students who will not take such a course. After completing this derivation of the Black-Scholes option pricing formula students should also seek out other derivations in the literature for the purposes of comparison.

Chapter Eight introduces some of the commonly discussed partial derivatives of the Black-Scholes option pricing formula. These partial derivatives help the reader to understand the sensitivity of option prices to movements in the underlying security's value, the risk-free interest rate, and the volatility of the underlying security's value. The collection of partial derivatives introduced in this chapter is commonly referred to as "the Greeks" by many financial practitioners. The Greeks are used in the ninth chapter on hedging strategies for portfolios. Hedging strategies are used to protect the value of a portfolio against movements in the underlying security's value, the risk-free interest rate, and the volatility of the underlying security's value. Mathematically the hedging strategies remove some of the low order terms from the Black-Scholes option pricing formula making it less sensitive to changes in the variables upon which it depends. Chapter Nine will discuss and illustrate several examples of hedging strategies.

Chapter Ten extends the ideas introduced in Chapter Nine by modeling the effects of correlated movements in the values of investments. The tenth chapter discusses several different notions of optimality in selecting portfolios of investments. Some of the classical models of portfolio selection are introduced in this chapter including the Capital Assets Pricing Model (CAPM) and the Minimum Variance Portfolio.

It is the author's hope that students will find this book a useful introduction to financial mathematics and a springboard to further study in this area. Writing this book has been hard, but intellectually rewarding work.

During the summer of 2005 a draft version of this manuscript was used by the author to teach a course in financial mathematics. The author is indebted to the students of that class for finding numerous typographical errors in that earlier version which were corrected before the camera ready copy was sent to the publisher. The author wishes to thank Jill Bachstadt, Jason Buck, Mark Elicker, Kelly Flynn, Jennifer Gomulka, Nicole Hundley, Alicia Kasif, Stephen Kluth, Patrick McDevitt, Jessica Paxton, Christopher Rachor, Timothy Refi, Pamela Wentz, Joshua Wise, and Michael Zrncic.

A list of errata and other information related to this book can be found at a web site I created:

http://banach.millersville.edu/~bob/book/

Please feel free to share your comments, criticism, and (I hope) praise for this work through the email address that can be found at that site.

J. Robert Buchanan
Lancaster, PA, USA
October 31, 2005

Contents

Chapter 1

The Theory of Interest

One of the first types of investments that people learn about is some variation on the savings account. In exchange for the temporary use of an investor's money, a bank or other financial institution agrees to pay **interest**, a percentage of the amount invested, to the investor. There are many different schemes for paying interest. In this chapter we will describe some of the most common types of interest and contrast their differences. Along the way the reader will have the opportunity to renew their acquaintanceship with exponential functions and the geometric series. Since an amount of capital can be invested and earn interest and thus numerically increase in value in the future, the concept of **present value** will be introduced. Present value provides a way of comparing values of investments made at different times in the past, present, and future. As an application of present value, several examples of saving for retirement and calculation of mortgages will be presented. Sometimes investments pay the investor varying amounts of money which change over time. The concept of **rate of return** can be used to convert these payments in effective interest rates, making comparison of investments easier.

1.1 Simple Interest

In exchange for the use of a depositor's money, banks pay a fraction of the account balance back to the depositor. This fractional payment is known as **interest**. The money a bank uses to pay interest is generated by investments and loans that the bank makes with the depositor's money. Interest is paid in many cases at specified times of the year, but nearly always the fraction of the deposited amount used to calculate the interest is called the **interest rate** and is expressed as a percentage paid per year.

1

For example a credit union may pay 6% annually on savings accounts. This means that if a savings account contains $100 now, then exactly one year from now the bank will pay the depositor $6 (which is 6% of $100) provided the depositor maintains an account balance of $100 for the entire year.

In this chapter and those that follow, interest rates will be denoted symbolically by r. To simplify the formulas and mathematical calculations, when r is used it will be converted to decimal form even though it may still be referred to as a percentage. The 6% annual interest rate mentioned above would be treated mathematically as $r = 0.06$ per year. The initially deposited amount which earns the interest will be called the **principal amount** and will be denoted P. The sum of the principal amount and any earned interest will be called the **capital** or the **amount due**. The symbol A will be used to represent the amount due. The reader may even see the amount due referred to as the **compound amount**, though this use of the adjective "compound" is independent of its use in the term "compound interest" to be explored in Section 1.2. The relationship between P, r, and A for a single year period is

$$A = P + Pr = P(1 + r).$$

In general if the time period of the deposit is t years then the amount due is expressed in the formula

$$A = P(1 + rt). \tag{1.1}$$

This implies that the average account balance for the period of the deposit is P and when the balance is withdrawn (or the account is closed), the principal amount P plus the interest earned Prt is returned to the investor. No interest is credited to the account until the instant it is closed. This is known as the **simple interest** formula.

Some financial institutions credit interest earned by the account balance at fixed points in time. Banks and other financial institutions "pay" the depositor by adding the interest to the depositor's account. The interest, once paid to the depositor, is the depositor's to keep. Unless the depositor withdraws the interest or some part of the principal, the process begins again for another interest earning period. If P is initially deposited, then after one year, the amount due according the Eq. (1.1) with $t = 1$ would be $P(1 + r)$. This amount can be thought of as the principal amount for the account at the beginning of the second year. Thus two years after the initial deposit the amount due would be

$$A = P(1 + r) + P(1 + r)r = P(1 + r)^2. \qquad (1.2)$$

Continuing in this way we can see that t years after the initial deposit of an amount P, the capital A will grow to

$$A = P(1 + r)^t \qquad (1.3)$$

A mathematical "purist" may wish to establish Eq. (1.2) using the principle of induction.

Banks and other interest-paying financial institutions often pay interest more than a single time per year. The yearly interest formula given in Eq. (1.2) must be modified to track the compound amount for interest periods of other than one year.

1.2 Compound Interest

The typical interest bearing savings or checking account will be described by an investor as earning a nominal annual interest rate compounded some number of times per year. Investors will often find interest compounded semi-annually, quarterly, monthly, weekly, or daily. In this section we will compare and contrast compound interest to the simple interest case of the previous section. Whenever interest is allowed to earn interest itself, an investment is said to earn **compound interest**. In this situation, part of the interest is paid to the depositor once or more frequently per year. Once paid, the interest begins earning interest. We will let n denote the number of compounding periods per year. For example for interest "compounded monthly" $n = 12$. Only two small modifications to the simple interest formula (1.2) are needed to calculate the compound interest. First, it is now necessary to think of the interest rate per compounding period. If the annual interest rate is r, then the interest rate per compounding period is r/n. Second, the elapsed time should be thought of as some number of compounding periods rather than years. Thus with n compounding periods per year, the number of compounding periods in t years is nt. Therefore the formula for compound interest is

$$A = P \left(1 + \frac{r}{n}\right)^{nt}. \qquad (1.4)$$

Eq. (1.4) simplifies to the formula for the amount due given in Eq. (1.2) when $n = 1$.

Example 1.1 Suppose an account earns 5.75% annually compounded monthly. If the principal amount is $3104 then after three and one-half years the amount due will be

$$A = 3104 \left(1 + \frac{0.0575}{12}\right)^{(12)(3.5)} = 3794.15.$$

The reader should verify using Eq. (1.1) that if the principal in the previous example earned only *simple* interest at an annual rate of 5.75% then the amount due after 3.5 years would be only $3728.68. Thus happily for the depositor, compound interest builds capital faster than simple interest. Frequently it is useful to compare an annual interest rate with compounding to an equivalent simple interest, *i.e.* to the simple annual interest rate which would generate the same amount of interest as the annual compound rate. This equivalent interest rate is called the **effective interest rate**. For the rate mentioned in the previous example we can find the effective interest rate by solving the equation

$$\left(1 + \frac{0.0575}{12}\right)^{12} = 1 + r_e$$
$$0.05904 = r_e$$

Thus the nominal annual interest rate of 5.75% compounded monthly is equivalent to an effective annual rate of 5.90%.

In general if the nominal annual rate r is compounded n times per year the equivalent effective annual rate r_e is given by the formula:

$$r_e = \left(1 + \frac{r}{n}\right)^n - 1. \tag{1.5}$$

Intuitively it seems that more compounding periods per year implies a higher effective annual interest rate. In the next section we will explore the limiting case of frequent compounding going beyond semiannually, quarterly, monthly, weekly, daily, hourly, *etc.* to continuously.

1.3 Continuously Compounded Interest

Mathematically when considering the effect on the compound amount of more frequent compounding, we are contemplating a limiting process. In symbolic form we would like to find the compound amount A which satisfies the equation

$$A = \lim_{n \to \infty} P \left(1 + \frac{r}{n}\right)^{nt}. \tag{1.6}$$

Fortunately there is a simple expression for the value of the limit on the right-hand side of Eq. (1.6). We will find it by working on the limit

$$\lim_{n \to \infty} \left(1 + \frac{r}{n}\right)^{n}.$$

This limit is indeterminate of the form 1^{∞}. We will evaluate it through a standard approach using the natural logarithm and l'Hôpital's Rule. The reader should consult an elementary calculus book such as [Smith and Minton (2002)] for more details. We see that if $y = (1 + r/n)^n$, then

$$\begin{aligned} \ln y &= \ln \left(1 + \frac{r}{n}\right)^{n} \\ &= n \ln(1 + r/n) \\ &= \frac{\ln(1 + r/n)}{1/n} \end{aligned}$$

which is indeterminate of the form $0/0$ as $n \to \infty$. To apply l'Hôpital's Rule we take the limit of the derivative of the numerator over the derivative of the denominator. Thus

$$\begin{aligned} \lim_{n \to \infty} \ln y &= \lim_{n \to \infty} \frac{\frac{d}{dn}\left(\ln(1 + r/n)\right)}{\frac{d}{dn}(1/n)} \\ &= \lim_{n \to \infty} \frac{r}{1 + r/n} \\ &= r \end{aligned}$$

Thus $\lim_{n \to \infty} y = e^r$. Finally we arrive at the formula for **continuously compounded interest**,

$$A = Pe^{rt}. \tag{1.7}$$

This formula may seem familiar since it is often presented as the exponential growth formula in elementary algebra, precalculus, or calculus. The quantity A has the property that A changes with time t at a rate proportional to A itself.

Example 1.2 Suppose \$3585 is deposited in an account which pays interest at an annual rate of 6.15% compounded continuously. After two and one half years the principal plus earned interest will have grown to

$$A = 3585e^{(0.0615)(2.5)} = 4180.82.$$

The effective simple interest rate is the solution to the equation

$$e^{0.0615} = 1 + r_e$$

which implies $r_e \approx 6.34\%$.

1.4   Present Value

One of the themes we will see many times in the study of financial mathematics is the comparison of the value of a particular investment at the present time with the value of the investment at some point in the future. This is the comparison between the **present value** of an investment versus its **future value**.   We will see in this section that present and future value play central roles in planning for retirement and determining loan payments. Later in this book present and future values will help us determine a fair price for stock market derivatives.

The future value t years from now of an invested amount P subject to an annual interest rate r compounded continuously is

$$A = Pe^{rt}.$$

Thus by comparison with Eq. (1.7), the future value of P is just the compound amount of P monetary units invested in a savings account earning interest r compounded continuously for t years. By contrast the present value of A in an environment of interest rate r compounded continuously for t years is

$$P = Ae^{-rt}.$$

In other words if an investor wishes to have A monetary units in savings t years from now and they can place money in a savings account earning interest at an annual rate r compounded continuously, the investor should deposit P monetary units now. There are also formulas for future and present value when interest is compounded at discrete intervals, not continuously. If the interest rate is r annually with n compounding periods per year then the future value of P is

$$A = P\left(1 + \frac{r}{n}\right)^{nt}.$$

Compare this equation with Eq. (1.4). Simple algebra shows then the present value of P earning interest at rate r compounded n times per year for t years is

$$P = A\left(1 + \frac{r}{n}\right)^{-nt}.$$

Example 1.3 Suppose an investor will receive payments at the end of the next six years in the amounts shown in the table.

Year	1	2	3	4	5	6
Payment	465	233	632	365	334	248

If the interest rate is 3.99% compounded monthly, what is the present value of the investments? Assuming the first payment will arrive one year from now, the present value is the sum

$$465\left(1 + \frac{0.0399}{12}\right)^{-12} + 233\left(1 + \frac{0.0399}{12}\right)^{-24} + 632\left(1 + \frac{0.0399}{12}\right)^{-36}$$
$$+ 365\left(1 + \frac{0.0399}{12}\right)^{-48} + 334\left(1 + \frac{0.0399}{12}\right)^{-60} + 248\left(1 + \frac{0.0399}{12}\right)^{-72}$$
$$= 2003.01$$

Notice that the present value of the payments from the investment is different from the sum of the payments themselves (which is 2277).

Unless the reader is among the very fortunate few who can always pay cash for all purchases, you may some day apply for a loan from a bank or other financial institution. Loans are always made under the assumptions of a prevailing interest rate (with compounding), an amount to be borrowed, and the lifespan of the loan, *i.e.* the time the borrower has to repay the loan. Usually portions of the loan must be repaid at regular intervals (for example, monthly). Now we turn our attention to the question of using the amount borrowed, the length of the loan, and the interest rate to calculate the loan payment.

A very helpful mathematical tool for answering questions regarding present and future values is the **geometric series**. Suppose we wish to find the sum

$$S = 1 + a + a^2 + \cdots + a^n \tag{1.8}$$

where n is a positive whole number. If both sides of Eq. (1.8) are multiplied by a and then subtracted from Eq. (1.8) we have

$$S - aS = 1 + a + a^2 + \cdots + a^n - (a + a^2 + a^3 + \cdots + a^{n+1})$$
$$S(1 - a) = 1 - a^{n+1}$$
$$S = \frac{1 - a^{n+1}}{1 - a}$$

provided $a \neq 1$.

Now we will apply this tool to the task of finding out the monthly amount of a loan payment. Suppose someone borrows P to purchase a new car. The bank issuing the automobile loan charges interest at the annual rate of r compounded n times per year. The length of the loan will be t years. The monthly installment can be calculated if we apply the principle that the present value of all the payments made must equal the amount borrowed. Suppose the payment amount is the constant x. If the first payment must be made at the end of the first compounding period, then the present value of all the payments is

$$x(1 + \frac{r}{n})^{-1} + x(1 + \frac{r}{n})^{-2} + \cdots + x(1 + \frac{r}{n})^{-nt}$$
$$= x(1 + \frac{r}{n})^{-1} \frac{1 - (1 + \frac{r}{n})^{-nt}}{1 - (1 + \frac{r}{n})^{-1}}$$
$$= x \frac{1 - (1 + \frac{r}{n})^{-nt}}{\frac{r}{n}}$$

Therefore the relationship between the interest rate, the compounding frequency, the period of the loan, the principal amount borrowed, and the payment amount is expressed in the following equation.

$$P = x \frac{n}{r} \left(1 - \left[1 + \frac{r}{n} \right]^{-nt} \right) \tag{1.9}$$

Example 1.4 If a person borrows $25000 for five years at an interest rate of 4.99% compounded monthly and makes equal monthly payments, the payment amount will be

$$x = 25000(0.0499/12) \left(1 - [1 + (0.0499/12)]^{-(12)(5)} \right)^{-1} = 471.67.$$

Similar reasoning can be used when determining how much to save for retirement. Suppose a person is 25 years of age now and plans to retire at age 65. For the next 40 years they plan to invest a portion of their monthly income in securities which earn interest at the rate of 10% compounded monthly. After retirement the person plans on receiving a monthly payment (an annuity) in the absolute amount of \$1500 for 30 years. The amount of money the person should invest monthly while working can be determined by equating the present value of all their deposits with the present value of all their withdrawals. The first deposit will be made one month from now and the first withdrawal will be made 481 months from now. The last withdrawal will be made 840 months from now. The monthly deposit amount will be be denoted by the symbol x. The present value of all the deposits made into the retirement fund is

$$x \sum_{i=1}^{480} \left(1 + \frac{0.10}{12}\right)^{-i} = x \left(1 + \frac{0.10}{12}\right)^{-1} \frac{1 - \left(1 + \frac{0.10}{12}\right)^{-480}}{1 - \left(1 + \frac{0.10}{12}\right)^{-1}}$$
$$\approx 117.765x.$$

Meanwhile the present value of all the annuity payments is

$$1500 \sum_{i=481}^{840} \left(1 + \frac{0.10}{12}\right)^{-i} = 1500 \left(1 + \frac{0.10}{12}\right)^{-481} \frac{1 - \left(1 + \frac{0.10}{12}\right)^{-360}}{1 - \left(1 + \frac{0.10}{12}\right)^{-1}}$$
$$\approx 3182.94.$$

Thus $x \approx 27.03$ dollars per month. This seems like a small amount to invest, but such is the power of compound interest and starting a savings plan for retirement early. If the person waits ten years (*i.e.*, until age 35) to begin saving for retirement, but all other factors remain the same, then

$$x \sum_{i=1}^{360} \left(1 + \frac{0.10}{12}\right)^{-i} \approx 113.951x$$

$$1500 \sum_{i=361}^{720} \left(1 + \frac{0.10}{12}\right)^{-i} \approx 8616.36$$

which implies the person must invest $x \approx 75.61$ monthly. Waiting ten years to begin saving for retirement nearly triples the amount which the future retiree must set aside for retirement.

The initial amounts invested are of course invested for a longer period of time and thus contribute a proportionately greater amount to the future value of the retirement account.

Example 1.5 Suppose two persons will retire in twenty years. One begins saving immediately for retirement but due to unforeseen circumstances must abandon their savings plan after four years. The amount they put aside during those first four years remains invested, but no additional amounts are invested during the last sixteen years of their working life. The other person waits four years before putting any money into a retirement savings account. They save for retirement only during the last sixteen years of their working life. Let us explore the difference in the final amount of retirement savings that each person will possess. For the purpose of this example we will assume that the interest rate is $r = 0.05$ compounded monthly and that both workers will invest the same amount x, monthly. The first worker has upon retirement an account whose present value is

$$x \sum_{i=1}^{48} \left(1 + \frac{0.05}{12}\right)^{-i} \approx 43.423x.$$

The present value of the second worker's total investment is

$$x \sum_{i=49}^{240} \left(1 + \frac{0.05}{12}\right)^{-i} \approx 108.102x.$$

Thus the second worker retires with a larger amount of retirement savings; however, the ratio of their retirement balances is only $43.423/108.102 \approx 0.40$. The first worker saves, in only one fifth of the time, approximately 40% of what the second worker saves.

The discussion of retirement savings makes no provision for rising prices. The economic concept of **inflation** is the phenomenon of the decrease in the purchasing power of a unit of money relative to a unit amount of goods or services. The rate of inflation (usually expressed as an annual percentage rate, similar to an interest rate) varies with time and is a function of many factors including political, economic, and international factors. While the causes of inflation can be many and complex, inflation is generally described as a condition which results from an increase in the amount of money in circulation without a commensurate increase in the amount of available goods. Thus relative to the supply of goods, the value of the currency is decreased. This can happen when wages are arbitrarily increased without an equal increase in worker productivity.

We now focus on the effect that inflation may have on the worker planning to save for retirement. If the interest rate on savings is r and the inflation rate is i we can calculate the **inflation-adjusted rate** or as it is sometimes called, the **real rate of interest**. This derivation will test your understanding of the concepts of present and future value discussed earlier in this chapter. We will let the symbol r_i denote the inflation-adjusted interest rate [Broverman (2004)]. Suppose at the current time one unit of currency will purchase one unit of goods. Invested in savings, that one unit of currency has a future value (in one year) of $1 + r$. In one year the unit of goods will require $1 + i$ units of currency for purchase. The difference

$$(1 + r) - (1 + i) = r - i$$

will be the real rate of growth in the unit of currency invested now. However, this return on saving will not be earned until one year from now. Thus we must adjust this rate of growth by finding its present value under the inflation rate. This leads us to the following formula for the inflation-adjusted interest rate.

$$r_i = \frac{r - i}{1 + i} \tag{1.10}$$

Note that when inflation is low (i is small), $r_i \approx r - i$ and this latter approximation is sometimes used in place of the more accurate value expressed in Eq. (1.10).

Returning to the earlier example of the worker saving for retirement, consider the case in which $r = 0.10$, the worker will save for 40 years and live on a monthly annuity whose inflation adjusted value will be $1500 for 30 years, and the rate of inflation will be $i = 0.03$ for the entire lifespan of the worker/retiree. Thus $r_i \approx 0.0680$. Assuming the worker will make the first deposit in one month the present value of all deposits to be made is

$$x \sum_{i=1}^{480} \left(1 + \frac{0.068}{12}\right)^{-i} = x \left(1 + \frac{0.068}{12}\right)^{-1} \frac{1 - \left(1 + \frac{0.068}{12}\right)^{-480}}{1 - \left(1 + \frac{0.068}{12}\right)^{-1}}$$

$$\approx 164.756x.$$

The present value of all the annuity payments is given by

$$1500 \sum_{i=481}^{840} \left(1 + \frac{0.068}{12}\right)^{-i} = 1500 \left(1 + \frac{0.068}{12}\right)^{-481} \frac{1 - \left(1 + \frac{0.068}{12}\right)^{-360}}{1 - \left(1 + \frac{0.068}{12}\right)^{-1}}$$

$$\approx 15273.80.$$

Thus the monthly deposit amount is approximately $92.71. This is roughly four times the monthly investment amount when inflation is ignored. However, since inflation does tend to take place over the long run, ignoring a 3% inflation rate over the lifetime of the individual would mean that the present purchasing power of the last annuity payment would be

$$1500\left(1+\frac{0.03}{12}\right)^{-840} \approx 184.17.$$

This is not much money to live on for an entire month. Retirement planning should include provisions for inflation, varying interest rates, the period of retirement, the period of savings, and desired monthly annuity during retirement.

1.5   Rate of Return

The present value of an item is one way to determine the absolute worth of the item and to compare its worth to that of other items. Another way to judge the value of an item which an investor may own or consider purchasing is known as the **rate of return**.   If a person invests an amount P now and receives an amount A one time unit from now, the rate of return can be thought of as the interest rate per time unit that the invested amount would have to earn so that the present value of the payoff amount is equal to the invested amount. Since the rate of return is going to be thought of as an equivalent interest rate, it will be denoted by the symbol r. Then by definition

$$P = A(1+r)^{-1} \quad \text{or equivalently} \quad r = \frac{A}{P} - 1.$$

Example 1.6   If you loan a friend \$100 today with the understanding that they will pay you back \$110 in one year's time, then the rate of return is $r = 0.10$ or 10%.

In a more general setting a person may invest an amount P now and receive a sequence of positive payoffs $\{A_1, A_2, \ldots, A_n\}$ at regular intervals. In this case the rate of return per period is the interest rate such that the present value of the sequence of payoffs is equal to the amount invested. In this case

$$P = \sum_{i=1}^{n} A_i(1+r)^{-i}.$$

It is not clear from this definition that r has a unique value for all choices of P and payoff sequences. Defining the function $f(r)$ to be

$$f(r) = -P + \sum_{i=1}^{n} A_i(1+r)^{-i} \qquad (1.11)$$

we can see that $f(r)$ is continuous on the open interval $(-1, \infty)$. In the limit as r approaches -1 from the right, the function values approach positive infinity. On the other hand as r approaches positive infinity, the function values approach $-P < 0$ asymptotically. Thus by the Intermediate Value Theorem (p. 108 of [Smith and Minton (2002)]) there exists r^* with $-1 < r^* < \infty$ such that $f(r^*) = 0$. The reader is encouraged to show that r^* is unique in the exercises.

Rates of return can be either positive or negative. If $f(0) > 0$, *i.e.*, the sum of the payoffs is greater than the amount invested then $r^* > 0$ since $f(r)$ changes sign on the interval $[0, \infty)$. If the sum of the payoffs is less than the amount invested then $f(0) < 0$ and the rate of return is negative. In this case the function $f(r)$ changes sign on the interval $(-1, 0]$.

Example 1.7 Suppose you loan a friend \$100 with the agreement that they will pay you at the end of each year for the next five years amounts $\{21, 22, 23, 24, 25\}$. The rate of return per year is the solution to the equation,

$$-100 + \frac{21}{1+r} + \frac{22}{(1+r)^2} + \frac{23}{(1+r)^3} + \frac{24}{(1+r)^4} + \frac{25}{(1+r)^5} = 0.$$

Newton's Method (Sec. 3.2 of [Smith and Minton (2002)]) can be used to approximate the solution $r^* \approx 0.047$.

1.6 Exercises

(1) Suppose that \$3659 is deposited in a savings account which earns 6.5% simple interest. What is the amount due after five years?
(2) Suppose that \$3993 is deposited in an account which earns 4.3% interest. What is the compound amount after two years if the interest is compounded

  (a) monthly?
  (b) weekly?
  (c) daily?

(d) continuously?

(3) Find the effective annual interest rate which is equivalent to 8% interest compounded quarterly.

(4) You are preparing to open a bank which will accept deposits into savings accounts and which will pay interest compounded monthly. In order to be competitive you must meet or exceed the interest paid by another bank which pays 5.25% compounded daily. What is the minimum interest rate you can pay and remain competitive?

(5) Suppose you have $1000 to deposit in one of two types of savings accounts. One account pays interest at an annual rate of 4.75% compounded daily, while the other pays interest at an annual rate of 4.75% compounded continuously. How long would it take for the compound amounts to differ by $1?

(6) Many textbooks determine the formula for continuously compounded interest through an argument which avoids the use of l'Hôpital's Rule (for example [Goldstein *et al.* (1999)]). Beginning with Eq. (1.6) let $h = r/n$. Then

$$P\left(1 + \frac{r}{n}\right)^{nt} = P(1 + h)^{(1/h)rt}$$

and we can focus on finding the $\lim_{h \to 0}(1 + h)^{1/h}$. Show that

$$(1 + h)^{1/h} = e^{(1/h)\ln(1+h)}$$

and take the limit of both sides as $h \to 0$. Hint: you can use the definition of the derivative in the exponent on the right-hand side.

(7) Which of the two investments described below is preferable? Assume the first payment will take place exactly one year from now and further payments are spaced one year apart. Assume the continually compounded annual interest rate is 2.75%.

Year	1	2	3	4
Investment A	200	211	198	205
Investment B	198	205	211	200

(8) Suppose you wish to buy a house costing $200000. You will put a down payment of 20% of the purchase price and borrow the rest from a bank for 30 years at a fixed interest rate r compounded monthly. If you wish your monthly mortgage payment to be $1500 or less, what is the maximum annual interest rate for the mortgage loan?

(9) If the effective annual interest rate is 5.05% and the rate of inflation is 2.02%, find the nominal annual real rate of interest compounded quarterly.

(10) Use the Mean Value Theorem (p. 235 of [Stewart (1999)]) to show the rate of return defined by the root of the function in Eq. (1.11) is unique.

(11) Suppose for an investment of $10000 you will receive payments at the end of each of the next four years in the amounts $\{2000, 3000, 4000, 3000\}$. What is the rate of return per year?

(12) Suppose you have the choice of investing $1000 in just one of two ways. Each investment will pay you an amount listed in the table below at the end of each year for the next five years.

Year	1	2	3	4	5
Investment A	225	215	250	225	205
Investment B	220	225	250	250	210

(a) Using the present value of the investment to make the decision, which investment would you choose? Assume the annual interest rate is 4.33%.

(b) Using the rate of return per year of the investment to make the decision, which investment would you choose?

Chapter 2

Discrete Probability

Since the number and interactions of forces driving the values of investments are so large and complex, development of a deterministic mathematical model of a market is likely to be impossible. In this book a probabilistic or stochastic model of a market will be developed instead. This chapter presents some elementary concepts of probability and statistics. Here the reader will find explanations of discrete events and their outcomes. A discrete outcome can take on only one of a finite number of values. For example the outcome of a roll of a fair die can be only one of the six values in the set $\{1, 2, 3, 4, 5, 6\}$. No one ever rolls a die and discovers the outcome to be π for example. Basic methods for determining the probabilities of outcomes will be presented. The concept of the **random variable**, a numerical quantity whose value is not known until an experiment is conducted, will be explained. There are many different kinds of discrete random variables, but one that frequently arises in financial mathematics is the **binomial random variable**. While statistics is a field of study unto itself, two important descriptive statistics will be introduced in this chapter, **expected value** (or mean) and **standard deviation**. The expected value provides a number which is representative of typical values of a random variable. The standard deviation is a number which provides a measure related to the width of an interval centered at the mean into which values of the random variable are likely to fall. As will be seen when discussing specific experiments, the standard deviation measures the degree to which values of the random variable are "spread out" around the mean.

2.1   Events and Probabilities

To the layman an event is something that happens. To the statistician, an **event** is an outcome or set of outcomes of an experiment. This brings up the question of what is an experiment? For our purposes an **experiment** will be any activity that generates an observable outcome. Some simple examples of experiments include flipping a coin, rolling a pair of dice, and drawing cards from a deck. An outcome of each of these experiments could be "heads", 7, or the ace of hearts respectively. For the example of the coin flip or drawing a card from the deck, the example outcomes given can be thought of as "atomic" in the sense that they cannot be further broken down into simpler events. The outcome of achieving a 7 on a roll of a pair of dice could be thought of as consisting of a pair of outcomes, one for each die. For example the 7 could be the result of 2 on the first die and 5 on the second. Having a flipped coin land heads up cannot be similarly decomposed. An event can also be thought of as a collection of outcomes rather than just a single outcome. For example the experiment of drawing a card from a standard deck, the events could be segregated into hearts, diamonds, spades, or clubs depending on the suit of the card drawn. Then any of the atomic events $2, 3, \ldots, 10$, jack, queen, king, or ace of hearts would be a "heart" event for the experiment of drawing a card and observing its suit.

In this chapter the outcomes of experiments will be thought of as **discrete** in the sense that the outcomes will be from a set whose members are isolated from each other by gaps. The discreteness of a coin flip, a roll of a pair of dice, and card draw are apparent due to the condition that there is no outcome between "heads" and "tails", or between 6 and 7, or between the two of clubs and the three of clubs respectively. Also in this chapter the number of different outcomes of an experiment will be either finite or countable (meaning that the outcomes can be put into one-to-one correspondence with a subset of the natural numbers).

The **probability** of an event is a real number measuring the likelihood of that event occurring as the outcome of an experiment. To begin the more formal study of events and probabilities, let the symbol A represent an event. The probability of event A will be denoted $P(A)$. By convention, probabilities are always real numbers in the interval $[0, 1]$, that is, $0 \leq P(A) \leq 1$. If A is an event for which $P(A) = 0$, then A is said to be an **impossible** event. If $P(A) = 1$, then A is said to be a **certain** event. Impossible events never occur, while certain events always occur. Events with probabilities closer to 1 are more likely to occur than events whose probabilities are closer to 0.

There are two approaches to assigning a probability to an event, the **classical** approach and the **empirical** approach. Adopting the empirical approach requires an investigator to conduct (or at least simulate) the experiment N times (where N is usually taken to be as large as practical). During the N repetitions of the experiment the investigator counts the number of times that event A occurred. Suppose this number is x. Then the probability of event A is estimated to be $P(A) = x/N$. The classical approach is a more theoretical exercise. The investigator must consider the experiment carefully and determine the total number of different outcomes of the experiment (call this number M), assume that each outcome is equally likely, and then determine the number of outcomes among the total in which event A occurs (suppose this number is y). The probability of event A is then assigned the value $P(A) = y/M$. In practice the two methods closely agree, especially when N is very large.

Some experiments involve events which can be thought of as the result of two or more outcomes occurring simultaneously. For example, suppose a red coin and a green coin will be flipped. One compound outcome of the experiment is the red coin lands on "heads" and the green coin lands on "heads" also. The next section contains some simple rules for handling the probabilities of these compound events.

2.2 Addition Rule

Suppose A and B are two events which may occur as a result of conducting an experiment. An investigator may wish to know the probability that A or B occurs. Symbolically this would be represented as $P(A \vee B)$. If an investigator rolls a pair of fair dice they may want to know the probability that a total of 2 or 12 results. Let event A be the outcome of 2 and B be the outcome of 12. Since $P(A) = 1/36$ and $P(B) = 1/36$ and the two events are **mutually exclusive**, that is, they cannot both simultaneously occur, $P(A \vee B) = P(A) + P(B) = 1/18$. Suppose instead that the investigator wants to know the probability that a total of less than 6 or an odd total results. We can let event A be the outcome of a total less than 6 (that is, a total of 2, 3, 4, or 5) and let event B be the outcome of an odd total (specifically 3, 5, 7, 9, or 11). We see this time that events A an B are not mutually exclusive, there are outcomes which overlap both events, namely the odd numbers 3 and 5 are less than 6. To adjust the calculation the probabilities of the non-exclusive events should be counted only once. Thus $P(A \vee B) = P(A) + P(B) - P(A \wedge B)$ where $P(A \wedge B)$ is the probability that one of the events in the overlapping non-exclusive set of outcomes occurs. Hence

$$P(A \vee B) = \left(\frac{1}{36} + \frac{1}{18} + \frac{1}{12} + \frac{1}{9}\right) + \left(\frac{1}{18} + \frac{1}{9} + \frac{1}{6} + \frac{1}{9} + \frac{1}{18}\right) - \left(\frac{1}{18} + \frac{1}{9}\right)$$
$$= \frac{11}{18}.$$

Thus the calculation of the probability of event A or event B occurring is different depending on whether A and B are mutually exclusive.

The concept outlined above is known as the **Addition Rule** for Probabilities and can be stated in the form of a theorem.

Theorem 2.1   *(Addition Rule)* *For events A and B, the probability of A or B occurring is*

$$P(A \vee B) = P(A) + P(B) - P(A \wedge B). \tag{2.1}$$

If A and B are mutually exclusive events then $P(A \wedge B) = 0$ and the Addition Rule simplifies to

$$P(A \vee B) = P(A) + P(B).$$

Determining the probability of the occurrence of A or B rests on determining the probability that *both* A and B occur. This topic is explored in the next section.

2.3   Conditional Probability and Multiplication Rule

During the past decade a very famous puzzle involving probability has come to be known as the "Monty Hall Problem". This paradox of probability was published in a different but equivalent form in Martin Gardner's "Mathematical Games" feature of *Scientific American* [Gardner (1959)] in 1959 and in the *American Statistician* [Selvin (1975)] in 1975. In 1990 it appeared in its present form in the "Ask Marilyn" column of *Parade Magazine* [vos Savant (1990)].

A game show host hides a prize behind one of three doors. A contestant must guess which door hides the prize. First, the contestant announces the door they have chosen. The host will then open one of the two doors, not chosen, in order to reveal the prize is not behind it. The host then tells the contestant they may keep their original choice or switch to the other unopened door. Should the contestant switch doors?

At first glance when faced with two identical unopened doors, it may seem that there is no advantage to switching doors; however, if the contestant switches they will win with probability 2/3. When the contestant makes the first choice they have a 1/3 chance of being correct and a 2/3 chance of being incorrect. When the host reveals the non-winning, unchosen door, the contestant's first choice still has a 1/3 chance of being correct, but now the unchosen unopened door has a 2/3 probability of being correct, so the contestant should switch.

This example illustrates the concept known as **conditional probability**. Essentially the decision the contestant faces is "given that I have seen that one of the doors I did not choose is not the winning door, should I alter my choice?" The probability that one event occurs given that another event has occurred is called conditional probability. The probability that event A occurs given that event B has occurred is denoted $P(A|B)$. One of the classical thought experiments of discrete probability involves selecting balls from an urn. Suppose an urn contains 20 balls, 6 of which are blue and the remaining 14 are green. Two balls will be drawn, the second will be drawn without replacing the first. The question "what is the probability that the second ball is green, given that the first ball was green?" could be asked. The answer to this question will motivate the statement of the multiplication rule of probability. One approach to the answer involves determining the probability that when two balls are drawn without replacement they are both green. The probability that both selections are green would be the number of two green ball outcomes divided by the total number of outcomes. There are 20 candidates for the first ball selected and there are 19 candidates for the second ball selected. Thus the total number of outcomes is 380. Of those outcomes $(14)(13) = 182$ are both green balls. Thus the probability that both balls are green is $182/380 = 91/190$. The reader may be asking what this situation has to do with the question originally posed. The outcome in which both balls are green is a subset of all the outcomes in which the first ball is green. Consider the diagram in Fig. 2.1. Thus the probability that both balls are green is the product of the probability that the first ball is green multiplied by the probability the second ball is green. Let event A be the set of outcomes in which the first ball is green and event

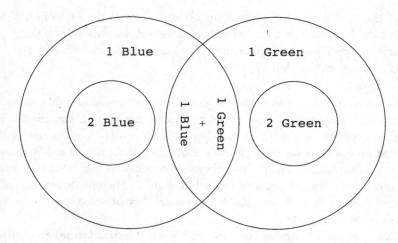

Fig. 2.1   The sets of outcomes of drawing one or two balls from an urn containing blue and green balls.

B be the set of outcomes in which the second ball is green. Numerically $P(A) = 7/10$ and symbolically

$$P(A \wedge B) = P(A) P(B|A)$$

Thus $P(B|A) = P(A \wedge B)/P(A) = (91/190)/(7/10) = 13/19$.

The concept illustrated above is known as the **Multiplication Rule** for Probabilities and can be stated in the form of a theorem.

Theorem 2.2   *(Multiplication Rule) For events A and B, the probability of A and B occurring is*

$$P(A \wedge B) = P(A) P(B|A). \qquad\qquad (2.2)$$

Equation 2.2 can be used to find $P(B|A)$ directly

$$P(B|A) = \frac{P(A \wedge B)}{P(A)}.$$

This expression is meaningful only when $P(A) > 0$.

Example 2.1   One type of roulette wheel, known as the American type, has 38 potential outcomes represented by the integers 1 through 36 and two special outcomes 0 and 00. The positive integers are placed on alternating red and black backgrounds while 0 and 00 are on green backgrounds. What is the probability that the outcome is less than 10 and more than 3 given that the outcome is an even number?

Let event A be the set of outcomes in which the number is even. $P(A) = 10/19$ if 0 and 00 are treated as even numbers. Let B be the set of outcomes in which the number is greater than 3 and less than 10. Then $P(A \land B) = 3/38$ and

$$P(B|A) = \frac{3/38}{10/19} = 3/20.$$

To expand on the previous example, suppose the roulette wheel will be spun twice. One could ask what is the probability that both spins have a red outcome. If event A is the outcome of red on the first spin and event B is the outcome of red on the second spin, then we have as before $P(A \land B) = P(A) P(B|A)$. However there is no reason to believe that the wheel somehow "remembers" the outcome of the first spin while it is being spun the second time. The first outcome has no effect on the second outcome. In any experiment, if event A has no effect on event B then A and B are said to be **independent**. In this situation $P(B|A) = P(B)$. Thus for independent events the Multiplication Rule can be modified to

$$P(A \land B) = P(A) P(B).$$

Therefore the probability that both spins will have red outcomes is $P(A \land B) = (9/19)(9/19) = 81/361$.

2.4 Random Variables and Probability Distributions

The outcome of an experiment is not known until after the experiment is performed. For example the number of people who vote in an election is not known until the election is concluded. In a more formal sense we can describe a **random variable** as a function which maps the set of outcomes of an experiment to some subset of the real numbers. In the election example (assuming the number of registered voters is N and that there will be no fraudulent voting) the sample space of outcomes of voter turnout is the set $S = \{0, 1, 2, \ldots, N\}$. Symbolically a random variable for the voter turnout is the function $X : S \to \mathbb{R}$. Often X is thought of as the eventual numeric result of the experiment.

A **probability distribution** (or **probability function**) is a function which assigns a probability to each element in the sample space of outcomes

Table 2.1   The genders of four
children born to the same set of
parents.

Child			
1	2	3	4
B	B	B	B
G	B	B	B
B	G	B	B
B	B	G	B
B	B	B	G
G	G	B	B
G	B	G	B
G	B	B	G
B	G	G	B
B	G	B	G
B	B	G	G
B	G	G	G
G	B	G	G
G	G	B	G
G	G	G	B
G	G	G	G

of an experiment. If S is the set of outcomes of an experiment and f is the associated probability function, then f maps each element in S to a unique real number in the interval $[0, 1]$. If x is a potential outcome of an experiment with sample space S then $f(x) = P(X = x)$, in other words $f(x)$ is the probability that x occurs as the outcome of the experiment. Since a probability function maps an outcome to a probability then the following two characteristics are true of the function.

(1) If x_i is one of the N outcomes of an experiment then $0 \le f(x_i) \le 1$.
(2) The sum of the values of the probability function is unity, *i.e.*

$$1 = \sum_{i=1}^{N} f(x_i).$$

Example 2.2   Consider a family with four children. The random variable X will represent the number of children who are male. The sample space for this experiment (having four children and counting the number of boys) is the set $S = \{0, 1, 2, 3, 4\}$. The gender of each child is independent from the genders of their siblings, so assuming that the probability of a child being male is $1/2$ then the 16 events shown in table 2.1 are equally likely. There is one outcome in which there are no male children, thus $f(0) =$

$P(X = 0) = 1/16$. There are four cases in which there is a single male child and hence $f(1) = 1/4$. The reader can readily determine from the table that $f(2) = 3/8$, $f(3) = 1/4$, and $f(4) = 1/16$.

Several common types of random variables and their associated probability distributions will be important to the study of financial mathematics. The binomial random variable will be discussed in the next section. It will be seen to be related to the **Bernoulli random variable** which takes on only one of two possible values, often thought of as true or false (or sometimes as success and failure). It is mathematically convenient to designate the outcomes as 0 and 1. The probability function of a Bernoulli random variable is particularly simple, $f(1) = P(X = 1) = p$ where $0 \le p \le 1$ and $f(0) = 1 - p$.

2.5 Binomial Random Variables

Returning to the last example of the previous section, we can think of the births of four children as four independent events. The gender of one child in no way influences the genders of children born before or after. Since the gender of a child can take on only one of two values it is possible to think of the birth of each child as a Bernoulli event. The probability of having a male or female child does not change between births. Thus the experiment of producing four children in a family is the same as repeating the experiment of having a single child four times or, repeating a Bernoulli experiment four times. This is the idea of the **binomial random variable** defined next. A binomial random variable X is the number of successful outcomes out of n independent Bernoulli random trials.

A binomial random variable is parameterized by the number of repetitions of the Bernoulli experiment (referred to as trials from here on) and by the probability of success on a single trial. If the number of trials is n and the probability of success on a single trial is p, then the set of possible outcomes of the binomial experiment is the set $\{0, 1, \ldots, n\}$. The number of combinations of x successes out of n trials is

$$\binom{n}{x} = \frac{n!}{x!(n-x)!}.$$

The probability of x successes out of n independent trials in a specified combination is, according to the Multiplication Rule, $p^x(1-p)^{n-x}$. Since the various combinations are mutually exclusive, by the Addition Rule the probability of x successes out of n trials is given by the function

$$P(X = x) = \binom{n}{x} p^x (1-p)^{n-x} = \frac{n!}{x!(n-x)!} p^x (1-p)^{n-x} \qquad (2.3)$$

Thus if the probability of an individual child being born male or female is $1/2$, the probability that a family with four children will have two female children is

$$P(X = 2) = \frac{4!}{2!\,(4-2)!} \left(\frac{1}{2}\right)^2 \left(\frac{1}{2}\right)^{4-2} = \frac{3}{8}.$$

This result agrees with the result of the more cumbersome method used to determine this probability in the previous section.

Example 2.3 The probability that a computer memory chip is defective is 0.02. A SIMM (single in-line memory module) contains 16 chips for data storage and a 17th chip for error correction. The SIMM can operate correctly if one chip is defective, but not if two or more are defective. The probability that a SIMM will not function is

$$P(X \geq 2) = \sum_{x=2}^{17} \binom{17}{x} (0.02)^x (0.98)^{17-x} \approx 0.044578.$$

2.6 Expected Value

When faced with experimental data, summary statistics are often useful for making sense of the data. In this context "statistics" refers to numbers which can be calculated from the data rather than the means and algorithms by which these numbers are calculated. In financial mathematics the statistical needs are somewhat more specialized than in a general purpose course in statistics. Here we wish to answer the hypothetical question, "if an experiment was to be performed an infinite number of times, what would be the typical outcome?" Thus we will introduce only the statistical concepts to be used later in this text. A reader interested in a broader, deeper, and more rigorous background in statistics should consult one of the many textbooks devoted to the subject for example [Ross (2006)].

To take an example, if a fair die was rolled an infinite number of times, what would be the typical result? The notion to be explored in this section is that of **expected value**. In some ways expected value is synonymous with the mean or average of a list of numerical values; however, it can differ in at least two important ways. First, the expected value usually refers to the typical value of a random variable whose outcomes are not necessarily equally likely whereas the mean of a list of data treats each observation as equally likely. Second, the expected value of a random variable is the typical outcome of an experiment performed an infinite number of times whereas the statistical mean is calculated based on a finite collection of observations of the outcome of an experiment.

If X is a discrete random variable with probability distribution $P(X)$ then the expected value of X is denoted $E[X]$ and defined as

$$E[X] = \sum_X (X \cdot P(X)). \tag{2.4}$$

It is understood that the summation is taken over all values that X may assume. In the case that X takes on only a finite number of values with nonzero probability, then this sum is well-defined. If X may assume an infinite number of values with probabilities greater than zero, we will assume that the sum converges. Since each value of X is multiplied by its corresponding probability, the expected value of X is a weighted average of the variable X. Returning to the question posed in the previous paragraph as to the typical outcome achieved when rolling a fair die an infinite number of times, we may determine this number from the formula for expected value. Since $X \in \{1,2,3,4,5,6\}$ and $P(X) = 1/6$ for all possible values of X, then the expected value of X is

$$E[X] = \sum_{X=1}^{6} \frac{X}{6} = \frac{1}{6}\sum_{X=1}^{6} X = \frac{1}{6} \cdot \frac{(6)(7)}{2} = \frac{7}{2}.$$

Thus the average outcome of rolling a fair die is 3.5.

Example 2.4 Let random variable X represent the number of female children in a family of four children. Assuming that births of males and females are equally likely and that all births are independent events, what is the $E[X]$? The sample space of X is the set $\{0,1,2,3,4\}$. Using the binomial probability formula $P(0) = 1/16$, $P(1) = 1/4$, $P(2) = 3/8$, $P(3) = 1/4$, and $P(4) = 1/16$, we have

$$E[X] = 0 \cdot \frac{1}{16} + 1 \cdot \frac{1}{4} + 2 \cdot \frac{3}{8} + 3 \cdot \frac{1}{4} + 4 \cdot \frac{1}{16} = 2.$$

In families having four children, typically there are two female children and consequently two male children.

The notion of the expected value of a random variable X can be extended to the expected value of a function of X. Thus we say that if F is a function applied to X, then

$$E[F(X)] = \sum_X F(X) P(X).$$

When the function F is merely multiplication by a constant then the expected value takes on a simple form.

Theorem 2.3   *If X is a random variable and a is a constant, then $E[aX] = aE[X]$.*

Proof.   By the definition of expected value

$$E[aX] = \sum_X ((aX) \cdot P(X)) = a \sum_X (X \cdot P(X)) = aE[X].$$

$\square$

Later in this work sums of random variables will become important. Thus some attention must be given to the expected value of the sum of random variables. However, this requires that the probability of two or more random variables be considered simultaneously. The reader should already be familiar with one example of this situation, namely the rolling of a pair of dice. Suppose that the two dice can be distinguished from one another (imagine that one of them is red while the other is green). Let X be the random variable denoting the outcome of the green die while Y is a random variable denoting the outcome of the red die. If the experiment to be performed is rolling the pair of dice and considering the total of the upward faces then the random variable denoting the outcome of this experiment is $X + Y$. This naturally leads us to the issue of describing the probabilities associated with various values of the random variable $X + Y$. The **joint probability function** is denoted $P(X, Y)$ and we will understand it to mean $P(X, Y) = P(X \wedge Y)$. Thus $P(1, 3)$ symbolizes the probability that the outcome of the red die is 1 while the outcome of the green die is 3. If the individual dice are independent then $P(1, 3) = P(1) P(3) = 1/36$ according to the multiplication rule. A couple of additional comments are in order. First, joint probabilities exist even for random events which are not independent. Second, realize that in general $P(X + Y) \neq P(X) + P(Y)$. This is an abuse of notation, but is not likely to cause confusion in what follows. $P(X + Y)$ refers to the probability of the sum $X + Y$ which depends on the joint probabilities of X and Y. $P(X)$ and $P(Y)$ refer respectively to the individual probabilities of random variable X and Y. The following is true, if we wish to know the probability that the sum of the discrete random variables X and Y is m then by using the addition rule for probabilities

$$P\left(X+Y=m\right) = \sum_{X+Y=m} P\left(X,Y\right).$$

The summation is taken over all combinations of X and Y such that $X + Y = m$. Returning to the dice example introduced earlier in the paragraph we see that the probability that the sum of the dice is 4 is

$$P\left(X+Y=4\right) = \sum_{X+Y=4} P\left(X,Y\right)$$
$$= P\left(1,3\right) + P\left(2,2\right) + P\left(3,1\right)$$
$$= \frac{1}{36} + \frac{1}{36} + \frac{1}{36} = \frac{1}{12}.$$

The joint probability distribution of a pair of random variables possesses many of the same properties that the probability distribution of a single random variable possesses. For example $0 \leq P\left(X,Y\right) \leq 1$ for all X and Y in the discrete sample space. It is also true that

$$\sum_{X}\sum_{Y} P\left(X,Y\right) = \sum_{Y}\sum_{X} P\left(X,Y\right) = 1.$$

An important property will be used in the proof of the next theorem. The sum of the joint probability of X and Y where Y is allowed to take on each of its possible values is called the **marginal probability of X**. Without confusion we will denote the marginal probability of X as $P\left(X\right)$ and realize that

$$P\left(X\right) = \sum_{Y} P\left(X,Y\right).$$

Similarly the marginal probability of Y is denoted $P\left(Y\right)$ and defined as

$$P\left(Y\right) = \sum_{X} P\left(X,Y\right).$$

Conveniently the expected value of a sum of random variables is the sum of the expected values of the random variables. This notion is made more precise in the following theorem.

Theorem 2.4 *If $X_1, X_2, \ldots, X_k$ are random variables then*

$$E\left[X_1 + X_2 + \cdots X_k\right] = E\left[X_1\right] + E\left[X_2\right] + \cdots + E\left[X_k\right].$$

Proof.　If $k = 1$ then the proposition is certainly true. If $k = 2$ then

$$
\begin{aligned}
\mathrm{E}\left[X_1 + X_2\right] &= \sum_{X_1, X_2} \left((X_1 + X_2)\mathrm{P}\left(X_1, X_2\right)\right) \\
&= \sum_{X_1} \sum_{X_2} \left((X_1 + X_2)\mathrm{P}\left(X_1, X_2\right)\right) \\
&= \sum_{X_1} \sum_{X_2} X_1 \mathrm{P}\left(X_1, X_2\right) + \sum_{X_2} \sum_{X_1} X_2 \mathrm{P}\left(X_1, X_2\right) \\
&= \sum_{X_1} X_1 \sum_{X_2} \mathrm{P}\left(X_1, X_2\right) + \sum_{X_2} X_2 \sum_{X_1} \mathrm{P}\left(X_1, X_2\right) \\
&= \sum_{X_1} X_1 \mathrm{P}\left(X_1\right) + \sum_{X_2} X_2 \mathrm{P}\left(X_2\right) \\
&= \mathrm{E}\left[X_1\right] + \mathrm{E}\left[X_2\right].
\end{aligned}
$$

For a finite value of $k > 2$ the result is true by induction. Suppose the result is true for $n < k$ where $k > 2$, then

$$
\begin{aligned}
\mathrm{E}\left[X_1 + \cdots + X_{k-1} + X_k\right] &= \mathrm{E}\left[X_1 + \cdots + X_{k-1}\right] + \mathrm{E}\left[X_k\right] \\
&= \mathrm{E}\left[X_1\right] + \cdots + \mathrm{E}\left[X_{k-1}\right] + \mathrm{E}\left[X_k\right]
\end{aligned}
$$

The last step is true by the induction hypothesis.　□

We can use Theorem 2.4 to determine the expected value of a binomial random variable. Along the way we will also find the expected value of a Bernoulli random variable. Suppose n trials of a Bernoulli experiment will be conducted for which the probability of success on a single trial is $0 \le p \le 1$. Random variable X represents the number of successes out of n trials. By assumption the trials are independent of one another and the outcomes are mutually exclusive. The result of the binomial experiment can be thought of as the sum of the results of n Bernoulli experiments. Let the random variable X_i be the number of successes of the ith Bernoulli trial, then

$$
\mathrm{E}\left[X\right] = \mathrm{E}\left[X_1 + \cdots + X_n\right] = \mathrm{E}\left[X_1\right] + \cdots + \mathrm{E}\left[X_n\right] = p + \cdots + p = np.
$$

If functions are applied to random variables a corollary to Theorem 2.4 can be stated.

Corollary 2.1　Let X_1, X_2, ..., X_k be random variables and let F_i be a function defined on X_i for $i = 1, 2, \ldots, k$ then

$$
\begin{aligned}
&\mathrm{E}\left[F_1(X_1) + F_2(X_2) + \cdots + F_k(X_k)\right] \\
&= \mathrm{E}\left[F_1(X_1)\right] + \mathrm{E}\left[F_2(X_2)\right] + \cdots + \mathrm{E}\left[F_k(X_k)\right]
\end{aligned}
$$

Proof. We will prove this result for the case when $k = 2$ and leave it to the reader to apply the principle of mathematical induction to extend the result to the case when $k > 2$.

$$
\begin{aligned}
E\left[F_1(X_1) + F_2(X_2)\right] &= \sum_{X_1, X_2} \left((F_1(X_1) + F_2(X_2)) P\left(X_1, X_2\right)\right) \\
&= \sum_{X_1} F_1(X_1) \sum_{X_2} P\left(X_1, X_2\right) + \sum_{X_2} F_2(X_2) \sum_{X_1} P\left(X_1, X_2\right) \\
&= \sum_{X_1} F_1(X_1) P\left(X_1\right) + \sum_{X_2} F_2(X_2) P\left(X_2\right) \\
&= E\left[F_1(X_1)\right] + E\left[F_2(X_2)\right].
\end{aligned}
$$

$\square$

Later we will have need to calculate the expected value of a product of random variables. This situation is not as straightforward as the case of a sum of random variables.

Theorem 2.5 *Let $X_1, X_2, \ldots, X_k$ be pairwise independent random variables, then*

$$
E\left[X_1 X_2 \cdots X_k\right] = E\left[X_1\right] E\left[X_2\right] \cdots E\left[X_k\right].
$$

Proof. Naturally we see that when $k = 1$ the theorem is true. Next we will consider the case when $k = 2$. Let X_1 and X_2 be independent random variables with joint probability distribution $P\left(X_1, X_2\right)$. Since the random variables are assumed to be independent then $P\left(X_1, X_2\right) = P\left(X_1\right) P\left(X_2\right)$. Once again we are lax in our use of notation, since in the previous equation the symbol P is used in three senses ((1) the joint probability distribution of X_1 and X_2, (2) the probability distribution of X_1, and (3) the probability distribution of X_2); however, there is little chance of confusion in this elementary proof.

$$
\begin{aligned}
E\left[X_1 X_2\right] &= \sum_{X_1, X_2} X_1 X_2 P\left(X_1, X_2\right) \\
&= \sum_{X_1} \sum_{X_2} X_1 X_2 P\left(X_1\right) P\left(X_2\right) \\
&= \sum_{X_1} X_1 P\left(X_1\right) \sum_{X_2} X_2 P\left(X_2\right) \\
&= E\left[X_1\right] E\left[X_2\right]
\end{aligned}
$$

For a finite value of $k > 2$ the result is true by induction. Suppose the result is true for $n < k$ where $k > 2$, then

$$\mathrm{E}\,[X_1 \cdots X_{k-1} X_k] = \mathrm{E}\,[X_1 \cdots X_{k-1}]\,\mathrm{E}\,[X_k]$$
$$= \mathrm{E}\,[X_1] \cdots \mathrm{E}\,[X_{k-1}]\,\mathrm{E}\,[X_k]$$

The last step is true by the induction hypothesis.   □

A corollary to Theorem 2.5 holds for functions of pairwise independent random variables as well.

Corollary 2.2   *Let X_1, X_2, ..., X_k be pairwise independent random variables and let F_i be a function defined on X_i for $i = 1, 2, \ldots, k$ then*

$$\mathrm{E}\,[F_1(X_1) F_2(X_2) \cdots F_k(X_k)] = \mathrm{E}\,[F_1(X_1)]\,\mathrm{E}\,[F_2(X_2)] \cdots \mathrm{E}\,[F_k(X_k)].$$

Proof.   Once again we will prove this result for the case when $k = 2$ and leave it to the reader to extend the result to the case when $k > 2$.

$$\mathrm{E}\,[F_1(X_1) F_2(X_2)] = \sum_{X_1, X_2} F_1(X_1) F_2(X_2) \mathrm{P}\,(X_1, X_2)$$
$$= \sum_{X_1} F_1(X_1) \mathrm{P}\,(X_1) \sum_{X_2} F_2(X_2) \mathrm{P}\,(X_2)$$
$$= \mathrm{E}\,[F_1(X_1)]\,\mathrm{E}\,[F_2(X_2)]$$

□

If the reader is interested in more properties of the expected value of sum and products of random variables, consult a textbook on probability such as [Ross (2003)].

The expected value of a random variable specifies the average outcome of an infinite number of repetitions of an experiment. In the next section the notions of variance and standard deviation are introduced. They specify measures of the spread of the outcomes from the expected value.

2.7   Variance and Standard Deviation

The **variance** of a random variable is a measure of the spread of values of the random variable about the expected value of the random variable. The variance is defined as

$$\mathrm{Var}\,(X) = \mathrm{E}\,\left[(X - \mathrm{E}\,[X])^2\right]. \tag{2.5}$$

As the reader can see from Eq. (2.5), the variance is always non-negative. The expression $X - \mathrm{E}\,[X]$ is the signed deviation of X from its expected value. The variance may be interpreted as the average of the squared deviation of a random variable from its expected value. An alternative formula for the variance is sometimes more convenient in calculations.

Theorem 2.6 *Let X be a random variable, then the variance of X is* $\mathrm{E}\left[X^2\right] - \mathrm{E}\left[X\right]^2$.

Proof. By definition,

$$
\begin{aligned}
\mathrm{Var}\left(X\right) &= \mathrm{E}\left[(X - \mathrm{E}\left[X\right])^2\right] \\
&= \mathrm{E}\left[X^2\right] - \mathrm{E}\left[2X\mathrm{E}\left[X\right]\right] + \mathrm{E}\left[\mathrm{E}\left[X\right]^2\right] \\
&= \mathrm{E}\left[X^2\right] - 2\mathrm{E}\left[X\right]\mathrm{E}\left[X\right] + \mathrm{E}\left[X\right]^2 \\
&= \mathrm{E}\left[X^2\right] - \mathrm{E}\left[X\right]^2.
\end{aligned}
$$

The third and fourth steps of this derivation made use of theorems 2.4 and 2.3 respectively. $\qquad\square$

Returning to the previous example of the hypothetical family with four children, we can now investigate the variance in the number of female children. We already know that $\mathrm{E}\left[X\right] = 2$. If we make use of the result of Theorem 2.6 then

$$
\begin{aligned}
\mathrm{Var}\left(X\right) &= \mathrm{E}\left[X^2\right] - \mathrm{E}\left[X\right]^2 \\
&= (0^2)(1/16) + (1^2)(1/4) + (2^2)(3/8) + (3^2)(1/4) + (4^2)(1/16) - 2^2 \\
&= 1.
\end{aligned}
$$

Before investigating the variance of a binomial random variable, we should determine the variance of a Bernoulli random variable. If the probability of success is $0 \le p \le 1$ then according to Eq. (2.5),

$$
\begin{aligned}
\mathrm{Var}\left(X\right) &= \mathrm{E}\left[(X - \mathrm{E}\left[X\right])^2\right] \\
&= \mathrm{E}\left[(X - p)^2\right] \\
&= (1 - p)^2 p + (0 - p)^2(1 - p) \\
&= p(1 - p).
\end{aligned}
$$

The following theorem provides an easy formula for calculating the variance of *independent* random variables.

Theorem 2.7 *Let $X_1, X_2, \ldots, X_k$ be pairwise independent random variables, then*

$$
\mathrm{Var}\left(X_1 + X_2 + \cdots + X_k\right) = \mathrm{Var}\left(X_1\right) + \mathrm{Var}\left(X_2\right) + \cdots + \mathrm{Var}\left(X_k\right).
$$

Proof. If $k = 1$ then the result is trivially true. Take the case when $k = 2$. By the definition of variance,

$$\begin{aligned}
\text{Var}\,(X_1 + X_2) &= \text{E}\left[((X_1 + X_2) - \text{E}\,[X_1 + X_2])^2\right] \\
&= \text{E}\left[((X_1 - \text{E}\,[X_1]) + (X_2 - \text{E}\,[X_2]))^2\right] \\
&= \text{E}\left[(X_1 - \text{E}\,[X_1])^2\right] + \text{E}\left[(X_2 - \text{E}\,[X_2])^2\right] \\
&\quad + 2\text{E}\left[(X_1 - \text{E}\,[X_1])(X_2 - \text{E}\,[X_2])\right] \\
&= \text{Var}\,(X_1) + \text{Var}\,(X_2) + 2\text{E}\left[(X_1 - \text{E}\,[X_1])(X_2 - \text{E}\,[X_2])\right]
\end{aligned}$$

Since we are assuming that random variables X_1 and X_2 are independent, then by Theorem 2.5

$$\begin{aligned}
\text{E}\left[(X_1 - \text{E}\,[X_1])(X_2 - \text{E}\,[X_2])\right] &= \text{E}\,[X_1 - \text{E}\,[X_1]]\,\text{E}\,[X_2 - \text{E}\,[X_2]] \\
&= (\text{E}\,[X_1] - \text{E}\,[X_1])(\text{E}\,[X_2] - \text{E}\,[X_2]) \\
&= 0,
\end{aligned}$$

and thus

$$\text{Var}\,(X_1 + X_2) = \text{Var}\,(X_1) + \text{Var}\,(X_2).$$

The result can be extended to any finite value of k by induction. Suppose the result has been shown true for $n < k$ with $k > 2$. Then

$$\begin{aligned}
\text{Var}\,(X_1 + \cdots + X_{k-1} + X_k) &= \text{Var}\,(X_1 + \cdots + X_{k-1}) + \text{Var}\,(X_k) \\
&= \text{Var}\,(X_1) + \cdots + \text{Var}\,(X_{k-1}) + \text{Var}\,(X_k)
\end{aligned}$$

where the last equality is justified by the induction hypothesis. $\square$

Readers should think carefully about the validity of the claim that $X_1 - \text{E}\,[X_1]$ and $X_2 - \text{E}\,[X_2]$ are independent in light of the assumption that X_1 and X_2 are independent.

Example 2.5 Suppose a binomial experiment is characterized by n independent repetitions of a Bernoulli trial for which the probability of success on a single trial is $0 \le p \le 1$. Random variable X denotes the total number of successes accrued over the n trials.

$$\text{Var}\,(X) = \text{Var}\left(\sum_{j=1}^{n} X_j\right)$$

$$= \sum_{j=1}^{n} \text{Var}\,(X_j) \quad \text{(since trials are independent)}$$

$$= \sum_{j=1}^{n} p(1-p)$$

$$= np(1-p)$$

So far we have made no mention of the other topic in the heading for this section, namely **standard deviation**. There is little more that must be said since by definition the standard deviation is the square root of the variance. Standard deviation of a random variable X is denoted by $\sigma\,(X)$ and thus

$$\sigma\,(X) = \sqrt{\text{Var}\,(X)}.$$

The reader may also be left wondering about the possible existence of a result regarding the variance of a product of random variables. The general result for the variance of a product would take us too far afield, but we can state and prove a result for the product of pairwise independent random variables.

Theorem 2.8 *Let $X_1, X_2, \ldots, X_k$ be pairwise independent random variables, then*

$$\text{Var}\,(X_1 X_2 \cdots X_k) = \text{E}\left[X_1^2\right]\text{E}\left[X_2^2\right]\cdots\text{E}\left[X_k^2\right] - (\text{E}\left[X_1\right]\text{E}\left[X_2\right]\cdots\text{E}\left[X_k\right])^2.$$

Proof. The case when $k = 1$ follows from Theorem 2.6. Take the case when $k > 1$.

$$\begin{aligned}
\text{Var}\,(X_1 X_2 \cdots X_k) &= \text{E}\left[(X_1 X_2 \cdots X_k)^2\right] - (\text{E}\left[X_1 X_2 \cdots X_k\right])^2 \\
&= \text{E}\left[X_1^2 X_2^2 \cdots X_k^2\right] - (\text{E}\left[X_1\right]\text{E}\left[X_2\right]\cdots\text{E}\left[X_k\right])^2 \\
&= \text{E}\left[X_1^2\right]\text{E}\left[X_2^2\right]\cdots\text{E}\left[X_k^2\right] - (\text{E}\left[X_1\right]\text{E}\left[X_2\right]\cdots\text{E}\left[X_k\right])^2
\end{aligned}$$

The last equation holds as a result of Corollary 2.2. $\qquad\qquad\square$

2.8 Exercises

(1) Suppose the four sides of a regular tetrahedron are labeled 1 through 4. If the tetrahedron is rolled like a die, what is the probability of it landing on 3?

(2) Use the classical approach and the assumption of a fair die to find the probabilities of the outcomes obtained by rolling a pair of dice and summing the dots shown on the the upward faces.

(3) If the probability that a batter strikes out in the first inning of a baseball game is 1/3 and the probability that the batter strikes out in the fifth inning is 1/4, and the probability that the batter strikes out in both innings is 1/10, then what is the probability that the batter strikes out in either inning?

(4) Part of a well-known puzzle involves three people entering a room. As each person enters, at random either a red or a blue hat is placed on the person's head. The probability that an individual receives a red hat is 1/2. No person can see the color of their own hat, but they can see the color of the other two persons' hats. The three will split a prize if at least one person guesses the color of their own hat correctly and no one guesses incorrectly. A person may decide to pass rather than to guess. The three people are not allowed to confer with one another once the hats have been placed on their heads, but they are allowed to agree on a strategy prior to entering the room. At the risk of spoiling the puzzle, one strategy the players may follow instructs a player to pass if they see the other two persons wearing mis-matched hats and to guess the opposite color if their friends are wearing matching hats. Why is this a good strategy and what is the probability of winning the game?

(5) Suppose cards will be drawn without replacement from a standard 52-card deck. What is the probability that the first two cards will be aces?

(6) Suppose cards will be drawn without replacement from a standard 52-card deck. What is the probability that the second card drawn will be an ace and that the first card was not an ace?

(7) Suppose cards will be drawn without replacement from a standard 52-card deck. What is the probability that the fourth card drawn will be an ace given that the first three cards drawn were all aces?

(8) Suppose cards will be drawn without replacement from a standard 52-card deck. On which draw is the card mostly likely to be the first ace

drawn?

(9) On the last 100 spins of an American style roulette wheel, the outcome has been black. What is the probability of the outcome being black on the 101st spin?

(10) On the last 5000 spins of an American style roulette wheel, the outcome has been 00. What is the probability of the outcome being 00 on the 5001st spin?

(11) Suppose that a random variable X has a probability distribution function f so that $f(x) = c/x$ for $x = 1, 2, \ldots, 10$ and is zero otherwise. Find the appropriate value of the constant c.

(12) Suppose that a box contains 15 black balls and 5 white balls. Three balls will be selected without replacement from the box. Determine the probability function for the number of black balls selected.

(13) Quality control for a manufacturer of integrated circuits is done by randomly selecting 25 chips from the previous days manufacturing run. Each of the 25 chips is tested. If two or more chips are faulty, then the entire run is discarded. Previously gathered evidence indicates that the defect rate for chips is 0.0016. What is the probability that a manufacturing run of chips will be discarded?

(14) The probabilities of a child being born male or female are not exactly equal to 1/2. Typically there are nearly 105 live male births per 100 live female births. Determine the expected number of female children in a family of 6 total children using these birth ratios and ignoring infant mortality.

(15) Suppose a standard deck of 52 cards is well shuffled and one card at a time will be drawn without replacement from the deck. What is the expected value of the first ace drawn (in other words, of the first, second, third, *etc* cards drawn, on average which will be the first ace drawn)?

(16) Show that for constants a and b and discrete random variable X that $E[aX + b] = aE[X] + b$.

(17) For the situation described in exercise 15 determine the variance in the occurrence of the first ace drawn.

(18) Show that for constants a and b and discrete random variable X that $\text{Var}(aX + b) = a^2\text{Var}(X)$.

Chapter 3

Normal Random Variables and Probability

Whereas in Chapter 2 random variables could take on only a finite number of values taken from a set with gaps between the values, in the present chapter, **continuous** random variables will be described. A continuous random variable can take on an infinite number of different values from a range without gaps. Calculus-based methods for determining the expected value and standard deviation of a continuous random variable will be described. A very useful type of continuous random variable for the study of financial mathematics is the **normal** random variable. We will see that when many random factors and influences come together (as in the complex situation of a financial market), the sum of the influences can be modeled by a normal random variable. Finally we will examine some stock market data and detect the presence of normal randomness in the fluctuations of stock prices.

3.1 Continuous Random Variables

Our understanding of probability must change if we are to understand the differences between discrete and continuous random variables. Suppose we consider the interval $[0, 1]$. If we think of only the integers contained in this interval, then only the discrete set $\{0, 1\}$ need be considered. If the probability of selecting either integer from this set is $1/2$ then we remain in the realm of discrete probability, random variables, and distributions. If we think of selecting, with equal likelihood, a number from the interval $[0, 1]$ of the form $k/10$ where $k \in \{0, 1, \ldots, 10\}$, then once again our approach to probability remains discrete. The $\mathrm{P}\left(X = k/10\right) = \frac{1}{11}$. Continuing in this way we see that if we consider only the real numbers in $[0, 1]$ of the form k/n where $k \in \{0, 1, \ldots, n\}$, then $\mathrm{P}\left(X = k/n\right) = \frac{1}{n+1}$. So long as n is finite we are dealing with the familiar concept of discrete probability. What happens as $n \to \infty$? In one sense we can think of this limiting case as the case of **continuous probability**. We say continuous because in the limit, the gaps between the outcomes in the sample space disappear. However, if we retain the old notion of probability, the likelihood of choosing a particular number from the interval $[0, 1]$ becomes

$$\lim_{n \to \infty} \frac{1}{n+1} = 0.$$

This is correct but brings to mind a paradox. Does this imply that the probability of choosing any number from $[0, 1]$ is 0? Surely we must choose some number. What is needed is a shift in our notion of probability when dealing with continuous random variables. Instead of determining the probability that a continuous random variable equals some real number, the meaningful determination is the likelihood that the random variable lies in a set, usually an interval or finite union of intervals on the real line. These notions are defined next.

A random variable X has a **continuous distribution** if there exists a non-negative function $f : \mathbb{R} \to \mathbb{R}$ such that for an interval $[a, b]$ the

$$P(a \leq X \leq b) = \int_a^b f(x)\, dx. \tag{3.1}$$

The function f which is known as the **probability distribution function** or **probability density function** must, in addition to satisfying $f(x) \geq 0$, have the following property,

$$\int_{-\infty}^{\infty} f(x)\, dx = 1. \tag{3.2}$$

These properties are the analogues of properties of discrete random variables and their distributions. The probabilities of individual values of the random variable are non-negative. In fact, as explained above $P(X = x) = 0$ for a continuous random variable provided $f(t)$ is continuous at $t = x$. By contrast $P(X = x)$ can be greater than zero for a discrete random variable. The sum, expressed now as an integral over the real number line, of the values of the probability distribution function must be one. Interpreted graphically, if X represents a continuous random variable then $P(a \leq X \leq b)$ represents the area of the region bounded by the graph of the probability distribution function, the x-axis, and the lines $x = a$ and $x = b$. See Fig. 3.1. The total area under the graph of the probability

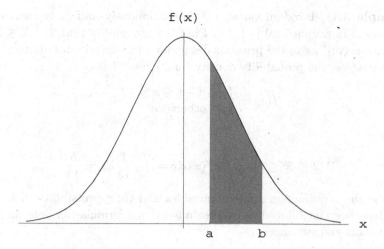

Fig. 3.1 An example of a probability distribution function. The shaded area represents the $P(a \leq X \leq b)$.

density function is unity.

Perhaps the most elementary of the continuous random variables is the **uniformly distributed continuous random variable**. A continuous random variable X is uniformly distributed in the interval $[a, b]$ (with $b > a$) if the probability that X belongs to any subinterval of $[a, b]$ is equal to the length of the subinterval divided by $b - a$. The definition of the continuous uniform random variable allows us to determine the probability density function for X. Since the length of the subinterval is proportional to the probability that X lies in the subinterval then the probability density function $f(x)$ must be constant. Suppose this constant is k. From the property expressed in Eq. (3.2) we know that

$$1 = \int_{-\infty}^{\infty} f(x)\, dx = \int_{a}^{b} k\, dx = \int_{a}^{b} \frac{1}{b-a}\, dx.$$

We are assuming here that the probability distribution function vanishes outside of the interval $[a, b]$. This simplifies the evaluation of the improper integral. Thus the probability density function for a continuously uniformly distributed random variable on interval $[a, b]$ is the piecewise-defined function

$$f(x) = \begin{cases} \frac{1}{b-a} & \text{if } a \leq x \leq b \\ 0 & \text{otherwise.} \end{cases}$$

Example 3.1 Random variable X is continuously uniformly randomly distributed in the interval $[-1, 12]$. Find the probability that $2 \leq X \leq 7$.

Since according to the previous discussion of uniformly distributed random variables, the probability density function of X is

$$f(x) = \begin{cases} \frac{1}{13} & \text{if } -1 \leq x \leq 12 \\ 0 & \text{otherwise,} \end{cases}$$

we have

$$P(2 \leq X \leq 7) = \int_2^7 f(x)\, dx = \int_2^7 \frac{1}{13}\, dx = \frac{5}{13}.$$

Now that continuous random variables and their probability distributions have been introduced we can begin to discuss formulae for determining their means and variances.

3.2   Expected Value of Continuous Random Variables

By definition the **expected value** or **mean** of a continuous random variable X with probability density function $f(x)$ is

$$E[X] = \int_{-\infty}^{\infty} x f(x)\, dx. \tag{3.3}$$

This equation is analogous to Eq. (2.4) for the case of a discrete random variable. In the continuous case the product of the value of the random variable and its probability density function is summed (this time through the use of a definite integral) over all values of the random variable. The expected value is only meaningful in cases in which the improper integral in (3.3) converges.

Example 3.2 Find the expected value of X if X is a continuously uniformly distributed random variable on the interval $[-50, 75]$.

Using Eq. (3.3) we have

$$E[X] = \int_{-\infty}^{\infty} \frac{x}{75 - (-50)}\, dx = \frac{1}{125} \int_{-50}^{75} x\, dx = \frac{1}{250} \left(75^2 - (-50)^2\right) = \frac{25}{2}.$$

The results which follow in this section will be similar to the results that were presented for discrete random variables. In most cases the notion of a discrete sum need only be replaced by a definite integral to justify these new results. To keep the exposition as brief as possible, many of the proofs of results in this section will be left to the reader. The expected value of a function g of a continuously distributed random variable X which has probability distribution function f is defined as

$$\mathrm{E}\left[g(X)\right] = \int_{-\infty}^{\infty} g(x)f(x)\,dx, \tag{3.4}$$

provided the improper integral converges absolutely, *i.e.*, $\mathrm{E}\left[g(X)\right]$ is defined if and only if

$$\int_{-\infty}^{\infty} |g(x)|f(x)\,dx < \infty.$$

Joint probability distributions for continuous random variables are defined similarly to those for discrete random variables. A joint probability distribution for a pair of random variables, X and Y, is a non-negative function $f(x,y)$ for which

$$\int_{-\infty}^{\infty}\int_{-\infty}^{\infty} f(x,y)\,dx\,dy = 1.$$

Consequently the expected value of a function $g : \mathbb{R}^2 \to \mathbb{R}$ of the two random variables X and Y is defined as

$$\mathrm{E}\left[g(X,Y)\right] = \int_{-\infty}^{\infty}\int_{-\infty}^{\infty} g(x,y)f(x,y)\,dx\,dy,$$

again, provided the integral is absolutely convergent.

Before turning our attention to sums and products of continuous random variables we will define the notion of a marginal distribution. If X and Y are continuous random variables with joint distribution $f(x,y)$ then the **marginal distribution** for X is defined as the function

$$f_X(x) = \int_{-\infty}^{\infty} f(x,y)\,dy.$$

The marginal distribution $f_Y(y)$ for random variable Y is defined similarly. Recall that for jointly distributed discrete random variables A and B we said that A and B are independent if and only if $\mathrm{P}\left(A \wedge B\right) = \mathrm{P}\left(A\right)\mathrm{P}\left(B\right)$. For continuous random variables we will use a similar definition. Two continuous random variables are **independent** if and only if the joint probability distribution function factors into the product of the marginal distributions of X and Y. In other words X and Y are independent if and only if

$$f(x,y) = f_X(x)f_Y(y)$$

for all real numbers x and y.

Example 3.3   Consider the jointly distributed random variables $(X, Y) \in [1, \infty) \times [-1, 2]$ whose distribution is the function $f(x, y) = \frac{2}{3x^3}$. We can find the mean of $X + Y$ as follows.

$$
\begin{aligned}
\mathrm{E}\left[X + Y\right] &= \int_{-\infty}^{\infty} \int_{-\infty}^{\infty} (x + y) \frac{2}{3x^3}\, dx\, dy \\
&= \int_{-1}^{2} \int_{1}^{\infty} (x + y) \frac{2}{3x^3}\, dx\, dy \\
&= \int_{-1}^{2} \int_{1}^{\infty} \frac{2x}{3x^3}\, dx\, dy + \int_{-1}^{2} \int_{1}^{\infty} \frac{2y}{3x^3}\, dx\, dy \\
&= \int_{1}^{\infty} \int_{-1}^{2} \frac{2x}{3x^3}\, dy\, dx + \int_{-1}^{2} \int_{1}^{\infty} \frac{2y}{3x^3}\, dx\, dy \\
&= 3 \int_{1}^{\infty} \frac{2}{3x^2}\, dx + \int_{-1}^{2} \frac{y}{3}\, dy \\
&= 2 + \frac{1}{2} = \frac{5}{2}
\end{aligned}
$$

If we examine the example above we recognize that $\mathrm{E}\left[X + Y\right] = \mathrm{E}\left[X\right] + \mathrm{E}\left[Y\right]$. This is true in general, not just for the previous example. The additivity property of the expected value of continuous random variables is stated in the following theorem.

Theorem 3.1   *If $X_1, X_2, \ldots, X_k$ are continuous random variables with joint probability distribution $f(x_1, x_2, \ldots, x_k)$ then $\mathrm{E}\left[X_1 + X_2 + \cdots X_k\right] = \mathrm{E}\left[X_1\right] + \mathrm{E}\left[X_2\right] + \cdots + \mathrm{E}\left[X_k\right]$.*

Proof.   See exercise 6.   □

The reader should again note that, just as was the case in the previous chapter, the previous theorem is true for random variables which are dependent or independent.

Theorem 3.2   *Let $X_1, X_2, \ldots, X_k$ be pairwise independent random variables with joint distribution $f(x_1, x_2, \ldots, x_k)$, then*

$$
\mathrm{E}\left[X_1 X_2 \cdots X_k\right] = \mathrm{E}\left[X_1\right] \mathrm{E}\left[X_2\right] \cdots \mathrm{E}\left[X_k\right].
$$

Proof.   See exercise 7.   □

Just as was the case for discrete random variables in the previous chapter, the expected value of a continuous random variable can be thought of as the average value of the outcome of an infinite number of experiments. The variance and standard deviation of a continuous random variable again mirror the concept earlier defined for discrete random variables.

3.3 Variance and Standard Deviation

The variance of a continuous random variable is a measure of the spread of values of the random variable about the expected value of the random variable. The variance is defined as

$$\text{Var}(X) = \text{E}\left[(X - \mu)^2\right] = \int_{-\infty}^{\infty} (x - \mu)^2 f(x)\, dx, \tag{3.5}$$

where $\mu = \text{E}[X]$ and $f(x)$ is the probability distribution function of X. The variance may be interpreted as the squared deviation of a random variable from its expected value. An alternative formula for the variance is given in the following theorem.

Theorem 3.3 *Let X be a random variable with probability distribution f and mean μ, then the variance of X is $\text{E}\left[X^2\right] - \mu^2$.*

Example 3.4 Find the variance of the continuous random variable whose probability distribution is given by

$$f(x) = \begin{cases} 2x & \text{if } 0 \le x \le 1, \\ 0 & \text{otherwise.} \end{cases}$$

The reader may readily check that $\mu = \text{E}[X] = 2/3$. Now by use of Theorem 3.3 we have

$$\begin{aligned} \text{Var}(X) &= \int_{-\infty}^{\infty} x^2 f(x)\, dx - \left(\frac{2}{3}\right)^2 \\ &= \int_0^1 2x^3\, dx - \frac{4}{9} \\ &= \frac{1}{2} - \frac{4}{9} = \frac{1}{18}. \end{aligned}$$

In the exercises the reader will be asked to prove the following two theorems which extend to continuous random variables results we have already seen for discrete random variables.

Theorem 3.4 *Let X be a continuous random variable with probability distribution $f(x)$ and let $a, b \in \mathbb{R}$, then*

$$\text{Var}\,(aX + b) = a^2 \text{Var}\,(X).$$

Theorem 3.5 *Let X_1, X_2, ..., X_k be pairwise independent continuous random variables with joint probability distribution $f(x_1, x_2, \ldots, x_k)$, then*

$$\text{Var}\,(X_1 + X_2 + \cdots + X_k) = \text{Var}\,(X_1) + \text{Var}\,(X_2) + \cdots + \text{Var}\,(X_k).$$

By definition the standard deviation of a continuous random variable is the square root of its variance. The standard deviation is sometimes denoted by σ. On occasion we will denote the variance of a random variable as σ^2.

3.4 Normal Random Variables

For our purposes one of the most important and useful continuous random variables will be the **normally distributed random variable**. A continuous random variable obeying a normal distribution is frequently said to follow the "bell curve". Many measurable quantities found in nature seem to have normal distributions, for example adult heights and weights. Statisticians, mathematicians, and physical scientists frequently assume any quantity subject to a large number of small independently acting forces (regardless of their distributions) is normally distributed. The normal distribution even finds its way into the financial arena via the assumption that movements in the price of an asset are subject to a large number of incompletely understood political, economic, and social forces, thus justifying the assumption that these changes in value are related to a normal random variable (this assumption will be explored further later).

The continuous normal random variable can also be thought of as the limiting behavior of the discrete binomial random variable introduced in Chapter 2. Recall the relatively simple probability function for a binomial random variable given in Eq. (2.3) and reproduced below.

$$P\,(X = x) = \frac{n!}{x!(n-x)!} p^x (1-p)^{n-x}$$

With suitable assumptions we can develop the probability density function for a continuous normally distributed random variable from the probability function of a binomial random variable. The remainder of this section is dedicated to this derivation and to some elementary properties of normal random variables. Before delving into the details of the development we should outline the steps to be taken. We start with a binomial experiment of n trials with outcomes $\pm\Delta x$ where $\Delta x > 0$. The probabilities of various outcomes of the experiment will be calculated using Eq. (2.3). To transition from the discrete case to the continuous realm we will take the limit of these probabilities as $\Delta x \to 0$, but this brings up the paradox mentioned at the beginning of Section 3.1. As $\Delta x \to 0$ the discrete probabilities will converge to zero unless the concept of a random variable equaling a specific value is replaced with the idea of a random variable lying in an interval around a specific value. We will also assume there exists a relationship between n and Δx in order to ensure the mean and variance of the random variable remain constant as $\Delta x \to 0$.

This derivation is outlined in [Bleecker and Csordas (1996), pg. 139]. We begin by supposing that a particle sits at the origin of the x-axis and may move to the left or to the right a distance Δx in each of n independent identically distributed trials. We will assume that $n(\Delta x)^2 = 2kt$, a constant. The particular form of constant is chosen to simplify the ultimate result. It is convenient to think of the time necessary to conduct the n trials is t and hence the time required by one trial is $\Delta t = t/n$. The (discrete) probability of moving left is $p = 1/2$ which also happens to be the probability of moving to the right.

We will show that if $n \in \mathbb{N}$ and $m \in \mathbb{Z}$ with $-n \leq m \leq n$ then the probability that the particle is at location $X = m\Delta x$ at time $t = n\Delta t$ is

$$\frac{n! \left(\frac{1}{2}\right)^n}{\left(\frac{1}{2}(n+m)\right)! \left(\frac{1}{2}(n-m)\right)!}. \tag{3.6}$$

We will assume that all the steps taken by the particle are independent and identically distributed. Let the total number of steps taken be n and the number of steps taken to the right be r. Consequently the number of steps taken to the left will be $n - r$. The position of the particle after n steps will be

$$(r - (n - r))\Delta x = (2r - n)\Delta x = m\Delta x,$$

where we have assigned $m = 2r - n$. Consequently $n + m$ is even (exercise 12 asks the reader to confirm that likewise $n - m$ is even). Using the binomial probability distribution given in Eq. (2.3) we can state that the probability that the particle is at position $m\Delta x$ after n steps is

$$
\begin{aligned}
P\left(X=m\Delta x\right) &= P\left(X=(2r-n)\Delta x\right) \\
&= \binom{n}{r}\left(\frac{1}{2}\right)^r\left(\frac{1}{2}\right)^{n-r} \\
&= \frac{n!}{r!(n-r)!}\left(\frac{1}{2}\right)^n \\
&= \frac{n!\left(\frac{1}{2}\right)^n}{\left(\frac{1}{2}(n+m)\right)!\left(\frac{1}{2}(n-m)\right)!}
\end{aligned}
$$

since $n+m=2r$. This is the same probability as in Eq. (3.6).

This claim can also be proved by induction on n. If $n=1$ then $m=\pm 1$ (the particle is not allowed to remain in place).

$$
P\left(X=-\Delta x\right)=\frac{1}{2}=\frac{1!\left(\frac{1}{2}\right)^1}{\left(\frac{1}{2}(1+(-1))\right)!\left(\frac{1}{2}(1-(-1))\right)!}
$$

$$
P\left(X=\Delta x\right)=\frac{1}{2}=\frac{1!\left(\frac{1}{2}\right)^1}{\left(\frac{1}{2}(1+1)\right)!\left(\frac{1}{2}(1-1)\right)!}
$$

Thus the claim is true for $n=1$. Now suppose the claim is true for $k\le n-1$. If the particle will move to $m\Delta x$ at time $n\Delta t$, then at time $(n-1)\Delta t$ the particle must be at either $(m-1)\Delta x$ or $(m+1)\Delta x$. Therefore at time $t=n\Delta t$

$$
\begin{aligned}
P\left(X=m\Delta x\right) &= \frac{1}{2}P\left(X=(m-1)\Delta x\right)+\frac{1}{2}P\left(X=(m+1)\Delta x\right) \\
&= \frac{1}{2}\left(\frac{(n-1)!\left(\frac{1}{2}\right)^{n-1}}{\left(\frac{1}{2}(n-1+m-1)\right)!\left(\frac{1}{2}(n-1-(m-1))\right)!}\right) \\
&\quad +\frac{1}{2}\left(\frac{(n-1)!\left(\frac{1}{2}\right)^{n-1}}{\left(\frac{1}{2}(n-1+m+1)\right)!\left(\frac{1}{2}(n-1-(m+1))\right)!}\right) \\
&= \frac{(n-1)!\left(\frac{1}{2}\right)^n}{\left(\frac{1}{2}(n-m-2)\right)!\left(\frac{1}{2}(n+m)\right)!} \\
&\quad +\frac{(n-1)!\left(\frac{1}{2}\right)^n}{\left(\frac{1}{2}(n-m)\right)!\left(\frac{1}{2}(n+m-2)\right)!} \\
&= \frac{n!\left(\frac{1}{2}\right)^n}{\left(\frac{1}{2}(n+m)\right)!\left(\frac{1}{2}(n-m)\right)!},
\end{aligned}
$$

which is the probability given in expression (3.6).

We can treat each step taken by the particle as a Bernoulli experiment with outcomes Δx and $-\Delta x$ with equal probabilities. Thus the mean of this Bernoulli "step" is 0. The variance in the outcome of this Bernoulli experiment is

$$\frac{1}{2}(-\Delta x)^2 + \frac{1}{2}(\Delta x)^2 = (\Delta x)^2.$$

Since the steps are independent and identically distributed then the mean position of the particle after n steps is the sum of the means of each of the steps, or again 0. Thus on average the particle will be at the origin at the end of the experiment. Calculating the variance directly using Theorem 2.6 is tedious, but if we think of the motion of the particle as the sum of the outcomes of n independent Bernoulli experiments, then the variance of the location of the particle is $\sigma^2 = n(\Delta x)^2$. Hence the constant $2kt$ can be interpreted as the variance in the final position of the particle.

Next we will make use of Stirling's Formula which approximates $n!$ when n is large.

$$n! \approx \sqrt{2\pi}e^{-n}n^{n+1/2} \tag{3.7}$$

If we apply this formula to every factorial present in expression (3.6) we obtain the following sequence of equivalent expressions.

$$\frac{\sqrt{2\pi}e^{-n}n^{n+1/2}\left(\frac{1}{2}\right)^n}{\sqrt{2\pi}e^{-(n+m)/2}\left(\frac{1}{2}(n+m)\right)^{(n+m+1)/2}\sqrt{2\pi}e^{-(n-m)/2}\left(\frac{1}{2}(n-m)\right)^{(n-m+1)/2}}$$

$$= \frac{1}{\sqrt{2\pi}}\frac{n^{n+1/2}\left(\frac{1}{2}\right)^n}{\left(\frac{1}{2}(n+m)\right)^{(n+m+1)/2}\left(\frac{1}{2}(n-m)\right)^{(n-m+1)/2}}$$

$$= \frac{2}{\sqrt{2\pi}}\frac{n^{n+1/2}}{(n+m)^{(n+m+1)/2}(n-m)^{(n-m+1)/2}}$$

$$= \frac{2}{\sqrt{2n\pi}}\frac{n^{(n+m+1)/2}n^{(n-m+1)/2}}{(n+m)^{(n+m+1)/2}(n-m)^{(n-m+1)/2}}$$

$$= \frac{2}{\sqrt{2n\pi}}\left(\frac{n}{n+m}\right)^{(n+m+1)/2}\left(\frac{n}{n-m}\right)^{(n-m+1)/2}$$

$$= \frac{2}{\sqrt{2n\pi}}\left(1+\frac{m}{n}\right)^{-(n+m+1)/2}\left(1-\frac{m}{n}\right)^{-(n-m+1)/2}$$

$$= \frac{2}{\sqrt{2n\pi}}\left(1+\frac{m}{n}\right)^{-m/2}\left(1-\frac{m}{n}\right)^{m/2}\left(1-\frac{m^2}{n^2}\right)^{-(n+1)/2}$$

Now we will make use of the fact that $m = x/\Delta x$ and $n = t/\Delta t$. The last expression above is then equal to

$$\frac{2\sqrt{\Delta t}}{\sqrt{2\pi t}}\left(1 + \frac{x\Delta t}{t\Delta x}\right)^{-\frac{x}{2\Delta x}}\left(1 - \frac{x\Delta t}{t\Delta x}\right)^{\frac{x}{2\Delta x}}\left(1 - \left[\frac{x\Delta t}{t\Delta x}\right]^2\right)^{-\frac{1+t/\Delta t}{2}}.$$

Earlier we assumed that n and Δx were related by the equation $n(\Delta x)^2 = 2kt = \sigma^2$, the constant variance in the final position of the particle. Thus we see that n and Δx are inversely related, as one becomes large (typically n), the other must become small *i.e.* Δx. We can also rewrite $n(\Delta x)^2 = 2kt$ as $(\Delta x)^2 = 2k\Delta t$. Using this relationship to replace the Δt's in the expression above yields

$$\frac{\Delta x}{\sqrt{k\pi t}}\left[1 + \frac{x}{2kt}\Delta x\right]^{-\frac{x}{2\Delta x}}\left[1 - \frac{x}{2kt}\Delta x\right]^{\frac{x}{2\Delta x}}\left[1 - \left(\frac{x}{2kt}\Delta x\right)^2\right]^{-\frac{kt}{(\Delta x)^2}-\frac{1}{2}}.$$

So far what we have derived is an approximation (which we will treat as an equality for large n or, what is equivalent, for small Δx) for the probability that the particle is located at position $x = m\Delta x$ at time $t = n\Delta t$. The reader should realize that these are still discrete probabilities. Next we would like to examine the case in which $\Delta x \to 0$. However, if we take the limit of the last expression we will obtain a probability of 0 just as in the elementary example discussed at the beginning of this chapter. In the limit the likelihood of the particle being at precisely any specified location is 0. What we must consider is the probability that the particle is in an interval, but what interval. Notice that for the discrete random random variable X

$$\mathrm{P}\left(X = m\Delta x\right) = \mathrm{P}\left((m-1)\Delta x < X < (m+1)\Delta x\right).$$

According to the Mean Value Theorem [Smith and Minton (2002)], the right-hand side of this equation is $2\Delta x$ times the probability density function of a continuous random variable. To derive a probability distribution function for this continuously distributed random variable we must divide the expression giving us $\mathrm{P}\left(X = m\Delta x\right)$ above by $2\Delta x$ and take the limit as $\Delta x \to 0$. Readers accustomed to numerical approximations for derivatives will recognize that we have used the centered difference formula for the first derivative (see exercise 13) in order to develop a difference quotient whose limit will be the probability distribution. Background information on finite difference approximations to derivatives can be found in [Burden and Faires (2005)]. We will make use of the result that $\lim_{\Delta x\to 0}(1 + a\Delta x)^{1/\Delta x} = e^a$.

Let us define $f(x,t)$ as

$f(x,t)$
$$= \frac{1}{2\sqrt{k\pi t}} \lim_{\Delta x \to 0} \left[1 + \frac{x\,\Delta x}{2kt}\right]^{\frac{-x}{2\Delta x}} \left[1 - \frac{x\,\Delta x}{2kt}\right]^{\frac{x}{2\Delta x}} \left[1 - \left(\frac{x\,\Delta x}{2kt}\right)^2\right]^{-\frac{kt}{(\Delta x)^2} - \frac{1}{2}}$$
$$= \frac{1}{2\sqrt{k\pi t}} \left(e^{\frac{x}{2kt}}\right)^{-\frac{x}{2}} \left(e^{-\frac{x}{2kt}}\right)^{\frac{x}{2}} \left(e^{-\frac{x^2}{4k^2 t^2}}\right)^{-kt}$$
$$= \frac{1}{2\sqrt{k\pi t}} e^{-\frac{x^2}{4kt}} e^{-\frac{x^2}{4kt}} e^{\frac{x^2}{4kt}}$$
$$= \frac{1}{2\sqrt{k\pi t}} e^{-\frac{x^2}{4kt}}.$$

We must verify that this expression satisfies the properties of a probability distribution function. For $k, t > 0$ the expression is non-negative. Since for $|x| \geq 1$ it is true that $0 < e^{-x^2} \leq e^{-|x|}$, then the integral of $f(x,t)$ over the entire real number line converges, suppose we write:

$$0 < \int_{-\infty}^{\infty} \frac{1}{2\sqrt{k\pi t}} e^{-\frac{x^2}{4kt}}\,dx = S < \infty.$$

Then we have

$$S^2 = \int_{-\infty}^{\infty} \frac{1}{2\sqrt{k\pi t}} e^{-\frac{x^2}{4kt}}\,dx \int_{-\infty}^{\infty} \frac{1}{2\sqrt{k\pi t}} e^{-\frac{y^2}{4kt}}\,dy$$
$$= \frac{1}{4k\pi t} \int_{-\infty}^{\infty} \int_{-\infty}^{\infty} e^{-(x^2+y^2)/4kt}\,dx\,dy.$$

Switching to polar coordinates by making the substitutions $x = r\cos\theta$, $y = r\sin\theta$, and $dx\,dy = r\,dr\,d\theta$ produces

$$S^2 = \frac{1}{4k\pi t} \int_0^{2\pi} \int_0^{\infty} re^{-r^2/4kt}\,dr\,d\theta$$
$$= \frac{1}{2kt} \int_0^{\infty} re^{-r^2/4kt}\,dr$$
$$= \int_0^{\infty} e^{-u}\,du$$
$$= \lim_{M \to \infty} \int_0^M e^{-u}\,du$$
$$= \lim_{M \to \infty} (-e^{-M} + 1) = 1.$$

Therefore $S = 1$. Notice in the third step we made use of the substitution $u = r^2/4kt$. The reader may consult a book on complex analysis such as [Marsden and Hoffman (1987)] for a more formal proof of this condition. Hence we see that the function given above satisfies the non-negativity condition and unit area condition of a probability distribution for a continuous random variable. Thus we have derived the probability distribution function for the particle. For a fixed value of t the probability density function has the familiar bell shape as seen in Fig. 3.2. The definition of expected

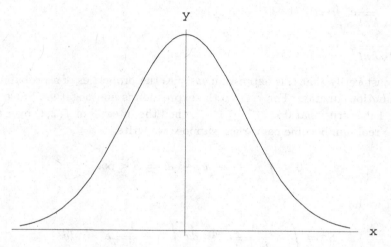

Fig. 3.2   For a fixed value of $t > 0$ the function $f(x,t) = \frac{1}{2\sqrt{k\pi t}}e^{-\frac{x^2}{4kt}}$ will have a graph resembling the often discussed "bell curve."

value given in Eq. (3.3) can be used to determine the average position of the particle.

$$
\begin{aligned}
\mathrm{E}\,[X] &= \int_{-\infty}^{\infty} \frac{x}{2\sqrt{k\pi t}} e^{-\frac{x^2}{4kt}}\, dx \\
&= \lim_{M\to\infty} \int_{0}^{M} \frac{x}{2\sqrt{k\pi t}} e^{-\frac{x^2}{4kt}}\, dx + \lim_{N\to\infty} \int_{-N}^{0} \frac{x}{2\sqrt{k\pi t}} e^{-\frac{x^2}{4kt}}\, dx \\
&= \lim_{M\to\infty} \sqrt{\frac{kt}{\pi}}\left(1 - e^{-M^2/4kt}\right) + \lim_{N\to\infty} \sqrt{\frac{kt}{\pi}}\left(e^{-N^2/4kt} - 1\right) \\
&= 0
\end{aligned}
$$

Thus, as might be expected, the average location of the particle is at the origin independent of t. This is due to the assumption that the particle has no preference for movement to the left or right. In a similar manner we may determine the variance of the position of the particle.

$$
\begin{aligned}
\mathrm{Var}\,(X) &= \int_{-\infty}^{\infty} \frac{x^2}{2\sqrt{k\pi t}} e^{-\frac{x^2}{4kt}}\, dx - (\mathrm{E}\,[X])^2 \\
&= \frac{1}{\sqrt{k\pi t}} \lim_{M\to\infty} \int_0^M x^2 e^{-\frac{x^2}{4kt}}\, dx \\
&= \frac{1}{\sqrt{k\pi t}} \lim_{M\to\infty} (-2kt) \left(M e^{-M^2/4kt} - \int_0^M e^{-x^2/4kt}\, dx \right) \\
&= \frac{2kt}{\sqrt{k\pi t}} \int_0^{\infty} e^{-x^2/4kt}\, dx \\
&= 2kt \cdot \frac{1}{2\sqrt{k\pi t}} \int_{-\infty}^{\infty} e^{-x^2/4kt}\, dx \\
&= 2kt
\end{aligned}
$$

We were able to avoid evaluating the last improper integral by making use of the unit area property of the probability distribution. From the fact that $\sigma^2 = 2kt$ we know that the "spread" in the location of the particle increases with time. This can be readily seen in the surface plot shown in Fig. 3.3.

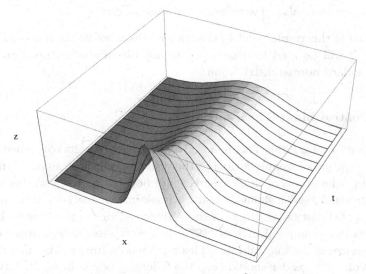

Fig. 3.3 For $t > 0$ the function $f(x,t) = \frac{1}{2\sqrt{k\pi t}} e^{-\frac{x^2}{4kt}}$ retains the bell-shaped cross section, but the profile flattens and broadens as t increases.

If the particle were initially placed at location μ but all other assumptions remained the same then the probability distribution would simply be shifted in the x direction by μ. Thus we may define the **normal probability distribution** with mean μ and variance σ^2 to be the function

$$f(x) = \frac{1}{\sigma\sqrt{2\pi}}e^{-\frac{(x-\mu)^2}{2\sigma^2}}. \tag{3.8}$$

When $\mu = 0$ and $\sigma = 1$, this is referred to as the **standard normal probability distribution**.

In the sequel the **cumulative distribution function** $\phi(x)$ where

$$\phi(x) = \mathrm{P}\,(X < x) = \int_{-\infty}^{x} \frac{1}{\sqrt{2\pi}}e^{-\frac{t^2}{2}}\,dt \tag{3.9}$$

will be frequently used in discussions involving probabilities of normal random variables. For example if X is a standard normal random variable then $\mathrm{P}\,(X < 0) = \phi(0) = 1/2$. It is frequently helpful when dealing with normally distributed random variables to perform a mathematical change of variable which produces a standard normal random variable.

Theorem 3.6 *If X is a normally distributed random variable with expected value μ and variance σ^2, then $Z = (X - \mu)/\sigma$ is normally distributed with an expected value of zero and a variance of one.*

The proof of this result is left to the reader in exercise 20. In the sequel the symbol Z will be used to denote a normally distributed random variable with standard normal distribution.

3.5 Central Limit Theorem

The name Central Limit Theorem is given to several results concerned with the distribution of sample means or the distribution of the sum of random variables. The proofs of these theorems are beyond the scope of this work. The interested reader should consult a book on the mathematical underpinnings of statistics for proofs of the various versions of the Central Limit Theorem (for example [DeGroot (1975)]). Nevertheless we can observe the consequences of the Central Limit Theorem on data from a random number simulation. The reader should keep the following points in mind. Given a random variable X which can be either discrete or continuous and which may have any probability distribution, we may collect a sample of size n

and denote the mean of that sample $\overline{X}_n$. If the process of collecting multiple samples and calculating their means is repeated then we can treat the sample means as random variables in their own right. One version of the Central Limit Theorem due to Lindeberg and Lévy implies that the sample means become normally distributed as the sample size becomes large.

Theorem 3.7 *If random variables $X_1, X_2, \ldots, X_n$ form a random sample of size n from a probability distribution with mean μ and standard deviation σ then for all x*

$$\lim_{n \to \infty} P\left(\frac{\sqrt{n}(\overline{X}_n - \mu)}{\sigma} \leq x\right) = \phi(x),$$

where $\overline{X}_n$ is the mean of a random sample of size n.

Example 3.5 The reader can replicate this example on most programmable computing devices. Collect 5000 samples of size n (where $n \in \{2, 5, 10, 20, 40\}$) of a uniformly distributed continuous random variable on the interval $[0, 1]$. Compute the means of each sample of size n and plot the frequency histogram of the means. As can be seen in Fig. 3.4 as n increases the histograms take on the appearance of a normal probability distribution function.

Another version of the Central Limit Theorem is due to Liapounov and concerns the asymptotic distribution of a sum of random variables. Suppose the random variables $X_1, X_2, \ldots, X_n$ are pairwise independent but not necessarily identically distributed. We will assume that for each $i \in \{1, 2, \ldots, n\}$, $E[X_i] = \mu_i$ and that $\text{Var}(X_i) = \sigma_i^2$. Now define a new random variable Y_n as

$$Y_n = \frac{\sum_{i=1}^n (X_i - \mu_i)}{\sqrt{\sum_{i=1}^n \sigma_i^2}}.$$

Using the assumption that the random variables are pairwise independent and Theorems 3.1, 3.4, and 3.5 we can determine that $E[Y_n] = 0$ and $\text{Var}(Y_n) = 1$. The following theorem establishes that as n becomes large Y_n is approximately normally distributed.

Theorem 3.8 *Suppose that the infinite collection $\{X_i\}_{i=1}^\infty$ of random variables are pairwise independent and that for each $i \in \mathbb{N}$ we have $E\left[|X_i - \mu_i|^3\right] < \infty$. If in addition,*

$$\lim_{n \to \infty} \frac{\sum_{i=1}^n E\left[|X_i - \mu_i|^3\right]}{\left(\sum_{i=1}^n \sigma_i^2\right)^{3/2}} = 0$$

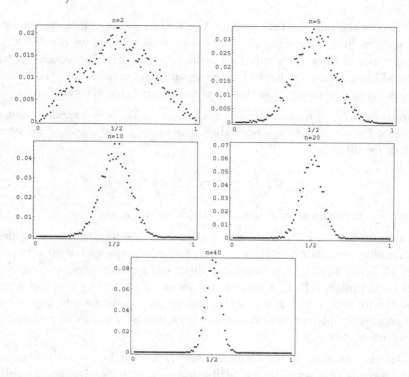

Fig. 3.4 An illustration of the Central Limit Theorem due to Lindeberg and Lévy. As the sample size increases the distribution of the sample means becomes more normal in appearance.

then for any $x \in \mathbb{R}$

$$\lim_{n \to \infty} P\left(Y_n \leq x\right) = \phi(x)$$

where random variable Y_n is defined as above.

Example 3.6 Once again, in place of a proof of Theorem 3.8 we will present a numerical simulation illustrating this result. The reader is encouraged to pursue their own similar numerical exploration. The data used in the following plots were generated by defining n continuous uniform distributions on $[\alpha_i, \beta_i]$ for $i = 1, 2, \ldots, n$ where each of α_i and β_i were randomly, uniformly distributed in $[-100, 100]$ and independently selected. The mean and variance of each uniform distribution were calculated. Then a random variable X_i was randomly chosen from each uniform distribution on $[\alpha_i, \beta_i]$ and Y_n was computed as above. A sample for Y_n of size 5000 was collected and frequency histograms were created. These are shown in Fig. 3.5. It is readily seen that as n increases the distribution of Y_n becomes more normal in appearance.

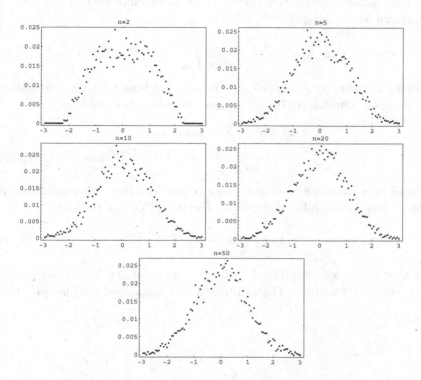

Fig. 3.5 An illustration of the Central Limit Theorem due to Liapounov. As the sample size increases the distribution of the sum of the random variables becomes more normal in appearance.

3.6 Lognormal Random Variables

While normal random variables are central to our discussion of probability and the upcoming Black-Scholes option pricing formula, of equal importance will be continuous random variables which are distributed in a lognormal fashion. A random variable X is a **lognormal random variable** with parameters μ and σ if $\ln X$ is a normally distributed random variable with mean μ and variance σ^2. When referring to a lognormal random variable, the parameters μ and σ are often called the **drift** and **volatility** respectively. A lognormal random variable is a continuous random variable which takes on values in the interval $(0, \infty)$. From Eq. (3.8) we see that for the lognormal random variable X, the probability that $\ln X < \ln x$ is expressed as the integral

$$P\left(\ln X < \ln x\right) = \frac{1}{\sigma\sqrt{2\pi}} \int_{-\infty}^{\ln x} e^{-(t-\mu)^2/2\sigma^2}\, dt.$$

Making the change of variable $t = \ln u$ we see then that the cumulative distribution function (CDF) for a lognormal random variable is

$$P\left(X < x\right) = P\left(\ln X < \ln x\right)$$
$$= \frac{1}{\sigma\sqrt{2\pi}} \int_{0}^{x} \frac{1}{u} e^{-(\ln u - \mu)^2/2\sigma^2}\, du. \qquad (3.10)$$

Therefore a lognormally distributed random variable with parameters μ and σ^2 has a probability distribution function (PDF) of the form

$$f(x) = \frac{1}{(\sigma\sqrt{2\pi})x} e^{-(\ln x - \mu)^2/2\sigma^2} \qquad (3.11)$$

for $0 < x < \infty$. See Fig. 3.6 for the graph of the lognormal probability density function. The introduction of lognormal random variables

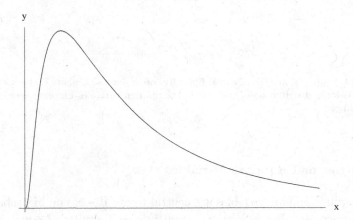

Fig. 3.6   The graph of the lognormal probability distribution function.

does not complicate matters much since the probability that a lognormally distributed random variable X is less than some $x > 0$ is then

$P(X < x) = \phi(\ln x)$ where ϕ is the cumulative probability distribution function for a normal random variable with mean zero and variance one. Similarly the $P(X > x) = 1 - \phi(\ln x)$ for a lognormally distributed random variable.

Using the definitions of expected value and variance we can prove the following lemma.

Lemma 3.1 *If X is a lognormal random variable with parameters μ and σ then*

$$E[X] = e^{\mu + \sigma^2/2} \tag{3.12}$$

$$Var(X) = e^{2\mu + \sigma^2}\left(e^{\sigma^2} - 1\right) \tag{3.13}$$

Proof. According to the definition of the expected value of a continuous random variable and using the probability density function found in Eq. (3.11)

$$E[X] = \frac{1}{\sigma\sqrt{2\pi}} \int_0^\infty x\left(\frac{1}{x}e^{-(\ln x - \mu)^2/2\sigma^2}\right) dx$$

$$= \frac{1}{\sigma\sqrt{2\pi}} \int_{-\infty}^\infty e^t e^{-(t-\mu)^2/2\sigma^2} dt$$

$$= e^{\mu + \sigma^2/2} \frac{1}{\sigma\sqrt{2\pi}} \int_{-\infty}^\infty e^{-(t-(\mu+\sigma^2))^2/2\sigma^2} dt$$

$$= e^{\mu + \sigma^2/2}$$

In the second line of the equation we have made the substitution $t = \ln x$. The last equality is true since the integral represents the area under the probability distribution curve for a normal random variable with mean $\mu + \sigma^2$ and variance σ^2.

Likewise by Eq. (3.5)

$$Var(X) = E[X^2] - (E[X])^2$$

$$= \frac{1}{\sigma\sqrt{2\pi}} \int_0^\infty x^2\left(\frac{1}{x}e^{-(\ln x - \mu)^2/2\sigma^2}\right) dx - \left(e^{\mu+\sigma^2/2}\right)^2$$

$$= \frac{1}{\sigma\sqrt{2\pi}} \int_{-\infty}^\infty e^{2t} e^{-(t-\mu)^2/2} dt - e^{2\mu+\sigma^2}$$

$$= e^{2(\mu+\sigma^2)} \frac{1}{\sigma\sqrt{2\pi}} \int_{-\infty}^\infty e^{-(t-(\mu+2\sigma))^2/2} dt - e^{2\mu+\sigma^2}$$

$$= e^{2\mu+\sigma^2}\left(e^{\sigma^2} - 1\right)$$

Between the second and third lines of the equation we used the substitution $t = \ln x$. $\square$

Now we will apply the concept of the lognormally distributed random variable to the situation of the price of a stock or other security. Suppose that the selling price of a security will be measured daily and that the starting measurement will be denoted $S(0)$. For $n \geq 1$, we will let $S(n)$ denote the selling price on day n. To a good approximation the ratios of consecutive days' selling prices are lognormal random variables, *i.e.* the expressions $X(n) = S(n)/S(n-1)$ for $n \geq 1$ are lognormally distributed. It is also generally assumed that the ratios $X(n)$ and $X(m)$ are identically distributed and independent when $n \neq m$. If a sufficient number of measurements have been made that a financial analyst estimates the parameters of the random variable X are $\mu = 0.0155$ and $\sigma = 0.0750$, then questions regarding the likelihood of future prices of the security can be answered.

Example 3.7   First, what is the probability that the selling price of the stock on the next day will be higher than the price on the present day?

$$
\begin{aligned}
P\left(X(n) > 1\right) &= P\left(\ln X(n) > 0\right) \\
&= P\left(z > \frac{0 - \mu}{\sigma}\right) \\
&= P\left(z > -0.206667\right) \\
&= 1 - \phi(-0.206667) \\
&\approx 0.582
\end{aligned}
$$

Thus for a security with the parameters measured and reported above, there is a better than one half probability of the selling price increasing on any given day.

Second, we may ask what is the probability that the selling price two days hence will be higher than the present selling price? At first glance the reader may be tempted to use the fact that each observation of the random variable is independent and hence declare that the sought after probability is $0.582^2 \approx 0.316$. This implies the mistaken assumption that the price of the security increased on each of the two days. This ignores the possibility that the price of the security could decrease on either (but not both) of the two days and still make a net gain over the two-day period. To correctly approach the problem we must make use of the Central Limit Theorem.

$$\begin{aligned}
P\left(S(n+2)/S(n) > 1\right) &= P\left(\frac{S(n+2)}{S(n+1)} \cdot \frac{S(n+1)}{S(n)} > 1\right) \\
&= P\left(X(n+1)X(n) > 1\right) \\
&= P\left(\ln X(n+1) + \ln X(n) > 0\right) \\
&= P\left(z > \frac{0 - 2(0.0155)}{0.0750\sqrt{2}}\right) \\
&= P\left(z > -0.292271\right) \\
&= 1 - \phi(-0.292271) \\
&\approx 0.615
\end{aligned}$$

Thus the probability of a gain over a two-day period is nearly twice as large as may have been first suspected.

The assumption that ratios of selling prices sampled at regular intervals for securities are lognormally distributed will be explored further in Chapter 5. Appendix A contains a sample of closing prices for Sony Corporation stock. Readers can collect their own data set from many corporations' websites or from finance sites such as `finance.yahoo.com`. The eager reader may want to collect data for a particular stock to further test the lognormal hypothesis.

3.7 Properties of Expected Value

In later chapters the reader will encounter frequent references to the positive part of the difference of two quantities. For instance if an investor may purchase a security worth X for price K, then the excess profit (if any) of the transaction is the positive part of the difference $X - K$. This is typically denoted $(X - K)^+ = \max\{0, X - K\}$. In this section a theorem and several corollaries are presented which will enable the reader to calculate the expected value of $(X - K)^+$ when X is a continuous random variable and K is a constant.

Theorem 3.9 *Let X be a continuous random variable with probability distribution function $f(x)$ and finite variance. If K is a constant then*

$$E\left[(X - K)^+\right] = \int_K^\infty \left(\int_x^\infty f(t)\,dt\right) dx. \qquad (3.14)$$

Proof. Starting with the definition of the expected value of a function of the continuous random variable X with probability density function $f(x)$, we have

$$\mathrm{E}\left[(X-K)^+\right] = \int_{-\infty}^{\infty} (x-K)^+ f(x)\,dx$$

$$= \int_{K}^{\infty} (x-K)f(x)\,dx$$

$$= \lim_{M\to\infty} \int_{K}^{M} (x-K)f(x)\,dx.$$

The last definite integral can be evaluated using integration by parts with

$$u = x - K \qquad v = \int_{-\infty}^{x} f(t)\,dt$$
$$du = dx \qquad dv = f(x)\,dx$$

This yields

$$\mathrm{E}\left[(X-K)^+\right] = \lim_{M\to\infty} \left(\left[\int_{-\infty}^{x} f(t)\,dt\right](x-K)\Big|_{K}^{M} - \int_{K}^{M} \int_{-\infty}^{x} f(t)\,dt\,dx \right)$$

$$= \lim_{M\to\infty} \left(\left[\int_{-\infty}^{M} f(t)\,dt\right](M-K) - \int_{K}^{M}\left[1 - \int_{x}^{\infty} f(t)\,dt\right]dx \right)$$

$$= \lim_{M\to\infty} \left(\left[1 - \int_{M}^{\infty} f(t)\,dt\right](M-K) - \int_{K}^{M}\left[1 - \int_{x}^{\infty} f(t)\,dt\right]dx \right)$$

$$= \lim_{M\to\infty} \left(\int_{K}^{M}\int_{x}^{\infty} f(t)\,dt\,dx - (M-K)\int_{M}^{\infty} f(t)\,dt \right)$$

$$= \int_{K}^{\infty}\left(\int_{x}^{\infty} f(t)\,dt\right)dx.$$

The reader is asked in exercise 14 to provide the details used in evaluating the final limit. □

Corollary 3.1 is a special case of Theorem 3.9.

Corollary 3.1 *If X is normal random variable with mean μ and variance σ^2 and K is a constant, then*

$$\mathrm{E}\left[(X-K)^+\right] = \frac{\sigma}{\sqrt{2\pi}} e^{-(\mu-K)^2/2\sigma^2} + (\mu-K)\phi\left(\frac{\mu-K}{\sigma}\right). \qquad (3.15)$$

Proof. The expected value of $(X - K)^+$ is

$$
\begin{aligned}
\mathrm{E}\left[(X - K)^+\right] &= \frac{1}{\sqrt{2\pi}\sigma} \int_{-\infty}^{\infty} (x - K)^+ e^{-(x-\mu)^2/2\sigma^2}\, dx \\
&= \frac{1}{\sqrt{2\pi}\sigma} \int_{K}^{\infty} (x - K) e^{-(x-\mu)^2/2\sigma^2}\, dx
\end{aligned}
$$

If we make the substitution $t = (x - \mu)/\sigma$ then we have

$$
\begin{aligned}
\mathrm{E}\left[(X - K)^+\right] &= \frac{1}{\sqrt{2\pi}} \int_{(K-\mu)/\sigma}^{\infty} (\mu - K + t\sigma) e^{-t^2/2}\, dt \\
&= \frac{\mu - K}{\sqrt{2\pi}} \int_{(K-\mu)/\sigma}^{\infty} e^{-t^2/2}\, dt + \frac{\sigma}{\sqrt{2\pi}} \int_{(K-\mu)/\sigma}^{\infty} t e^{-t^2/2}\, dt \\
&= (\mu - K)\phi\left(\frac{\mu - K}{\sigma}\right) + \frac{\sigma}{\sqrt{2\pi}} e^{(K-\mu)^2/2\sigma^2}.
\end{aligned}
$$

The first integral on the right-hand side follows from the definition of the cumulative distribution function for the standard normal random variable. The reader is asked to evaluate the second improper integral in exercise 15. $\square$

For the purpose of completing the mean-variance analysis encountered later in Chapter 11 we must also be able to evaluate the expected value of the positive part of $(X - K)$ when X is a lognormally distributed continuous random variable.

Corollary 3.2 *If X is a lognormally distributed random variable with parameters μ and σ^2 and $K > 0$ is a constant then*

$$
\mathrm{E}\left[(X - K)^+\right] = e^{\mu + \sigma^2/2}\phi\left(\frac{\mu - \ln K}{\sigma} + \sigma\right) - K\phi\left(\frac{\mu - \ln K}{\sigma}\right) \qquad (3.16)
$$

Proof. By definition

$$
\begin{aligned}
\mathrm{E}\left[(X - K)^+\right] &= \frac{1}{\sqrt{2\pi}\sigma} \int_{0}^{\infty} (x - K)^+ \left(\frac{1}{x} e^{-(\ln x - \mu)^2/2\sigma^2}\right) dx \\
&= \frac{1}{\sqrt{2\pi}\sigma} \int_{K}^{\infty} (x - K) \left(\frac{1}{x} e^{-(\ln x - \mu)^2/2\sigma^2}\right) dx.
\end{aligned}
$$

Making the substitution $t = (\ln x - \mu)/\sigma$ allows us to write

$$
\begin{aligned}
\mathrm{E}\left[(X - K)^+\right] &= \frac{1}{\sqrt{2\pi}} \int_{(\ln K - \mu)/\sigma}^{\infty} (e^{\sigma t + \mu} - K)e^{-t^2/2}\, dt \\
&= \frac{1}{\sqrt{2\pi}} \left(e^\mu \int_{(\ln K - \mu)/\sigma}^{\infty} e^{\sigma t} e^{-t^2/2}\, dt - K \int_{(\ln K - \mu)/\sigma}^{\infty} e^{-t^2/2}\, dt \right) \\
&= \frac{1}{\sqrt{2\pi}} e^{\mu + \sigma^2/2} \int_{(\ln K - \mu)/\sigma}^{\infty} e^{-(t-\sigma)^2/2}\, dt - K\phi\left(\frac{\mu - \ln K}{\sigma} \right)
\end{aligned}
$$

The remaining integral can be evaluated using another substitution, $z = t - \sigma$. Therefore we have

$$
\begin{aligned}
\mathrm{E}\left[(X - K)^+\right] &= e^{\mu + \sigma^2/2} \frac{1}{\sqrt{2\pi}} \int_{\frac{\ln K - \mu}{\sigma} - \sigma}^{\infty} e^{-z^2/2}\, dz - K\phi\left(\frac{\mu - \ln K}{\sigma} \right) \\
&= e^{\mu + \sigma^2/2} \phi\left(\frac{\mu - \ln K}{\sigma} + \sigma \right) - K\phi\left(\frac{\mu - \ln K}{\sigma} \right).
\end{aligned}
$$
$\square$

3.8   Properties of Variance

In this section we will derive properties of the variance for the positive part of the difference of a continuous random variable and a constant. These results will be of use in later sections of this text.

Lemma 3.2   *Let X be a normally distributed random variable with mean μ and variance σ^2. If K is a constant then*

$$
\begin{aligned}
&\mathrm{E}\left[((X - K)^+)^2\right] \\
&= \left((\mu - 2K)^2 + \sigma^2\right) \phi\left(\frac{\mu - 2K}{\sigma} \right) + \frac{(\mu - 2K)\sigma}{\sqrt{2\pi}} e^{-\frac{(\mu - 2K)^2}{2\sigma^2}}. \quad (3.17)
\end{aligned}
$$

Proof. Since X is normally distributed with mean μ and variance σ^2, then $X - K$ is normally distributed with mean $\mu - K$ and variance σ^2.

$$
\begin{aligned}
&\mathrm{E}\left[((X - K)^+)^2\right] \\
&= \frac{1}{\sqrt{2\pi}\sigma} \int_{-\infty}^{\infty} ((x - K)^+)^2 e^{-(x - (\mu - K))^2/2\sigma^2}\, dx \\
&= \frac{1}{\sqrt{2\pi}\sigma} \int_{K}^{\infty} (x - K)^2 e^{-(x - (\mu - K))^2/2\sigma^2}\, dx \\
&= \frac{1}{\sqrt{2\pi}} \int_{(2K - \mu)/\sigma}^{\infty} (\sigma t + \mu - 2K)^2 e^{-t^2/2}\, dt
\end{aligned}
$$

upon making the substitution $t = (x - (\mu - K))/\sigma$. Expanding the square in the last integrand and integrating produces

$$
\begin{aligned}
\mathrm{E}&\left[((X - K)^+)^2\right] \\
&= (\mu - 2K)^2 \phi\left(\frac{\mu - 2K}{\sigma}\right) + \frac{2(\mu - 2K)\sigma}{\sqrt{2\pi}} \int_{(2K-\mu)/\sigma}^{\infty} t e^{-t^2/2} \, dt \\
&\quad + \frac{\sigma^2}{\sqrt{2\pi}} \int_{(2K-\mu)/\sigma}^{\infty} t^2 e^{-t^2/2} \, dt \\
&= (\mu - 2K)^2 \phi\left(\frac{\mu - 2K}{\sigma}\right) + \frac{2(\mu - 2K)\sigma}{\sqrt{2\pi}} e^{-(\mu-2K)^2/2\sigma^2} \\
&\quad - \frac{(\mu - 2K)\sigma}{\sqrt{2\pi}} e^{-(\mu-2K)^2/2\sigma^2} + \sigma^2 \phi\left(\frac{\mu - 2K}{\sigma}\right) \\
&= \left((\mu - 2K)^2 + \sigma^2\right) \phi\left(\frac{\mu - 2K}{\sigma}\right) + \frac{(\mu - 2K)\sigma}{\sqrt{2\pi}} e^{-(\mu-2K)^2/2\sigma^2}.
\end{aligned}
$$

$\square$

Now the following result is immediately established.

Corollary 3.3 *Let X be a normally distributed random variable with mean μ and variance σ^2. If K is a constant then*

$$
\begin{aligned}
\mathrm{Var}&\left((X - K)^+\right) \\
&= \left((\mu - 2K)^2 + \sigma^2\right) \phi\left(\frac{\mu - 2K}{\sigma}\right) + \frac{(\mu - 2K)\sigma}{\sqrt{2\pi}} e^{-(\mu-2K)^2/2\sigma^2} \\
&\quad - \left(\frac{\sigma}{\sqrt{2\pi}} e^{-(\mu-K)^2/2\sigma^2} + (\mu - K)\phi\left(\frac{\mu - K}{\sigma}\right)\right)^2.
\end{aligned} \tag{3.18}
$$

Now we must establish a similar result for lognormally distributed random variables.

Lemma 3.3 *Let X be a lognormally distributed random variable with parameters μ and σ^2. If $K > 0$ is a constant then*

$$
\begin{aligned}
\mathrm{E}&\left[((X - K)^+)^2\right] \\
&= e^{2(\mu+\sigma^2)}\phi(w + 2\sigma) - 2Ke^{\mu+\sigma^2/2}\phi(w + \sigma) + K^2\phi(w)
\end{aligned} \tag{3.19}
$$

where $w = (\mu - \ln K)/\sigma$.

Proof.

$$E\left[((X-K)^+)^2\right] = \frac{1}{\sqrt{2\pi}\sigma} \int_0^\infty ((x-K)^+)^2 \frac{1}{x} e^{-(\ln x-\mu)^2/2\sigma^2} \, dx$$

$$= \frac{1}{\sqrt{2\pi}\sigma} \int_K^\infty (x-K)^2 \frac{1}{x} e^{-(\ln x-\mu)^2/2\sigma^2} \, dx$$

The expression $\sigma z = \ln x - \mu$ will be substituted into the last integral to yield:

$$E\left[((X-K)^+)^2\right] = \frac{1}{\sqrt{2\pi}} \int_{(\ln K-\mu)/\sigma}^\infty (e^{\sigma z+\mu} - K)^2 e^{-z^2/2} \, dz$$

$$= \frac{e^{2(\mu+\sigma^2)}}{\sqrt{2\pi}} \int_{(\ln K-\mu)/\sigma}^\infty e^{-(z-2\sigma)^2/2} \, dz$$

$$- \frac{2K e^{\mu+\sigma^2/2}}{\sqrt{2\pi}} \int_{(\ln K-\mu)/\sigma}^\infty e^{-(z-\sigma)^2/2} \, dz$$

$$+ \frac{K^2}{\sqrt{2\pi}} \int_{(\ln K-\mu)/\sigma}^\infty e^{-z^2/2} \, dz$$

$$= e^{2(\mu+\sigma^2)} \phi\left(\frac{\mu - \ln K}{\sigma} + 2\sigma\right) + K^2 \phi\left(\frac{\mu - \ln K}{\sigma}\right)$$

$$- 2K e^{\mu+\sigma^2/2} \phi\left(\frac{\mu - \ln K}{\sigma} + \sigma\right).$$

$\square$

At last we have an expression for the variance of the positive part of the difference of a lognormal random variable and a constant.

Corollary 3.4 *Let X be a lognormally distributed random variable with parameters μ and σ^2. If $K > 0$ is a constant then*

$$\text{Var}\left((X-K)^+\right) = e^{2(\mu+\sigma^2)}\phi(w+2\sigma) - 2K e^{\mu+\sigma^2/2}\phi(w+\sigma) + K^2\phi(w)$$

$$- \left(e^{\mu+\sigma^2/2}\phi(w+\sigma) - K\phi(w)\right)^2 \qquad (3.20)$$

where $w = (\mu - \ln K)/\sigma$.

3.9 Exercises

(1) Random variable X is continuously uniformly distributed in the interval $[-4, 1]$. Find $P(X \geq 0)$.

(2) Random variable X is continuously distributed on the interval $(1, \infty)$, with probability distribution function

$$f(x) = \begin{cases} \frac{c}{x^3} & \text{if } 1 \leq x \\ 0 & \text{otherwise.} \end{cases}$$

Determine the value of c.

(3) Show using properties of the definite integral that for a continuous random variable X with probability distribution function $f(x)$,

$$P(X \geq a) = 1 - P(X < a).$$

(4) A random variable X has a continuous Cauchy distribution with probability density function

$$f(x) = \frac{1}{\pi(1 + x^2)}.$$

Show that the mean of this random variable does not exist.

(5) If X is a continuous random variable with probability density function $f(x)$, show that $\mathrm{E}[aX + b] = a\mathrm{E}[X] + b$ where $a, b \in \mathbb{R}$.

(6) Prove Theorem 3.1.

(7) Prove Theorem 3.2.

(8) Prove Theorem 3.3.

(9) Prove Theorem 3.4.

(10) Prove Theorem 3.5.

(11) Find the expected value and variance of the continuous random variable X whose probability distribution function is given by

$$f(x) = \begin{cases} \frac{2}{5}|x| & \text{if } -1 \leq x \leq 2, \\ 0 & \text{otherwise.} \end{cases}$$

(12) Show that if m and n are integers then $n - m$ is even if and only if $n + m$ is even.

(13) Suppose $f(x)$ is three times continuously differentiable at $x = x_0$. Use Taylor's Theorem [Stewart (1999)] to expand $f(x)$ about $x = x_0$. Then by using $f(x_0 + h)$ and $f(x_0 - h)$ show that

$$f'(x_0) \approx \frac{f(x_0 + h) - f(x_0 - h)}{2h}.$$

(14) Show that

$$\lim_{M \to \infty} (M - K) \int_M^\infty f(t)\, dt = 0$$

where f is a probability density function for a continuous random variable with finite variance.

(15) Show that

$$\frac{\sigma}{\sqrt{2\pi}} \int_{(K-\mu)/\sigma}^{\infty} te^{-t^2/2} \, dt.$$

(16) Evaluate the following indefinite integral.

$$\int \frac{x}{2\sqrt{k\pi t}} e^{-\frac{x^2}{4kt}} \, dx$$

(17) Use the technique of integration by parts to evaluate

$$\int \frac{x^2}{2\sqrt{k\pi t}} e^{-\frac{x^2}{4kt}} \, dx.$$

(18) Evaluate the following limit with $k > 0$ and $t > 0$.

$$\lim_{M \to \infty} 2ktMe^{-M^2/4kt}$$

(19) Using a computer algebra system, graphing calculator, or some numerical method evaluate the following probabilities for a standard normal random variable:

   (a) $P(-1 < X < 1)$
   (b) $P(-2 < X < 2)$
   (c) $P(-3 < X < 3)$
   (d) $P(1 < X < 3)$

(20) Prove Theorem 3.6.

(21) The annual rainfall amount in a geographical area is normally distributed with a mean of 14 inches and a standard deviation of 3.2 inches. What is the probability that the sum of the annual rainfalls in two consecutive years will exceed 30 inches?

(22) The ratio of selling prices of a security on consecutive days is a lognormally distributed random variable with parameters $\mu = 0.01$ and $\sigma = 0.05$. What is the probability of a one-day increase in the selling price? What is the probability of a one-day decrease in the selling price? What is the probability of a four-day decrease in the selling price of the security?

(23) Let X be a uniformly distributed continuous random variable on the interval $[a, b]$. If K is a constant, find an expression for $E[(X - K)^+]$.

(24) Show that for the standard normal random variable $1 - \phi(x) = \phi(-x)$.

(25) Fill in the details of the proof of Corollary 3.3.
(26) Fill in the details of the proof of Corollary 3.4.

Chapter 4

The Arbitrage Theorem

The concept known as **arbitrage** is subtle and can seem counter-intuitive. In its basic form arbitrage exists whenever two financial instruments are mis-priced relative to one another. Due to the mis-pricing, it becomes possible to make a financial gain. For example suppose bank A issues loans at a 5% interest rate and bank B offers a savings account which pays 6% interest. A person could take out a loan from bank A and place the loan into savings with bank B. When it becomes time to repay bank A for the loan, the person closes the savings account with bank B, repays the principal and interest and still has 1% of the loaned amount as profit. The assumption that financial markets are efficient prevents such obvious arbitrage opportunities from being commonplace. When arbitrage opportunities arise, investors wanting to make a profit flock to the mis-priced instruments and the financial market reacts by correcting the pricing of the instruments.

We will see that financial instruments such as options, bonds, and stocks must be priced so as to be "arbitrage free." It is this absence of arbitrage which forms the basis of the derivation of the Black-Scholes equation found in Chapter 7.

4.1 The Concept of Arbitrage

As mentioned previously one way to think about arbitrage is as the situation which arises when two financial instruments are mis-priced relative to one another. In this section we will refine and make precise our definition. Suppose there is a set of possible outcomes for some experiment and that wagers can be placed on those outcomes. The **Arbitrage Theorem** states that either the probabilities of the outcomes are such that all bets are fair, or there is a betting scheme which produces a positive gain independent of the outcome of the experiment. Since we are adopting the language of wagering, a simple example will illustrate the Arbitrage Theorem.

To simplify the process of placing and paying off bets, **odds** are used rather than probabilities, though the two concepts are related by a simple formula. Suppose the "odds" of a particular sporting events outcome are quoted as "2 : 1 against." We can think of this as implying that there are three outcomes to the experiment and in two of them the desired outcome does not occur and in one it does. Thus the probability of the desired outcome arising is 1/3. In general then if the "odds against" a particular outcome are $n : 1$ then the probability of the outcome is $1/(n+1)$. Odds simplify the paying off of bets in the following way. If the odds against an outcome are $n : 1$ then a unit bet will pay us n units if the outcome occurs. If the outcome does not occur, the unit bet is lost. The payoffs scale multiplicatively for non-unit bets. Odds against the occurrence of an event can also be expressed in the form $n : m$ where n and m are natural numbers. In this case the probability of the event occurring is $m/(n+m)$. Odds of the form $n : m$ are equivalent to the odds $\frac{n}{m} : 1$.

Now consider this example. Suppose the odds against player A defeating player B in a tennis match are 3 : 1 and the odds against player B defeating player A are 1 : 1. Converting these to probabilities we see that player A defeats player B with probability 1/4 while player B defeats player A with probability 1/2. There is obviously something wrong with these probabilities since they should add to one, but do not. We will see that the Arbitrage Theorem implies there is a betting strategy which generates a positive gain regardless of the outcome of the tennis match. Now suppose we wager 1 on player A and 2 on player B. If player A wins we will win 3 on the first bet and lose 2 on the second, producing a net gain of 1. If player B wins we will lose 1 on the first bet and win 2 on the second, again yielding a positive payoff of 1. Notice, no matter which player wins the tennis match, we have a positive gain. If this scenario were real we would borrow as much money as possible and place these wagers. Once we collect our winnings we could pay back the loan and retire. This example is hardly transparent, so study of the Arbitrage Theorem would be beneficial in avoiding arbitrage opportunities in more complex financial situations. Before a statement and proof of the Arbitrage Theorem is presented a preliminary result from linear programming is needed.

4.2   Duality Theorem of Linear Programming

In this section we will derive and discuss a result from which the Arbitrage Theorem easily follows. The **Duality Theorem** is a familiar result to people having studied linear programming and operations research. Linear programming is the name given to a branch of mathematics often applied in business and economics in which a linear function of some (usually large) number of variables must be optimized (either maximized or minimized) subject to a set of linear equations or inequalities.

To sample the types of linear programming problems solved by businesses consider this simple example. A bank may invest its deposits in loans which earn 6% interest per year and in the purchase of stocks which increase in value by 13% per year. The bank wishes to maximize the total return on its investments. Assume the bank can invest a proportion x in loans and proportion y in stocks. Any remaining proportion is simply held by the bank. The total return is therefore $0.06x + 0.13y$. Suppose that government regulations require that the bank invest no more than 60% its deposits in stocks. As a good business practice the bank wishes to devote at least 25% of its deposits to loans. These constraints impose some inequalities on the bank's investment strategy. The inequalities are $x \geq 0$, $y \geq 0$, $x + y \leq 1$ (non-negative proportions on the deposits are invested and the total amount invested is no greater than the total amount on deposit), $y \leq 0.6$ (government regulation), and $x \geq 0.25$ (the bank's business practice). An investment strategy can be represented by a vector $\langle x, y \rangle$. The bank's linear programming problem is in picking the vector which satisfies the constraints and maximizes the return. Since this problem is two dimensional a plot can reveal the solution. As seen in Fig. 4.1 the optimal solution occurs when $x = 0.4$ and $y = 0.6$.

In the remainder of this section we give a more formal, but still brief, introduction to linear programming and the Duality Theorem here following the style of [Strang (1986)]. Other accessible introductions to linear programming and the Duality Theorem can be found in [Franklin (1980)] and [Noble and Daniel (1988)].

A generic linear programming problem, or **linear program**, consists of a set of linear equations or inequalities (constraints), sign conditions on the solution (more constraints), and a linear expression which must be optimized (either maximized or minimized). We will use bold letters to represent vectors and the dot product (Euclidean inner product) to express linear expressions. Vectors will be thought of as matrices with a single

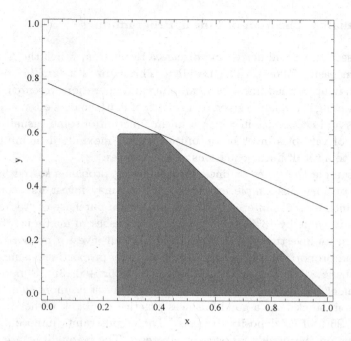

Fig. 4.1   The shaded region represents the set of possible solutions to the bank's investment decision. The total return is maximized at $\langle 0.4, 0.6 \rangle$.

column. Thus if $\mathbf{c}$ and $\mathbf{x}$ are vectors with n components each, the notation

$$\mathbf{c}^T\mathbf{x} = c_1 x_1 + c_2 x_2 + \cdots + c_n x_n$$

represents a linear expression. The notation $\mathbf{c}^T$ stands for the usual matrix transpose operation. Optimization problems may include **equality
constraints** or **inequality constraints**. We will choose to state every
constraint of every linear program in the form $\mathbf{a}^T\mathbf{x} \leq b$. Linear inequalities
of the form $\mathbf{a}^T\mathbf{x} \geq b$ are equivalent to inequalities of the form $(-\mathbf{a})^T\mathbf{x} \leq -b$.
Linear equation constraints are unnecessary since

$$\mathbf{a}^T\mathbf{x} = b \quad \Longleftrightarrow \quad \mathbf{a}^T\mathbf{x} \leq b \quad \text{and} \quad (-\mathbf{a})^T\mathbf{x} \leq -b.$$

For denoting the comparison of vectors we will adopt a convenient ex-
tension of the notion of inequality. We will say that vector $\mathbf{u}$ is less than
(less than or equal to) vector $\mathbf{v}$ if the vectors have the same number of
elements and $u_i < v_i$ ($u_i \leq v_i$) for all i. Similarly we will say that vector
$\mathbf{u}$ is greater than (greater than or equal to) vector $\mathbf{v}$ if $u_i > v_i$ ($u_i \geq v_i$)
for $i = 1, 2, \ldots, k$. These inequalities will be denoted as appropriate $\mathbf{u} < \mathbf{v}$,

$\mathbf{u} \leq \mathbf{v}$, $\mathbf{u} > \mathbf{v}$, or $\mathbf{u} \geq \mathbf{v}$. This new notation will simplify the statements of the theorems in this chapter.

We will assume that all the sign constraints on the solution to a linear program are of the form $\mathbf{x} \geq \mathbf{0}$, where $\mathbf{0}$ denoted the zero vector of the same dimension as $\mathbf{x}$. Any component, say x_i, of $\mathbf{x}$ which is not non-negative can be replaced by $x_i^+ - x_i^-$ where $x_i^+ \geq 0$ and $x_i^- \geq 0$. Lastly we adopt the convention that every optimization process is one of maximization, since maximizing $\mathbf{c}^T\mathbf{x}$ is equivalent to minimizing $(-\mathbf{c})^T\mathbf{x}$.

If there are several inequality constraints placed on a solution to a linear programming problem we may compactly describe them using matrix notation. Suppose there are m inequality constraints on the solution $\mathbf{x}$, a vector with n components.

$$\mathbf{a}_1^T\mathbf{x} \leq b_1$$
$$\mathbf{a}_2^T\mathbf{x} \leq b_2$$
$$\vdots$$
$$\mathbf{a}_m^T\mathbf{x} \leq b_m$$

This will be expressed as the vector inequality $A\mathbf{x} \leq \mathbf{b}$ where

$$A\mathbf{x} = \begin{bmatrix} a_{11} & a_{12} & \cdots & a_{1n} \\ a_{21} & a_{22} & \cdots & a_{2n} \\ \vdots & \vdots & & \vdots \\ a_{m1} & a_{m2} & \cdots & a_{mn} \end{bmatrix} \begin{bmatrix} x_1 \\ x_2 \\ \vdots \\ x_n \end{bmatrix} \leq \begin{bmatrix} b_1 \\ b_2 \\ \vdots \\ b_m \end{bmatrix} = \mathbf{b}.$$

Thus for our purposes the general form of a linear program will be one which can be stated as "maximize $\mathbf{c}^T\mathbf{x}$ subject to the constraints $A\mathbf{x} \leq \mathbf{b}$ and $\mathbf{x} \geq \mathbf{0}$". This choice of general form is arbitrary and there is not universal agreement among authors writing about linear programming as to which form should be considered the "general form". As long as we are consistent, no confusion is likely to result.

The vector $\mathbf{x}$ is **feasible** by definition if $\mathbf{x} \geq \mathbf{0}$ and $\mathbf{x}$ satisfies the set of constraint equations $A\mathbf{x} \leq \mathbf{b}$. The inner product $\mathbf{c}^T\mathbf{x}$ is referred to as the **cost function**. If the vector $\mathbf{x}$ is feasible and maximizes the cost function, then $\mathbf{x}$ is also called an **optimal solution**.

If m and n are not too large this optimization task is easily accomplished. For example suppose we try to maximize $5x_1 + 4x_2 + 8x_3$ subject to the constraints $x_1 + x_2 + x_3 \leq 1$ and $\mathbf{x} \geq \mathbf{0}$. The constraints require the solution to the optimization problem to lie in a subset of the positive orthant of $\mathbb{R}^3$. This subset is a tetrahedron. See Fig. 4.2. If the maximum

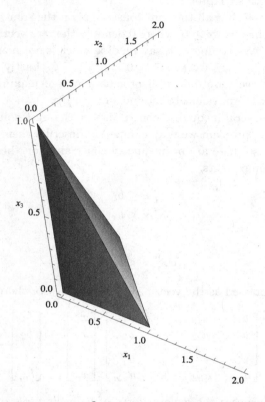

Fig. 4.2   The set of points in $\mathbf{R}^3$ where the cost function's maximum may occur.

of the cost function is the as yet unknown value k, then $5x_1 + 4x_2 + 8x_3 = k$ defines a plane. The largest value of k for which the level set of the cost function intersects the constrained set of points will be the maximum of the cost function. Thus the maximum of the cost function is 8 as can be seen in Fig. 4.3. The cost function is maximized at the point $(x_1, x_2, x_3) = (0, 0, 1)$.

We saw earlier that a constraint equation can be replaced by a pair constraint inequalities. Feasible sets of points in these types of optimization problems are always **convex sets**. A set is convex if for every pair of points P and Q contained in the set, the line segment connecting them also lies completely in the set. Inequality constraints can be converted to equality

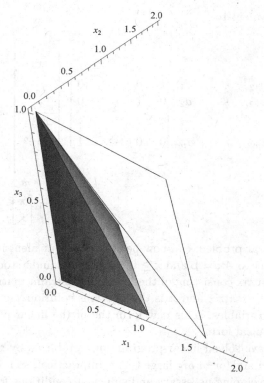

Fig. 4.3 The maximum of the cost function is the last level set of the cost function to intersect the tetrahedron of constrained solution points.

constraints by introducing **slack variables**. Thus

$$x_1 + x_2 + x_3 \leq 1 \quad \text{becomes} \quad x_1 + x_2 + x_3 + \hat{x}_4 = 1,$$

where slack variable, $\hat{x}_4 \geq 0$ "takes up the slack" to produce equality. No modification of the cost function is necessary, although it is always possible to think of the cost function in the form now of $5x_1 + 4x_2 + 8x_3 + 0\hat{x}_4$.

Now any general linear problem with inequality constraint $A\mathbf{x} \leq \mathbf{b}$ can be replaced with an equality constraint by the introduction of slack variables. Suppose A is an $m \times n$ matrix, $\mathbf{x}$ is a vector of n components, and $\mathbf{b}$ is a vector of m components, then by augmenting $\mathbf{x}$ with m slack variables and A with the $m \times m$ identity matrix the inequality constraint

$A\mathbf{x} \leq \mathbf{b}$ is equivalent to

$$
\begin{bmatrix}
a_{11} & a_{12} & \cdots & a_{1n} & 1 & 0 & \cdots & 0 \\
a_{21} & a_{22} & \cdots & a_{2n} & 0 & 1 & \cdots & 0 \\
\vdots & \vdots & & \vdots & \vdots & \vdots & & \vdots \\
a_{m1} & a_{m2} & \cdots & a_{mn} & 0 & 0 & \cdots & 1
\end{bmatrix}
\begin{bmatrix}
x_1 \\ x_2 \\ \vdots \\ x_n \\ \hline \hat{x}_{n+1} \\ \hat{x}_{n+2} \\ \vdots \\ \hat{x}_{n+m}
\end{bmatrix}
=
\begin{bmatrix}
b_1 \\ b_2 \\ \vdots \\ b_m
\end{bmatrix}
$$

$$
\begin{bmatrix} A | I_m \end{bmatrix} \begin{bmatrix} \mathbf{x} \\ \hat{\mathbf{x}} \end{bmatrix} = \mathbf{b}.
$$

The general linear problem can now be stated in equivalent form as "maximize $\mathbf{c}^T\mathbf{x}$ subject to $A\mathbf{x} = \mathbf{b}$ and $\mathbf{x} \geq \mathbf{0}$", where it is understood that A is an $m \times (n+m)$ matrix consisting of the original constraint matrix augmented with the identity matrix and $\mathbf{x}$ is the previous solution vector augmented with the slack variables. This new form the of the linear problem will be called the canonical form.

So far we have solved linear programming problems by graphical techniques. When n and/or m are large this is impractical, so now we turn our attention to developing necessary and sufficient conditions for determining if a linear programming problem has an optimal solution.

4.2.1 *Dual Problems*

In mathematics we frequently benefit from the ability to solve one problem by means of finding the solution to a related, but simpler, problem. This is certainly true of the linear programming problems. For every linear programming problem of the general form, there is an associated problem known as its **dual**. Henceforth the original problem will be known as the **primal**. These paired optimization problems are related in the following ways.

> **Primal:** Maximize $\mathbf{c}^T\mathbf{x}$ subject to $A\mathbf{x} \leq \mathbf{b}$ and $\mathbf{x} \geq \mathbf{0}$.
> **Dual:** Minimize $\mathbf{b}^T\mathbf{y}$ subject to $A^T\mathbf{y} \geq \mathbf{c}$ and $\mathbf{y} \geq \mathbf{0}$.

We should note that:

(1) the process of maximization in the primal is replaced with the process of minimization in the dual,

(2) the unknown of the dual is a vector $\mathbf{y}$ with m components,

(3) the vector $\mathbf{b}$ moves from the constraint of the primal to the cost function of the dual,

(4) the vector $\mathbf{c}$ moves from the cost of the primal to the constraint of the dual,

(5) the constraints of the dual are inequalities and there are n of them.

Given a primal or a dual problem it is a routine matter to construct its partner. The primal and the dual form a set of "fraternal twin" problems. The following theorem will shed some light on their relationship.

Theorem 4.1 *The dual of the dual is the primal.*

Proof. Starting with the dual problem,

$$\text{Minimize } \mathbf{b}^T \mathbf{y} \quad \text{subject to } A^T \mathbf{y} \geq \mathbf{c} \text{ and } \mathbf{y} \geq \mathbf{0}.$$

We can re-write the dual in general form,

$$\text{Maximize } (-\mathbf{b})^T \mathbf{y} \quad \text{subject to } (-A)^T \mathbf{y} \leq -\mathbf{c} \text{ and } \mathbf{y} \geq \mathbf{0}.$$

Now the dual of this problem (*i.e.*, the dual of the dual) is

$$\text{Minimize } (-\mathbf{c})^T \mathbf{x} \quad \text{subject to } ((-A)^T)^T \mathbf{x} \geq -\mathbf{b} \text{ and } \mathbf{x} \geq \mathbf{0}.$$

This problem is logically equivalent to the problem

$$\text{Maximize } \mathbf{c}^T \mathbf{x} \quad \text{subject to } A\mathbf{x} \leq \mathbf{b} \text{ and } \mathbf{x} \geq \mathbf{0},$$

which is the primal problem. □

Perhaps it is no surprise then that the solutions to the primal and dual are related.

Theorem 4.2 *(Weak Duality Theorem) If $\mathbf{x}$ and $\mathbf{y}$ are the feasible solutions of the primal and dual problems respectively, then $\mathbf{c}^T \mathbf{x} \leq \mathbf{b}^T \mathbf{y}$. If $\mathbf{c}^T \mathbf{x} = \mathbf{b}^T \mathbf{y}$ then these solutions are optimal for their respective problems.*

Proof. Feasible solutions to the primal and the dual problems must satisfy the constraints $A\mathbf{x} \leq \mathbf{b}$ with $\mathbf{x} \geq 0$ (for the primal problem) and $A^T \mathbf{y} \geq \mathbf{c}$ with $\mathbf{y} \geq \mathbf{0}$ (for the dual). We multiply the constraint in the dual by $\mathbf{x}^T$ to obtain

$$\mathbf{x}^T A^T \mathbf{y} \geq \mathbf{x}^T \mathbf{c} \quad \Longleftrightarrow \quad \mathbf{c}^T \mathbf{x} \leq \mathbf{y}^T A\mathbf{x}.$$

We multiply the constraint in the primal by $\mathbf{y}^T$ to find

$$\mathbf{y}^T A\mathbf{x} \le \mathbf{y}^T \mathbf{b} = \mathbf{b}^T \mathbf{y}.$$

Note that the directions of the inequalities are preserved because each component of $\mathbf{x}$ and $\mathbf{y}$ is non-negative. Combining these last two inequalities produces

$$\mathbf{c}^T \mathbf{x} \le \mathbf{y}^T A\mathbf{x} \le \mathbf{b}^T \mathbf{y}.$$

Therefore we have $\mathbf{c}^T \mathbf{x} \le \mathbf{b}^T \mathbf{y}$.

If there exist solutions $\mathbf{x}^*$ and $\mathbf{y}^*$ to the primal and dual problems respectively for which $\mathbf{c}^T \mathbf{x}^* = \mathbf{b}^T \mathbf{y}^*$ then $\mathbf{x}$ and $\mathbf{y}$ must be optimal for their respective problems. If $\mathbf{x}$ is any feasible solution to the primal problem, then

$$\mathbf{c}^T \mathbf{x} \le \mathbf{b}^T \mathbf{y}^* = \mathbf{c}^T \mathbf{x}^*$$

meaning $\mathbf{x}^*$ generated the maximum of the cost function for the primal problem. If $\mathbf{y}$ is any feasible solution to the dual problem, then

$$\mathbf{b}^T \mathbf{y} \ge \mathbf{c}^T \mathbf{x}^* = \mathbf{b}^T \mathbf{y}^*$$

meaning $\mathbf{y}^*$ generated the minimum of the cost function for the dual problem. □

One consequence of the Weak Duality Theorem is that if $\mathbf{x}$ is a feasible solution for the primal and $\mathbf{y}$ is a feasible solution for the dual, then the linear expression for the primal $\mathbf{c}^T \mathbf{x}$ is bounded above and the linear expression for the dual $\mathbf{b}^T \mathbf{y}$ is bounded below. Since we are trying the maximize (minimize) a linear expression for the primal (dual) problem, it is helpful to know that there exists an upper (a lower) bound on this expression.

Example 4.1　Consider the paired primal and dual problems:

> **Primal**: Maximize $4x_1 + 3x_2$ subject to $x_1 + x_2 \le 2$ and $x_1, x_2 \ge 0$.
> **Dual**: Minimize $2y_1$ subject to $y_1 \ge 3$ and $y_1 \ge 4$ and $y_1 \ge 0$.

While both problems may seem simple, the dual is trivial. The minimum value of y_1 subject to the constraints must be $y_1 = 4$. According to the Weak Duality Theorem then the minimum of the cost function of the primal must be at least 8. Applying the level set argument as before, the largest value of k for which the level set $4x_1 + 3x_2 = k$ intersects the set of feasible points for the primal is $k = 8$. See Fig. 4.4.

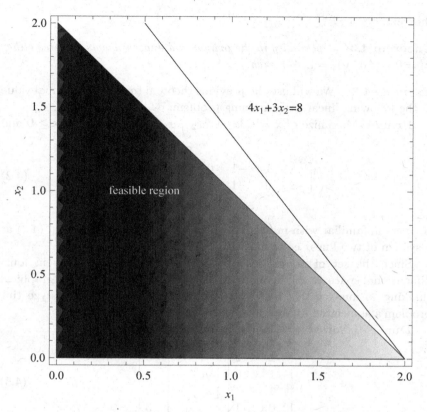

Fig. 4.4   The maximum of the cost function $4x_1+3x_2$ subject to the constraints $x_1+x_2 \leq$ 2 and $x_1, x_2 \geq 0$, occurs at the point with coordinates $(x_1, x_2) = (2, 0)$.

As a consequence of the Weak Duality Theorem we can determine that when $\mathbf{x}$ and $\mathbf{y}$ are optimal for their respective problems then

$$\mathbf{b}^T\mathbf{y} = \mathbf{y}^T A\mathbf{x} = \mathbf{c}^T\mathbf{x}$$
$$(\mathbf{y}^T A - \mathbf{c}^T)\mathbf{x} = 0$$
$$(A^T\mathbf{y} - \mathbf{c})^T\mathbf{x} = 0. \tag{4.1}$$

Since $\mathbf{x}$ is feasible then each component of that vector is non-negative. Likewise by assumption in the dual $A^T\mathbf{y} \geq \mathbf{c}$ which implies that each component of the vector $A^T\mathbf{y} - \mathbf{c}$ is non-negative. Using Eq. (4.1) we can conclude that vector $\mathbf{x}$ must be zero in every component for which vector $A^T\mathbf{y} - \mathbf{c}$ is positive and vice versa. Thus we have proved the following theorem.

Theorem 4.3 *Optimality in the primal and dual problems requires either* $x_j = 0$ *or* $(A^T y)_j = c_j$ *for each* $j = 1, \ldots, n$.

Example 4.2 We will use the previous theorem to find the optimal value of the following linear programming problem.

Primal: Maximize $\mathbf{c}^T \mathbf{x} = -3x_1 + 2x_2 - x_3 + 3x_4$ subject to $\mathbf{x} \geq \mathbf{0}$ and

$$
\begin{bmatrix} 1 & 1 & -1 & 0 \\ -2 & 0 & 1 & 1 \end{bmatrix} \begin{bmatrix} x_1 \\ x_2 \\ x_3 \\ x_4 \end{bmatrix} \leq \begin{bmatrix} 5 \\ 3 \end{bmatrix} \tag{4.2}
$$

Readers unfamiliar with matrix notation may wish to re-write Eq. (4.2) as a system of two linear equations with four unknowns.

Since the set of points described by the constraints exists in four-dimensional space, it will be more difficult to use geometrical and graphical thinking to analyze this problem. Fortunately we can also analyze this problem's associated dual problem.

Dual: Minimize $\mathbf{b}^T \mathbf{y} = 5y_1 + 3y_2$ subject to

$$
\begin{bmatrix} 1 & -2 \\ 1 & 0 \\ -1 & 1 \\ 0 & 1 \end{bmatrix} \begin{bmatrix} y_1 \\ y_2 \end{bmatrix} \geq \begin{bmatrix} -3 \\ 2 \\ -1 \\ 3 \end{bmatrix} \tag{4.3}
$$

The space of points on which the dual is to be optimized exists in two-dimensional space and thus is easily pictured. The constraints of the dual can be thought of as a system of inequalities.

$$
y_1 - 2y_2 \geq -3
$$
$$
y_1 \geq 2
$$
$$
-y_1 + y_2 \geq -1
$$
$$
y_2 \geq 3
$$

The solution to this set of inequalities can be pictured as the shaded region shown in Fig. 4.5. The minimum of the cost function for the dual occurs at the point with coordinates $(y_1, y_2) = (3, 3)$. Thus the minimum value is 24 which will also be the maximum value of the primal problem's cost function. At the optimal point for the dual problem, strict inequality is

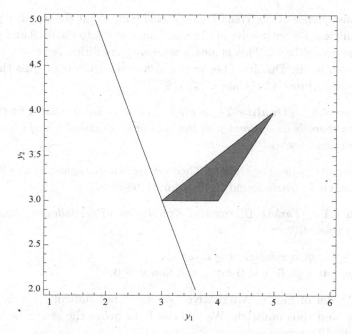

Fig. 4.5 The shaded region in the plot denotes the constrained set of points from Eq. (4.3) for the dual problem.

present in the second and third constraints since

$$y_1 = 3 > 2$$
$$-y_1 + y_2 = 0 > -1.$$

By the previous theorem then the second and third components of $\mathbf{x}$ in the primal problem must be zero. Therefore the primal can be recast as

Primal: Maximize $-3x_1 + 3x_4$ subject to $x_1 \geq 0$, $x_4 \geq 0$ and

$$\begin{bmatrix} 1 & 1 & -1 & 0 \\ -2 & 0 & 1 & 1 \end{bmatrix} \begin{bmatrix} x_1 \\ 0 \\ 0 \\ x_4 \end{bmatrix} = \begin{bmatrix} x_1 \\ -2x_1 + x_4 \end{bmatrix} \leq \begin{bmatrix} 5 \\ 3 \end{bmatrix} \qquad (4.4)$$

By inspection we must have $x_1 = 5$ and $x_4 = 13$. Consequently the maximum of the cost function for the primal is seen to be 24 and it occurs at the point with coordinates $(x_1, x_2, x_3, x_4) = (5, 0, 0, 13)$.

A consequence of the Weak Duality Theorem is that the equality $\mathbf{y} \cdot \mathbf{b} = \mathbf{c} \cdot \mathbf{x}$ is sufficient for optimality of the solutions $\mathbf{y}$ and $\mathbf{x}$ to the dual and primal problems respectively. This is also a necessary condition for optimality as will be seen in the Duality Theorem, which is stated below. This theorem was originally proved in [Gale *et al.* (1951)].

Theorem 4.4 *(Duality Theorem)* *If there is an optimal $\mathbf{x}$ in the primal, then there is an optimal $\mathbf{y}$ in the dual and the minimum of $\mathbf{c}^T \mathbf{x}$ equals the maximum of $\mathbf{y}^T \mathbf{b}$.*

In order to prove the Duality Theorem we will make use of the **Farkas Alternative Lemma** found in [Franklin (1980)].

Lemma 4.1 *(Farkas Alternative) Exactly one of the following two statements is true. Either*

(1) $A\mathbf{x} \leq \mathbf{b}$ has a solution $\mathbf{x} \geq \mathbf{0}$, or
(2) $A^T\mathbf{y} \geq \mathbf{0}$ with $\mathbf{b}^T y < 0$ has a solution $\mathbf{y} \geq \mathbf{0}$.

The proof of Farkas Alternative Lemma is not difficult, but is rather technical and thus omitted. We will use it to prove the strong version of the Duality Theorem

Proof. First consider the general form of the primal and dual problems.

Primal: Maximize $\mathbf{c}^T\mathbf{x}$ subject to $A\mathbf{x} \leq \mathbf{b}$ and $\mathbf{x} \geq \mathbf{0}$.
Dual: Minimize $\mathbf{b}^T\mathbf{y}$ subject to $A^T\mathbf{y} \geq \mathbf{c}$ and $\mathbf{y} \geq \mathbf{0}$.

Assuming there are feasible solutions to each problem then we can re-write the constraint of the dual as $(-A)^T\mathbf{y} \leq -\mathbf{c}$ with $\mathbf{y} \geq \mathbf{0}$. Thus according to the constraint on the primal, the re-written constraint on the dual, and the conclusion of the Weak Duality Theorem the following inequalities hold for $\mathbf{x}, \mathbf{y} \geq \mathbf{0}$.

$$A\mathbf{x} \leq \mathbf{b}$$
$$(-A)^T\mathbf{y} \leq -\mathbf{c}$$
$$\mathbf{c}^T\mathbf{x} - \mathbf{b}^T\mathbf{y} \leq 0$$

These can be written in the compound matrix form

$$\begin{bmatrix} A & 0 \\ 0 & -A^T \\ \mathbf{c}^T & -\mathbf{b}^T \end{bmatrix} \begin{bmatrix} \mathbf{x} \\ \mathbf{y} \end{bmatrix} \leq \begin{bmatrix} \mathbf{b} \\ -\mathbf{c} \\ 0 \end{bmatrix}.$$

According to the Farkas Alternative Lemma either this inequality has a solution $\langle \mathbf{x}, \mathbf{y} \rangle \geq \mathbf{0}$ or the alternative

$$\begin{bmatrix} A^T & 0 & \mathbf{c} \\ 0 & -A & -\mathbf{b} \end{bmatrix} \begin{bmatrix} \mathbf{u} \\ \mathbf{v} \\ \lambda \end{bmatrix} \geq \mathbf{0} \quad \text{and} \quad \begin{bmatrix} \mathbf{b}^T & -\mathbf{c}^T & 0 \end{bmatrix} \begin{bmatrix} \mathbf{u} \\ \mathbf{v} \\ \lambda \end{bmatrix} < 0$$

has a solution $\langle \mathbf{u}, \mathbf{v}, \lambda \rangle \geq \mathbf{0}$. We may decompose this compound matrix to derive the following system of inequalities:

$$A^T \mathbf{u} + \lambda \mathbf{c} \geq \mathbf{0}, \quad -A\mathbf{v} - \lambda \mathbf{b} \geq \mathbf{0}, \quad \mathbf{b}^T \mathbf{u} - \mathbf{c}^T \mathbf{v} < 0$$

with $\mathbf{u} \geq \mathbf{0}$, $\mathbf{v} \geq \mathbf{0}$, and $\lambda \geq 0$. If $\lambda > 0$ then this system of inequalities is equivalent to the following system.

$$A \left(\frac{1}{\lambda} \mathbf{v} \right) \leq -\mathbf{b}$$

$$A^T \left(\frac{1}{\lambda} \mathbf{u} \right) \geq -\mathbf{c}$$

$$-\mathbf{b}^T \left(\frac{1}{\lambda} \mathbf{u} \right) > -\mathbf{c}^T \left(\frac{1}{\lambda} \mathbf{v} \right)$$

Since $\mathbf{u} \geq \mathbf{0}$ and $\mathbf{v} \geq \mathbf{0}$ the vectors $\frac{1}{\lambda} \mathbf{u} \geq \mathbf{0}$ and $\frac{1}{\lambda} \mathbf{v} \geq \mathbf{0}$ as well. The first two inequalities above form a primal problem and its dual. If we apply the Weak Duality Theorem (Theorem 4.2), then it must be the case that $-\mathbf{b}^T \left(\frac{1}{\lambda} \mathbf{u} \right) \leq -\mathbf{c}^T \left(\frac{1}{\lambda} \mathbf{v} \right)$, a contradiction to the third inequality in the system above. Therefore we know that $\lambda = 0$. Thus the Farkas Alternative simplifies to the following system:

$$A\mathbf{v} \leq \mathbf{0}, \quad A^T \mathbf{u} \geq \mathbf{0}, \quad \text{and} \quad \mathbf{b}^T \mathbf{u} < \mathbf{c}^T \mathbf{v}$$

where $\mathbf{u} \geq \mathbf{0}$ and $\mathbf{v} \geq \mathbf{0}$. The last (strict) inequality implies that $\mathbf{b}^T \mathbf{u} < 0$ or $\mathbf{c}^T \mathbf{v} > 0$ (possibly both are true, see exercise 12). If $\mathbf{b}^T \mathbf{u} < 0$ then the primal problem $A\mathbf{x} \leq \mathbf{b}$ has no feasible solution $\mathbf{x} \geq \mathbf{0}$. To see this note that together the inequalities $\mathbf{x} \geq \mathbf{0}$, $A\mathbf{x} \leq \mathbf{b}$, and $\mathbf{b}^T \mathbf{u} < 0$ imply that

$$(A\mathbf{x})^T \leq \mathbf{b}^T$$
$$\mathbf{x}^T A^T \leq \mathbf{b}^T$$
$$\mathbf{x}^T \left(A^T \mathbf{u} \right) \leq \mathbf{b}^T \mathbf{u} < 0.$$

However, $\mathbf{x} \geq \mathbf{0}$ and $A^T \mathbf{u} \geq \mathbf{0}$ and thus $\mathbf{x}^T \left(A^T \mathbf{u} \right) \geq 0$, a contradiction.

If $\mathbf{c}^T\mathbf{v} > 0$ then the dual problem $A^T\mathbf{y} \geq \mathbf{c}$ has no feasible solution $\mathbf{y} \geq \mathbf{0}$. To see this note that together the inequalities $\mathbf{y} \geq \mathbf{0}$, $A^T\mathbf{y} \geq \mathbf{c}$, and $\mathbf{c}^T\mathbf{v} > 0$ imply that

$$\left(A^T\mathbf{y}\right)^T \geq \mathbf{c}^T$$
$$\mathbf{y}^T A \geq \mathbf{c}^T$$
$$\mathbf{y}^T\left(A\mathbf{v}\right) \geq \mathbf{c}^T\mathbf{v} > 0$$
$$\mathbf{y}^T\left(-A\mathbf{v}\right) < 0$$

However, $\mathbf{y} \geq \mathbf{0}$ and $-A\mathbf{v} \geq \mathbf{0}$ and thus $\mathbf{y}^T\left(-A\mathbf{v}\right) \geq 0$, a contradiction. Thus the strong version of the Duality Theorem is established. $\square$

Thus far in this chapter we have developed a great deal of background knowledge in linear programming and duality. Linear programming is a vast field of study in its own right. The previous material is merely an introduction intended to enable us to prove the Fundamental Theorem of Finance in the next section.

4.3   The Fundamental Theorem of Finance

Consider an experiment with m possible outcomes numbered 1 through m. Suppose we can place n wagers (numbered 1 through n) on the outcomes of the experiment. Let r_{ji} be the return for a unit bet on wager i when the outcome of the experiment is j. The vector $\mathbf{x} = (x_1, x_2, \ldots, x_n)$ is called a **betting strategy**. Component x_i is the amount placed on wager i. The composite return from this betting strategy when outcome j occurs is then $\sum_{i=1}^{n} x_i r_{ji}$. The Arbitrage Theorem states that the probabilities of the m outcomes of the experiment are such that for each bet the expected value of the payoff is zero, or there exists a betting strategy for which the payoff is positive regardless of the outcome of the experiment.

Theorem 4.5   *(Arbitrage Theorem) Exactly one of the following is true: either*

(1) there is a vector of probabilities $\mathbf{y} = (y_1, y_2, \ldots, y_m)$ for which

$$\sum_{j=1}^{m} y_j r_{ji} = 0, \quad \text{for each } i = 1, 2, \ldots, n, \text{ or}$$

(2) there is a betting strategy $\mathbf{x} = (x_1, x_2, \ldots, x_n)$ *for which*

$$\sum_{i=1}^{n} x_i r_{ji} > 0, \quad \text{for each } j = 1, 2, \ldots, m.$$

Proof. Let the n-tuple $(x_1, x_2, \ldots, x_n)$ be the betting strategy. Vector $\mathbf{x} = \langle x_1, x_2, \ldots, x_n, x_{n+1} \rangle$ where the $n + 1$st element x_{n+1} represents the amount of the payoff for the betting strategy. We would like to maximize x_{n+1} under the constraint that $\sum_{i=1}^{n} x_i r_{ji} \geq x_{n+1}$ for $j = 1, 2, \ldots, m$. This is equivalent to $\sum_{i=1}^{n} x_i(-r_{ji}) + x_{n+1} \leq 0$ for $j = 1, 2, \ldots, m$. The reader should recognize the primal problem being set up. Let the vector $\mathbf{c}$ be the vector of $n + 1$ components where the first n are 0 and the $n + 1$st is 1. Let the vector $\mathbf{b} = (0, \ldots, 0, 0)$ have m components. Then the maximization problem outlined above can be stated in matrix form as "Maximize $\mathbf{c}^T \mathbf{x} = x_{n+1}$ subject to $A\mathbf{x} \leq \mathbf{b}$ and $\mathbf{x} \geq 0$" where

$$A\mathbf{x} = \begin{bmatrix} -r_{11} & -r_{12} & \cdots & -r_{1n} & 1 \\ -r_{21} & -r_{22} & \cdots & -r_{2n} & 1 \\ \vdots & \vdots & & \vdots & \vdots \\ -r_{m1} & -r_{m2} & \cdots & -r_{mn} & 1 \end{bmatrix} \begin{bmatrix} x_1 \\ x_2 \\ \vdots \\ x_n \\ x_{n+1} \end{bmatrix} \leq \begin{bmatrix} 0 \\ 0 \\ \vdots \\ 0 \end{bmatrix}.$$

The dual of this primal problem can be stated as "Minimize $\mathbf{b}^T \mathbf{y} = 0$ subject to $A^T \mathbf{y} \geq \mathbf{c}$ and $\mathbf{y} \geq 0$." The unknown $\mathbf{y}$ is a vector of m components. Since $\mathbf{b}$ is the zero vector the cost function to be minimized is really just the constant 0. Therefore in matrix form the dual can be thought of as minimizing 0 subject to $\mathbf{y} \geq 0$ and

$$\begin{bmatrix} -r_{11} & -r_{21} & \cdots & -r_{m1} \\ -r_{12} & -r_{22} & \cdots & -r_{m2} \\ \vdots & \vdots & & \vdots \\ -r_{1n} & -r_{2n} & \cdots & -r_{mn} \\ 1 & 1 & \cdots & 1 \end{bmatrix} \begin{bmatrix} y_1 \\ y_2 \\ \vdots \\ y_{m-1} \\ y_m \end{bmatrix} = \begin{bmatrix} 0 \\ 0 \\ \vdots \\ 0 \\ 1 \end{bmatrix}.$$

This matrix equation is equivalent to the system of equations, $\sum_{j=1}^{m} y_j r_{ji} = 0$ for $i = 1, 2, \ldots, n$; $\sum_{j=1}^{m} y_j = 1$, and $y_j \geq 0$ for $j = 1, 2, \ldots, m$. The dual problem is feasible with a minimum value of zero if and only if $\mathbf{y}$ is a probability vector for which all bets have an expected return of zero. Suppose the dual problem is feasible, the primal problem is also feasible since $x_i = 0$ for $i = 1, 2, \ldots, n + 1$ satisfies the inequality constraints of the dual. According to the Duality Theorem then the maximum of the primal problem is zero which means no guaranteed profit is possible. In other words if (1) is true then (2) is false. Now suppose the dual problem is not feasible. According to the Duality Theorem the primal problem has no optimal solution. Therefore zero is not the maximum payoff, thus there is a betting strategy which produces a positive payoff. Thus we see that if (1) is false then (2) is true.   □

4.4 Exercises

(1) Suppose the odds against the three possible outcomes of an experiment are as given in the table below.

Outcome	Odds
A	2:1
B	3:1
C	1:1

Find a betting strategy which produces a positive net profit regardless of the outcome of the experiment.

(2) Sketch the region in the plane which satisfies the following inequalities:

$$x_1 + 3x_2 \geq 6$$
$$3x_1 + x_2 \geq 6$$
$$x_1 \geq 0$$
$$x_2 \geq 0$$

Is the region convex? What are the coordinates of the points at the corners of the region?

(3) Minimize the cost function $x_1 + x_2$ on the region described in exercise 2.

(4) Find inequality constraints which will describe the rectangular region with corners at $(0, 0)$, $(2, 0)$, $(2, 3)$, and $(0, 3)$.

(5) Introduce slack variables in the inequality constraints found in exercise 4 to produce equality constraints $A\mathbf{x} = \mathbf{b}$. What are A and $\mathbf{b}$?

(6) Minimize the cost function $x_1 + x_2 + 2x_3$ on the set where $x_1 + 2x_2 + 3x_3 \geq 15$ and $x_i \geq 0$ for $i = 1, 2, 3$.

(7) Minimize the cost function $7x_2 + 9x_3$ subject to the constraints $x_1 + x_2 + x_3 \geq 5$ and $x_2 + x_3 + 2x_4 \geq 1$ with $x_i \geq 0$ for $i = 1, 2, 3$.

(8) Write down the dual problem of the following linear programming problem: Minimize $x_1 + x_2 + x_3$ subject to $2x_1 + x_2 = 4$ and $x_3 \leq 6$ with $x_i \geq 0$ for $i = 1, 2, 3$.

(9) Find the solutions to the primal and dual problems of exercise 8.

(10) Write down the dual problem to the following linear programming problem: Maximize $2y_1 + 4y_3$ subject to $y_1 + y_2 \leq 1$ and $y_2 + 2y_3 \leq 1$.

(11) Find the solutions to the primal and dual problems of exercise 10.

(12) Show that if $a < b$ then either $a < 0$ or $b > 0$.

Chapter 5

Random Walks and Brownian Motion

In this chapter we will introduce and explain some of the concepts surrounding the probabilistic models used to capture the behavior of stock, security, option, and index prices. Random walks are related to properties of discrete random variables studied in Chap. 2. Brownian motion can be thought of as a continuous random process which is the limiting case of a random walk as the time and space steps become infinitesimally small. Topics covered here could be expanded into an entire book of their own. For our purposes we will explore just enough of the stochastic calculus to provide some justification for Itô's Lemma, the main result of this chapter.

5.1 Intuitive Idea of a Random Walk

Earlier in Chapter 3 we analyzed the motion of a particle taking a discrete step along the x-axis during every "tick" of the clock. At first it was revealed that the location of the particle relative to the origin followed a binomial distribution. As the step size was decreased, the probability of the particle lying in a particular interval was seen to obey the normal distribution. In this chapter we will extend this discussion to a more general setting. Imagine a person standing at the origin of the real number line. They will flip a coin. For every time the coin lands on heads, they will take a unit step to the right (the positive direction). For every time the coin lands on tails, they will move a unit step to the left (the negative direction). The evolution of this stochastic discrete dynamical system is called a **random walk**. A plot of the person's location versus the number of times the coin has been flipped might resemble the graph in Fig. 5.1. Random walks do not require that the probabilities of moving left or right be equal (*i.e.*, the coin does not have to be fair). The magnitude of the movement at each stage does

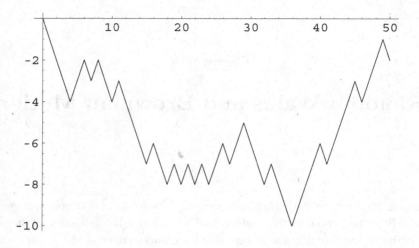

Fig. 5.1   A realization of a random walk on the real number line.

not have to be a single unit. The person could move left or right depending on the value of a random number chosen from a normal distribution for example. The bias in choosing to move left or right and the possibility of moving different distances during different steps are generalizations not present in the discussion presented in Chapter 3. In the remainder of this chapter we will apply the concept of the random walk to modeling the movement of the value of a stock or other security.

5.2   First Step Analysis

A simple first application of a random walk is the situation of a person owning a share of stock whose current value is denoted by S. At discrete intervals (for example, once per day) the stock's value can increase or decrease by one unit, in other words if its current value on day n is $S(n)$, then tomorrow the value will either be $S(n+1) = S(n)+1$ or $S(n+1) = S(n)-1$. To keep the following discussion simple, the probability of increase in value will be $p = 1/2$ and consequently the probability of decrease is the same. This assumption implies the random walk is **unbiased**. If we think of X_n as a random variable which takes on value 1 with probability $p = 1/2$ and value -1 with the same probability, we can think of the Nth state of the random walk as a partial sum where for $N > 0$,

$$S(N) = S(0) + X_1 + X_2 + \cdots + X_N. \tag{5.1}$$

The random variable X_n is a generalization of the Bernoulli random variable discussed earlier (recall that we had defined a Bernoulli random variable to have outcomes 0 and 1). $S(N)$ can be defined inductively by the formula $S(N) = S(N-1) + X_N$ for $N > 0$ where $S(0)$ is a specified initial state of the walk. If we assume that the random variables X_i and X_j have the same distribution and are independent for $i \neq j$, then we see that the transitions between states in the random walk, $S(N) - S(N-1) = X_N$ are independent random variables. If the initial state of the random walk is $S(0)$ and $n \geq 0$ steps are taken, where $0 \leq k \leq n$ of the steps are in the positive direction and $n - k \geq 0$ are in the negative direction, then

$$P\left(S(n) = S(0) + k - (n - k)\right) = P\left(S(n) - S(0) = 2k - n\right)$$
$$= \binom{n}{k} \left(\frac{1}{2}\right)^n.$$

The last expression is the familiar probability formula for a binomial random variable with $p = 1/2$.

If $S(0) \neq 0$ then we can perform a change of variable via $T(n) = S(n) - S(0)$ for $n = 0, 1, \ldots$ and define a related random walk where $T(0) = 0$. The transitions in the states of the new random walk are independent, identically distributed random variables just as before. Likewise we see that

$$P\left(T(n) = 2k - n\right) = \binom{n}{k} \left(\frac{1}{2}\right)^n. \tag{5.2}$$

Thus without loss of generality we will assume that $S(0) = 0$. This is known as the **spatial homogeneity** property of the random walk. Now that we have settled on a starting state for the random walk, what states can be visited in n steps? The next lemma provides the answer and the respective probabilities of reaching these states.

Lemma 5.1 *For the random walk defined in Eq. (5.1) with initial state $S(0) = 0$,*

(1) $P\left(S(n) = m\right) = 0$ if $|m| > n$,
(2) $P\left(S(n) = m\right) = 0$ if $n + m$ is odd,
(3) $P\left(S(n) = m\right) = \binom{n}{(n+m)/2} \left(\frac{1}{2}\right)^n$, otherwise.

Proof.

(1) According to Eq. (5.1), $-n \leq S(n) \leq n$, thus if $|m| > n$ the partial sum cannot attain this value.
(2) If we let $m = 2k - n$ in Eq. (5.2) then $n + m = 2k$ which implies $n + m$ is even, contradicting the assumption that $n + m$ is odd.
(3) We may assume that $n + m$ is even and thus the result is shown by Eq. (5.2).

$\square$

The state of the random walk after n steps can be summarized in the following theorem.

Theorem 5.1   *For the random walk defined in Eq. (5.1) with initial state* $S(0) = 0$,

$$\mathrm{E}\left[S(n)\right] = 0 \quad and \quad \mathrm{Var}\left(S(n)\right) = n.$$

Proof.

$$\mathrm{E}\left[S(n)\right] = \mathrm{E}\left[S(0)\right] + \mathrm{E}\left[X_1\right] + \mathrm{E}\left[X_2\right] + \cdots + \mathrm{E}\left[X_n\right]$$
$$= 0$$

by Theorem 2.4 since $\mathrm{E}\left[X_i\right] = 0$ for $i = 1, 2, \ldots, n$. According the assumption that X_i and X_j are independent when $i \neq j$ we have

$$\mathrm{Var}\left(S(n)\right) = \mathrm{Var}\left(S(0)\right) + \sum_{i=1}^{n} \mathrm{Var}\left(X_i\right) = n$$

by Theorem 2.7.

$\square$

Consider an un-restricted random walk $\{S(j)\}_{j=0}^{n}$ which possesses the property that $S(k) = 0$ for some k between 0 and n. An example is graphed in Fig. 5.2. If we reflect the path of the random walk across the j-axis for $k < j \leq n$, we obtain another random walk $\{\hat{S}(j)\}_{j=0}^{n}$ with the properties that

$$\hat{S}(j) = \begin{cases} S(j) & \text{for } j = 0, 1, \ldots, k \\ -S(j) & \text{for } j = k+1, k+2, \ldots, n. \end{cases}$$

The probability the original random walk ended up in state A equals the probability the reflected random walk ended up in state $-A$. To see this

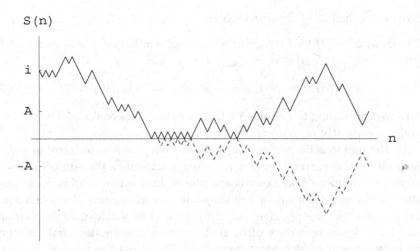

Fig. 5.2 The probabilities that an unbiased random walk will follow either the solid or dashed (reflected) path are equal.

suppose the original random walk starts at 0 (and hence the reflected random walk also starts at 0). Since downward steps occur with equal probability to upward steps, then

$$P\left(S(n) = A\right) = P\left(\hat{S}(n) = -A\right).$$

Random walks possess the **Markov property**, meaning that the history of movements of the random walk is irrelevant to the next random step. In other words the random walk cannot "remember" how it arrived at a particular state. The current state and only the current state influences the next state of the random walk. Therefore we can think of partitioning a random walk which crosses the j-axis into two segments, the initial segment from $j = 0$ to $j = k$ at which time the random walk is in state 0, and the final segment from $j = k+1$ to $j = n$. Hence by using the Markov property

$$P\left(S(n) = A\right) = P\left(\hat{S}(n) = -A\right)$$

$$P\left(S(k) = 0\right) P\left(T(n-k) = A\right) = P\left(\hat{S}(k) = 0\right) P\left(\hat{T}(n-k) = -A\right)$$

$$= P\left(S(k) = 0\right) P\left(\hat{T}(n-k) = -A\right)$$

$$P\left(T(n-k) = A\right) = P\left(\hat{T}(n-k) = -A\right).$$

The last equation is true by the special case discussed above. Hence we have established the following theorem.

Theorem 5.2   *If $\{S(j)\}_{j=0}^n$ is an unbiased random walk with initial state $S(0) = i$ and if $|A - i| \leq n$ and $|A + i| \leq n$ then*

$$P\left(S(n) = A \,|\, S(0) = i\right) = P\left(S(n) = -A \,|\, S(0) = i\right). \qquad (5.3)$$

Note that according to Lemma 5.1 these probabilities could be 0 if $n + A - i$ (and consequently $n - A - i$) are odd.

In the previous discussion the random walk was free to move in either the positive or negative direction any integer amount as the number of steps increased. Suppose that bounds are placed in the path the particle may follow. Once again, for the sake of simplicity, we will assume the initial state of the random walk is positive, *i.e.*, $S(0) > 0$ and we will assume the random walk has a lower boundary of 0. If the state of the random walk reaches the lower boundary in a finite number of steps then the state remains at that boundary value. In the study of random walks, this is known as an **absorbing boundary condition**. For a gambler, "going broke" can be thought of as an absorbing boundary condition. In a financial setting we may think of the value of a security as following a random walk (though in reality a much more complicated one than the type we are exploring here). If the value of the security drops to 0, the security becomes worthless to the owner. We must further generalize our discussion of random walks to include the absorbing boundary. There are two questions related to this situation that we wish to explore.

(1) What is the probability that the state of the random walk crosses a threshold value of $A > 0$ before it hits the boundary at 0 (and hence remains there)?
(2) What is the expected value of the number of steps which will elapse before the state of the random variable first crosses the $A > 0$ threshold?

Figure 5.3 shows a random walk of a stock whose value was initially 10. This situation has been so simplified and abstracted from the reality of the stock market that it is of little use in making investment decisions, but serves as an instructive example of the use of the concept of a random walk. The answers to the two primary questions will be found following the lines of reasoning laid out in [Kijima (2003)] and [Steele (2001)].

Since the boundary at 0 is absorbing we must keep track of the smallest value which $S(n)$ takes on. Thus we define $m_n = \min\{S(k) : 0 \leq k \leq n\}$.

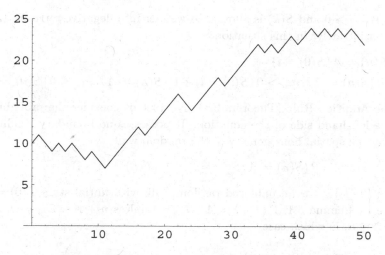

Fig. 5.3 A realization of a simple random walk attempting to capture the changes in the price of a stock.

If i is the smallest non-negative integer such that $m_i = 0$ (and hence that $S(i) = 0$) then by the absorbing boundary condition $S(k) = 0$ for all $k \geq i$. Our attention is now focused on developing an understanding of the following conditional probability:

$$P\left(S(n) = A \wedge m_n > 0 \mid S(0) = i\right).$$

We will assume that $A > 0$ and $i > 0$. This probability should depend on the three parameters A, i, and n. A formula for this conditional probability is derived in the proof of the following lemma.

Lemma 5.2 *Suppose a random walk has the form described in Eq. (5.1) in which the X_i for $i = 1, 2, \ldots$ are independent, identically distributed random variables taking on the values ± 1, each with probability $p = 1/2$. Suppose further that the boundary at 0 is absorbing, then if $A, i > 0$,*

$$P\left(S(n) = A \wedge m_n > 0 \mid S(0) = i\right)$$

$$= \left[\binom{n}{(n+A-i)/2} - \binom{n}{(n-A-i)/2}\right]\left(\frac{1}{2}\right)^n, \tag{5.4}$$

provided $|A - i| \leq n$, $|A + i| \leq n$, and $n + A - i$ is even.

Proof. In order to prove the lemma we will start by considering a random walk with no boundary, that is, the random variable $S(n)$ has an initial state of $S(0) = i > 0$ and $S(k)$ is allowed to wander into negative territory (and back) arbitrarily. In this situation

$$P\,(S(n) = A\,|\,S(0) = i)$$
$$= P\,(S(n) = A \wedge m_n > 0\,|\,S(0) = i) + P\,(S(n) = A \wedge m_n \leq 0\,|\,S(0) = i)$$

by the Addition Rule (Theorem 2.1). Now let us consider the probability on the left-hand side of the equation. It possesses no boundary condition and by the spatial homogeneity of the random walk

$$P\,(S(n) = A\,|\,S(0) = i) = P\,(T(n) = A - i)$$

where $\{T(j)\}_{j=0}^n$ is an unbiased random walk with initial state $T(0) = 0$. Hence by Lemma 5.1, $P\,(T(n) = A - i) = 0$ unless $n + A - i$ is even and $|A - i| \leq n$, in which case

$$P\,(S(n) = A\,|\,S(0) = i) = \binom{n}{(n + A - i)/2}\left(\frac{1}{2}\right)^n.$$

On the other hand if the random walk starts at a positive state i and finishes at $-A < 0$ then it is certain that $m_n \leq 0$. Consequently

$$P\,(S(n) = A \wedge m_n \leq 0\,|\,S(0) = i) = P\,(S(n) = -A\,|\,S(0) = i)$$
$$= \binom{n}{(n - A - i)/2}\left(\frac{1}{2}\right)^n$$

provided $|A + i| \leq n$ and $n - A - i$ is even. Finally we are able to determine that

$$P\,(S(n) = A \wedge m_n > 0\,|\,S(0) = i)$$
$$= \binom{n}{(n + A - i)/2}\left(\frac{1}{2}\right)^n - \binom{n}{(n - A - i)/2}\left(\frac{1}{2}\right)^n.$$

which is equivalent to the expression in Eq. (5.4). $\square$

The previous lemma now provides an answer to the first of the two questions posed.

Example 5.1   For an unbiased random walk with initial state $S(0) = 10$, what is the probability that $S(50) = 16$ and $S(n) > 0$ for $n = 0, 1, \ldots, 50$?

Making use of the formula found in Eq. (5.4) we have

$$P\,(S(50) = 16 \wedge m_{50} > 0\,|\,S(0) = 10) = \left[\binom{50}{28} - \binom{50}{12}\right]2^{-50}$$
$$\approx 0.0787178.$$

Since the current notation is so bulky, we will define the function

$$f_{A,i}(n) = \left[\binom{n}{(n+A-i)/2} - \binom{n}{(n-A-i)/2} \right] \left(\frac{1}{2} \right)^n$$

and use it from now on.

We will call the first time that the random walk $S(n)$ equals A the **stopping time**, denoted by Ω_A. The stopping time is a random variable. In the remainder of this section we will explore the probability that Ω_A takes on the various natural number values given the initial state of the random walk is $S(0) = i > 0$, parameter $A > 0$, and the boundary at 0 is absorbing. Suppose the initial state of the random walk $S(0) = i > A > 0$, then due to the spatial homogeneity of the random walk

$$P\left(S(n) = A \wedge m_{n-1} > A \,|\, S(0) = i \right) =$$
$$P\left(S(n) = 0 \wedge m_{n-1} > 0 \,|\, S(0) = i - A \right).$$

$\Omega_0 = n$ if and only if $S(n-1) = 1$, $m_{n-1} > 0$ and $X_n = -1$. Therefore

$$P\left(\Omega_0 = n \,|\, S(0) = i - A \right)$$
$$= P\left(X_n = -1 \wedge S(n-1) = 1 \wedge m_{n-1} > 0 \,|\, S(0) = i - A \right)$$
$$= \frac{1}{2} P\left(S(n-1) = 1 \wedge m_{n-1} > 0 \,|\, S(0) = i - A \right)$$
$$= \frac{1}{2} f_{1,(i-A)}(n-1).$$

Consequently $P\left(\Omega_A = n \,|\, S(0) = i \right) = \frac{1}{2} f_{1,(i-A)}(n-1)$ as well.

Next we take up the case in which the initial state of the random walk $0 < S(0) = i < A$. Since the stopping time is the *first* time that $S(n) = A$ then we can think of the spatial domain of the random walk as having two absorbing boundaries: the usual one at $S = 0$ and a new one at $S = A$. The analysis of this situation is modeled after the discussion in the first two chapters of [Redner (2001)].

The symbol $p_{i \to A}$ will denote any random walk $\{S(j)\}$ in the discrete interval $[0, A]$ starting at $i > 0$, terminating at A, and which avoids 0. The symbol $P_{p_{i \to A}}$ will denote the probability that the random walk starting at $S(0) = i$ follows $p_{i \to A}$. Finally the symbol $\mathcal{P}_A(i)$ will denote the probability that a random walk which starts at $S(0) = i$ will achieve state $S = A$ while avoiding the state $S = 0$. By the Addition Rule for probability

$$\mathcal{P}_A(i) = \sum_{p_{i \to A}} P_{p_{i \to A}}.$$

Since at each step of the random walk, the increment to the left or right is independent of the previous increments,

$$\mathcal{P}_A(i) = \mathrm{P}\left(S(1) = i - 1 \mid S(0) = i\right)\mathcal{P}_A(i - 1)$$
$$+ \mathrm{P}\left(S(1) = i + 1 \mid S(0) = i\right)\mathcal{P}_A(i + 1)$$
$$= \frac{1}{2}\mathcal{P}_A(i - 1) + \frac{1}{2}\mathcal{P}_A(i + 1)$$

and thus we have

$$\mathcal{P}_A(i - 1) - 2\mathcal{P}_A(i) + \mathcal{P}_A(i + 1) = 0. \tag{5.5}$$

Equation (5.5) is the discrete difference equation approximation to the second derivative [Burden and Faires (2005)]. It is also sometimes referred to as the discrete **Laplacian equation**. Since we know a priori that $\mathcal{P}_A(0) = 0$ and $\mathcal{P}_A(A) = 1$ then we may derive a system of linear equations from Eq. (5.5) whose solution will give us $\mathcal{P}_A(i)$ for $i = 1, 2, \ldots, A - 1$. In matrix form the system of equations resembles:

$$\begin{bmatrix} 1 & -2 & 1 & 0 & \cdots & 0 & 0 & 0 \\ 0 & 1 & -2 & 1 & \cdots & 0 & 0 & 0 \\ \vdots & & & & & & & \vdots \\ 0 & 0 & 0 & 0 & \cdots & 1 & -2 & 1 \end{bmatrix} \begin{bmatrix} \mathcal{P}_A(1) \\ \mathcal{P}_A(2) \\ \vdots \\ \mathcal{P}_A(A-1) \end{bmatrix} = \begin{bmatrix} 0 \\ 0 \\ \vdots \\ 0 \end{bmatrix}.$$

We can use a bit of intuition about Eq. (5.5) to hypothesize and then confirm a form of the solution to this matrix equation. The left-hand side of Eq. (5.5) represents an approximation to the second derivative. The right-hand side of the equation is zero, thus the solution is likely to be a linear function of i. If $\mathcal{P}_A(i) = ai + b$ then upon substituting this linear function in Eq. (5.5) we obtain

$$a(i - 1) + b - 2(ai + b) + a(i + 1) + b = 0$$

confirming that a linear function will express $\mathcal{P}_A(i)$. We can use the two boundary conditions to determine the proper values for the coefficients a and b.

$$\mathcal{P}_A(0) = b = 0$$
$$\mathcal{P}_A(A) = aA = 1 \qquad \Longrightarrow \qquad a = \frac{1}{A}$$

Consequently $\mathcal{P}_A(i) = i/A$. By the same type of argument we can also determine that the probability of a random walk finishing at state 0 while avoiding state $A > 0$ and starting at state $0 \leq i \leq A$ is $\mathcal{P}_0(i) = 1 - i/A$. We can summarize these results in the following theorem.

Theorem 5.3 *Suppose a random walk has the form described in Eq. (5.1) in which the X_i for $i = 1, 2, \ldots$ are independent, identically distributed random variables taking on the values ± 1, each with probability $p = 1/2$. Suppose further that the boundaries at 0 and A are absorbing, then if $0 \leq S(0) = i \leq A$*

(1) the probability that the random walk achieves state A without achieving state 0 is $\mathcal{P}_A(i) = i/A$,
(2). the probability that the random walk achieves state 0 without achieving state A is $\mathcal{P}_0(i) = 1 - i/A$.

Before answering the question as to the expected stopping time for the random walk attaining state A, we will explore the simpler issue of the stopping time of reaching either of the two absorbing boundaries. Let $B = \{0, A\}$ represent the discrete boundary set of the one-dimensional, discrete random walk. To help in the derivation we will define the following quantities:

$\omega_{p_{i \to B}}$: the exit time of the random walk which starts at $S(0) = i$, where $0 \leq i \leq A$ and which follows path $p_{i \to B}$.
$\Omega_B(i)$: the expected value of the exit time for a random walk which starts at $S(0) = i$, where $0 \leq i \leq A$.

By the definition of expected value,

$$\Omega_B(i) = \sum_{p_{i \to B}} P_{p_{i \to B}} \omega_{p_{i \to B}}.$$

Certainly it is the case that if the random walk starts on the boundary the expected stopping time is 0. Thus it is true that $\Omega_B(0) = \Omega_B(A) = 0$. There is also a recursive relationship between the stopping times at neighboring starting points in space. The path from $i \to B$ can be decomposed into paths from $(i - 1) \to B$ and $(i + 1) \to B$ with the addition of a single step. The expected value of the exit time of a random walk starting at i is one more than the expected value of a random walk starting at $i \pm 1$. Therefore

$$\Omega_B(i) = \frac{1}{2}\left(1 + \Omega_B(i - 1)\right) + \frac{1}{2}\left(1 + \Omega_B(i + 1)\right)$$

which is equivalent to the equation below.

$$\Omega_B(i-1) - 2\Omega_B(i) + \Omega_B(i+1) = -2 \tag{5.6}$$

Equation (5.6) is often called the **Poisson equation** for the stopping time. From the Poisson equation we can derive a system of linear equations for the stopping times which are parameterized by the various initial conditions of the random walk. In matrix form the system of equations has the form:

$$\begin{bmatrix} 1 & -2 & 1 & 0 & \cdots & 0 & 0 & 0 \\ 0 & 1 & -2 & 1 & \cdots & 0 & 0 & 0 \\ \vdots & & & & & & & \vdots \\ 0 & 0 & 0 & 0 & \cdots & 1 & -2 & 1 \end{bmatrix} \begin{bmatrix} \Omega_B(1) \\ \Omega_B(2) \\ \vdots \\ \Omega_B(A-1) \end{bmatrix} = \begin{bmatrix} -2 \\ -2 \\ \vdots \\ -2 \end{bmatrix}.$$

The matrix generated by the Poisson equation (as well as the earlier matrix generated by the Laplacian equation) are known as **tridiagonal matrices** due to the fact that all entries in the matrices are zero except for possibly the entries on the diagonal, the first sub-diagonal, and the first super-diagonal. There are many methods for solving tridiagonal linear systems.

Some of the simplest methods involve **Gaussian elimination** followed by **backwards substitution**. This technique as well as many others is explained in detail in [Golub and Van Loan (1989)].

An alternative to the matrix solution technique is to observe that the left-hand side of Eq. (5.6) is an approximation to the second derivative of Ω_B with respect to i. A reasonable hypothesis is that $\Omega_B(i)$ is a quadratic function of i, since the second derivative is the constant -2. Suppose $\Omega_B(i) = ai^2 + bi + c$ where a, b, and c are constants. Substituting this quadratic expression into Eq. (5.6) produces

$$-2 = a(i-1)^2 + b(i-1) + c - 2(ai^2 + bi + c) + a(i+1)^2 + b(i+1) + c$$
$$= 2a$$
$$a = -1$$

Recall that on the boundary $\Omega_B(0) = 0 = \Omega_B(A)$ and therefore

$$0 = c \quad \text{(the case } i = 0\text{)}$$
$$0 = -A^2 + bA \quad \text{(the case } i = A\text{)},$$

therefore $b = A$. Hence the solution to the Poisson equation for the expected value of the exit time is $\Omega_B(i) = i(A - i)$ for $i = 0, 1, \ldots, A$. The results are summarized in the following theorem.

Theorem 5.4   *Suppose a random walk has the form described in Eq. (5.1) in which the X_i for $i = 1, 2, \ldots$ are independent, identically distributed random variables taking on the values ± 1, each with probability $p = 1/2$. Suppose further that the boundaries at 0 and A are absorbing, then if $0 \leq S(0) = i \leq A$ the random walk intersects the boundary $(S = 0$ or $S = A)$ after a mean number of steps given by the formula*

$$\Omega_B(i) = i(A - i). \tag{5.7}$$

Example 5.2   Suppose an unbiased random walk takes place on the discrete interval $\{0, 1, 2, 3, 4\}$ for which the boundaries at 0 and 4 are absorbing. If $S(0) = i$ then the expected value of the stopping time of the random walk is

i	0	1	2	3	4
$\Omega_B(i)$	0	3	4	3	0

Theorem 5.4 yields a formula for the mean number of steps required for an unbiased random walk to reach either boundary point of the interval $[0, A]$ based on the starting point. Now we want to determine a related quantity, the conditional exit time. This is the average number of steps required for an unbiased random walk to reach the boundary at A while avoiding the other boundary at 0. The symbol $\Omega_A(i)$ will denote the expected value of the conditional exit time of a random walk in the initial state $S(0) = i$ which exits at boundary A while avoiding the boundary at 0. By definition this conditional exit time is

$$\Omega_A(i) = \frac{\displaystyle\sum_{p_{i \to A}} P_{p_{i \to A}} \omega_{p_{i \to A}}}{\displaystyle\sum_{p_{i \to A}} P_{p_{i \to A}}} = \frac{\displaystyle\sum_{p_{i \to A}} P_{p_{i \to A}} \omega_{p_{i \to A}}}{\mathcal{P}_A(i)}.$$

Note that the sums above are taken over all random walks which start at $S(0) = i$ and exit at A and avoid 0. This implies the conditional exit times and path probabilities are related by the equation:

$$\Omega_A(i)\mathcal{P}_A(i) = \sum_{p_{i \to A}} P_{p_{i \to A}} \omega_{p_{i \to A}}. \tag{5.8}$$

Once again the idea of decomposing a random walk into a first step and the remainder of the steps will be used on the right-hand side of Eq. (5.8). The conditional exit time of a random walk starting in state i will be one more than the conditional exit times of random walks starting in states $i \pm 1$. Therefore

$$\Omega_A(i) = 1 + \frac{\frac{1}{2}\Omega_A(i-1)\mathcal{P}_A(i-1) + \frac{1}{2}\Omega_A(i+1)\mathcal{P}_A(i+1)}{\frac{1}{2}\mathcal{P}_A(i-1) + \frac{1}{2}\mathcal{P}_A(i+1)}$$

$$= 1 + \frac{\frac{1}{2}\Omega_A(i-1)\mathcal{P}_A(i-1) + \frac{1}{2}\Omega_A(i+1)\mathcal{P}_A(i+1)}{\mathcal{P}_A(i)}$$

$$\Omega_A(i)\mathcal{P}_A(i) = \mathcal{P}_A(i) + \frac{1}{2}\Omega_A(i-1)\mathcal{P}_A(i-1) + \frac{1}{2}\Omega_A(i+1)\mathcal{P}_A(i+1)$$

$$\Omega_A(i)\frac{i}{A} = \frac{i}{A} + \frac{i-1}{2A}\Omega_A(i-1) + \frac{i+1}{2A}\Omega_A(i+1) \quad \text{(by Theorem 5.3)}.$$

The last equation is equivalent to

$$(i-1)\Omega_A(i-1) - 2i\Omega_A(i) + (i+1)\Omega_A(i+1) = -2i. \tag{5.9}$$

The hypothesis that the solution is a quadratic function of i was profitable once before, so we will try it again. If we let $\Omega_A(i) = ai^2 + bi + c$ and substitute this into Eq. (5.9) we obtain

$$-2i = (i-1)\left[a(i-1)^2 + b(i-1) + c\right] - 2i\left[ai^2 + bi + c\right]$$
$$+ (i+1)\left[a(i+1)^2 + b(i+1) + c\right]$$
$$0 = (3a+1)i + b.$$

Therefore $a = -\frac{1}{3}$ and $b = 0$. To evaluate the remaining coefficient c, we will use the boundary condition that $\Omega_A(A) = 0$ which implies.

$$0 = -\frac{1}{3}A^2 + c \implies c = \frac{1}{3}A^2.$$

Finally we have established the following theorem.

Theorem 5.5 *Suppose a random walk has the form described in Eq. (5.1) in which the X_i for $i = 1, 2, \ldots$ are independent, identically distributed random variables taking on the values ± 1, each with probability $p = 1/2$. Suppose further that the boundary at 0 is absorbing. The random walk that avoids state 0 will stop the first time that $S(n) = A$. The expected value of the stopping time is*

$$\Omega_A(i) = \frac{1}{3}\left(A^2 - i^2\right), \quad \text{for } i = 1, 2, \ldots, A. \tag{5.10}$$

If the random walk starts in state 0, since this state is absorbing the expected value of the exit time is infinity.

Example 5.3 Suppose an unbiased random walk takes place on the integer-valued number line. Suppose the boundary at 0 is absorbing. If $0 < S(0) = i \le 5$ then the expected values of the first passage, or stopping, times at which the random walk attains a value of 5 are given in the table below.

i	1	2	3	4	5
$\Omega_5(i)$	8	7	$\frac{16}{3}$	3	0

5.3 Intuitive Idea of a Stochastic Process

In this section we will bridge the gap between the discrete random walk and the continuous random walk. A fully rigorous treatment of this topic is quite beyond the undergraduate level, so this section will depend heavily on intuition. Readers interested in more of the details should consult any of the references on **stochastic calculus** such as [Lawler (2006)] or [Wilmott (2006)].

We now have the need for a more traditional mathematical description of these continuous random walks. An appropriate mathematical model must include some random component like the flipping coin of the previous sections. However a more useful random variable may be one which has a continuous distribution with a prescribed mean and standard deviation like the normal distribution encountered in Chap. 3.

Consider once again the unbiased random walk defined in Eq. (5.1). Suppose the n selections of the Bernoulli random variables X_j will take place equally spaced in time in the interval $[0, t]$. To keep the discussion simple, suppose as well that $S(0) = 0$. Our goal in this section is to derive a continuous random walk from the discrete random walk $\{S(jt/n)\}_{j=0}^n$ as $n \to \infty$. At each tick of the clock the random variable takes a step either to the left or to the right, each with probability $1/2$. All steps are identically distributed and pairwise independent. The size of the step is constrained to be $\sqrt{t/n}$. In other words for each j, $X_j = \pm\sqrt{t/n}$, each with probability $1/2$. Therefore the final state $S(t)$ is a binomial random variable with $\mathrm{E}[S(t)] = 0$ and $\mathrm{Var}(S(t)) = n(\sqrt{t/n})^2 = t$. By the Central Limit Theorem, as $n \to \infty$, $S(t)$ approaches a normal distribution, one we have engineered to have mean zero and standard deviation $\sqrt{t}$. A graphical interpretation of the limiting process is illustrated in Fig. 5.4. This limit is the continuous analogue of the random walk studied in Section 5.2. This mathematical model of continuous random motion is called **Brown-**

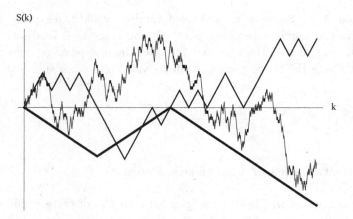

S(k)

k

Fig. 5.4 As more frequent, but smaller, steps are taken by a random walk, the process limits on a continuous stochastic process called Brownian motion.

ian motion, named after the Scottish botanist, Robert Brown, who first observed the random motion of particles of pollen suspended in a fluid. The time evolution of $S(t)$ is now described as a **stochastic process** due to its random behavior. The mathematical description of Brownian motion was developed by the mathematician, Norbert Wiener. In recognition of this contribution, this stochastic process is often called a **Wiener process** and is denoted by $W(t)$.

Brownian motion possesses several important properties, most of which require sophisticated real analysis and probability arguments to verify. First the reader may wonder about the choice of scaling ΔX with the square root of t. This is a practical choice, since (without proof) any other would produce a random walk which can become unbounded (*i.e.*, go to infinity) in finite time or would result in a constant stochastic process. Other properties of Brownian motion include:

(1) $W(t)$ is a continuous function of t,
(2) $W(0) = 0$ with probability one,
(3) Spatial homogeneity: if $W_0(t)$ represents a Wiener process for which the initial state is 0 and if $W_x(t)$ represents a Wiener process for which the initial state is x, then $W_x(t) = x + W_0(t)$.
(4) Markov property: for $t > s > 0$ the conditional distribution of $W(t)$ depends on the value of $W(s) + W(t - s)$.
(5) For each t, $W(t)$ is normally distributed with mean zero and variance t,

(6) The changes in W in non-overlapping intervals of t are independent random variables with means of zero and variances equal to the lengths of the time intervals.

This last condition is made clearer by considering four values of t satisfying $0 \le t_1 < t_2 \le t_3 < t_4$. The random variables

$$\Delta W_{[t_1,t_2]} = W(t_2) - W(t_1) \quad \text{and} \quad \Delta W_{[t_3,t_4]} = W(t_4) - W(t_3)$$

are independent. The condition that $t_2 = t_3$ does not constitute overlapping intervals. From the formulas for expected value and variance and the definition of the Wiener process we can readily see that when $t_2 > t_1 \ge 0$

$$\mathrm{E}\left[W(t_2) - W(t_1)\right] = \mathrm{E}\left[W(t_2)\right] - \mathrm{E}\left[W(t_1)\right] = 0 - 0 = 0. \tag{5.11}$$

The variance of the change in the Wiener process can also be found by making use of the fact that

$$
\begin{aligned}
t &= \mathrm{Var}\left(W(t)\right) \\
&= \mathrm{Var}\left(W(t) - W(0)\right) \quad \text{(by property (2))} \\
&= \mathrm{E}\left[(W(t) - W(0))^2\right] \quad \text{(by Eq. (5.11))} \\
&= \mathrm{E}\left[(W(t))^2\right].
\end{aligned}
$$

Hence

$$
\begin{aligned}
\mathrm{Var}\left(W(t_2) - W(t_1)\right) &= \mathrm{E}\left[(W(t_2) - W(t_1))^2\right] - \mathrm{E}\left[W(t_2) - W(t_1)\right]^2 \\
&= \mathrm{E}\left[(W(t_2))^2\right] + \mathrm{E}\left[(W(t_1))^2\right] - 2\mathrm{E}\left[W(t_1)W(t_2)\right] \\
&= t_2 + t_1 - 2\mathrm{E}\left[W(t_1)(W(t_2) - W(t_1) + W(t_1))\right] \\
&= t_2 + t_1 - 2\mathrm{E}\left[W(t_1)(W(t_2) - W(t_1))\right] \\
&\quad - 2\mathrm{E}\left[(W(t_1))^2\right] \\
&= t_2 + t_1 - 2t_1 \\
&= t_2 - t_1. \tag{5.12}
\end{aligned}
$$

The reader should examine each step in the derivation of the variance of the difference $W(t_2) - W(t_1)$ in Eq. (5.12) making note of where the various properties of the Wiener process were used.

If we let $\Delta t = t_2 - t_1$ where $t_2 > t_1 \ge 0$ and let $\Delta W = W(t_2) - W(t_1)$ then we may note that by Eq. (5.12) we have $\mathrm{E}\left[(\Delta W)^2\right] = \Delta t$. So far we have established these results for changes in a Wiener process over the interval $[t_1, t_2]$ in the discrete sense. We made use only of quantities evaluated at times t_1 and t_2.

(7) Quadratic variation: for a partition of the interval $[0, t]$ into n non-overlapping subintervals,

$$\lim_{n \to \infty} \sum_{k=1}^{n} [W(t_k) - W(t_{k-1})]^2 = t \qquad (5.13)$$

where the limit is taken in the **mean square** sense.

This quadratic variation property is important for the development of a **stochastic integral** and thus will be explained in more detail. Keep in mind that $W(s)$ is a random variable, so it is only meaningful to refer to its expected value and variance.

Before proceeding to the main result of this chapter we should explore where a continuous version of Eq. (5.12) holds. The following lemma (adapted from [Seydel (2002), Chap. 1]), while technical in nature, will establish the necessary result.

Lemma 5.3   *Let $\{P^{(n)}\}$ for $n \in \mathbb{N}$ be a sequence of partitions of the interval $[0, t]$ such that*

$$0 = t_0^{(n)} < t_1^{(n)} < \cdots < t_n^{(n)} = t.$$

For each i let $\Delta t_i^{(n)} = t_i^{(n)} - t_{i-1}^{(n)}$ and $\Delta W_i^{(n)} = W(t_i^{(n)}) - W(t_{i-1}^{(n)})$ and let $\delta_n = \max_i \{\Delta t_i^{(n)}\}$. Then

$$\mathrm{E}\left[\left(\sum_{i=1}^{n} ((\Delta W_i^{(n)})^2 - \Delta t_i^{(n)})\right)^2\right] \to 0 \quad as \quad \delta_n \to 0. \qquad (5.14)$$

Proof.   The proof will make use of the calculation that

$$\mathrm{E}\left[(\Delta W_i^{(n)})^4\right] = 3(\Delta t_i^{(n)})^2 \qquad (5.15)$$

see exercise 14. To simplify the notation we can drop the superscripts of the form $\cdot^{(n)}$. The reader may verify by expanding the product that

$$\left(\sum_{i=1}^{n} ((\Delta W_i)^2 - \Delta t_i)\right)^2 = \sum_{i=1}^{n} (\Delta W_i)^4 + \sum_{i=1}^{n} (\Delta t_i)^2 - 2 \sum_{i=1}^{n} \sum_{j=i}^{n} (\Delta W_i)^2 \Delta t_j$$

$$+ 2 \sum_{i=1}^{n-1} \sum_{j=i+1}^{n} (\Delta W_i)^2 (\Delta W_j)^2.$$

We can evaluate the following expected values:

$$E\left[\sum_{i=1}^{n}(\Delta W_i)^4\right] = 3\sum_{i=1}^{n}(\Delta t_i)^2 \quad \text{(by Eq. (5.15))}$$

$$E\left[\sum_{i=1}^{n}(\Delta t_i)^2\right] = \sum_{i=1}^{n}(\Delta t_i)^2$$

$$E\left[2\sum_{i=1}^{n}\sum_{j=i}^{n}(\Delta W_i)^2\Delta t_j\right] = 2\sum_{i=1}^{n}\sum_{j=i}^{n}(\Delta t_i)(\Delta t_j)$$

$$E\left[2\sum_{i=1}^{n-1}\sum_{j=i+1}^{n}(\Delta W_i)^2(\Delta W_j)^2\right] = 2\sum_{i=1}^{n-1}\sum_{j=i+1}^{n}(\Delta t_i)(\Delta t_j)$$

The last equation is true since the ΔW_i and ΔW_j are independent for $i \neq j$. Therefore

$$E\left[\left(\sum_{i=1}^{n}((\Delta W_i)^2 - \Delta t_i)\right)^2\right]$$

$$= 3\sum_{i=1}^{n}(\Delta t_i)^2 + \sum_{i=1}^{n}(\Delta t_i)^2 - 2\sum_{i=1}^{n}\sum_{j=i}^{n}(\Delta t_i)(\Delta t_j) + 2\sum_{i=1}^{n-1}\sum_{j=i+1}^{n}(\Delta t_i)(\Delta t_j)$$

$$= 4\sum_{i=1}^{n}(\Delta t_i)^2 - 2\sum_{i=1}^{n}\sum_{j=i}^{n}(\Delta t_i)(\Delta t_j) + 2\sum_{i=1}^{n-1}\sum_{j=i+1}^{n}(\Delta t_i)(\Delta t_j)$$

$$= 4\sum_{i=1}^{n}(\Delta t_i)^2 - 2\sum_{i=1}^{n}(\Delta t_i)^2$$

$$= 2\sum_{i=1}^{n}(\Delta t_i)^2 \qquad (5.16)$$

Now we can see that

$$2\sum_{i=1}^{n}(\Delta t_i)^2 = 2\left((\Delta t_1)^2 + (\Delta t_2)^2 + \cdots + (\Delta t_n)^2\right)$$

$$\leq 2\delta_n(\Delta t_1 + \Delta t_2 + \cdots + \Delta t_n)$$

$$= 2\delta_n t.$$

Finally since

$$0 \le \mathrm{E}\left[\left(\sum_{i=1}^{n}((\Delta W_i)^2 - \Delta t_i)\right)^2\right] \le 2\delta_n t,$$

the limit in Eq. (5.14) follows as a result of the Squeeze Theorem as $n \to \infty$ or equivalently as $\delta_n \to 0$.   □

Hence we may now pass to the limit as Δt becomes infinitesimally small and write

$$(dW(t))^2 = dt. \tag{5.17}$$

This result will be used to prove Itô's Lemma in the next section.

(8) The derivative dW/dt does not exist.

The final property addresses the "roughness" of the path followed by the Wiener process as seen in Fig. 5.4. Technically we should say that the probability that dW/dt exists is zero. Recall the limit definition of the derivative from calculus,

$$\frac{df}{dt} = \lim_{h \to 0} \frac{f(s+h) - f(s)}{h}.$$

Suppose $f(t)$ is a Wiener process $W(t)$. Since

$$\mathrm{E}\left[(W(s+h) - W(s))^2\right] = \mathrm{E}\left[|W(s+h) - W(s)|^2\right] = h$$

then on average $|W(s+h) - W(s)| \approx \sqrt{h}$, and therefore

$$\lim_{h \to 0} \frac{W(s+h) - W(s)}{h} \quad \text{does not exist.}$$

Of central importance in the remainder of the book will be the calculus of functions defined along a continuous random walk. We have already argued that the derivative of a Wiener process does not exist, but what about the integral? We define the stochastic integral of a function $f(\tau)$ defined on $[0, t]$ informally as

$$Z(t) = \int_0^t f(\tau)\, dW(\tau) = \lim_{n \to \infty} \sum_{k=1}^{n} f(t_{k-1}) \left(W(t_k) - W(t_{k-1})\right) \tag{5.18}$$

where $t_k = kt/n$. Of course this only makes sense if the limit exists. Note that the function f is always evaluated at the left-hand endpoint of the subinterval $[t_{k-1}, t_k]$. Think if this as ensuring that the function being integrated cannot use any information about the future movements of the random walk as it is being integrated. To give a perhaps more concrete example, suppose we are going to place continuous wagers defined by $f(\tau)$ during the interval $[0, t]$ on the outcome of the continuous random walk $W(\tau)$. The stochastic integral given in Eq. (5.18) would represent our net winnings.

The stochastic integral is sometimes written in a differential form obtained by informally differentiating Eq. (5.18).

$$dZ = f(t)dW(t) \tag{5.19}$$

However, beware that we are not allowed to divide both sides of Eq. (5.19) by dt since dW/dt is undefined. In general Eqs. (5.18) and (5.19) are considered to be synonymous.

With this informal introduction to stochastic processes completed we can now begin to generalize the discussion. A familiar place to start generalizing is with the deterministic process of exponential growth and decay. Most introductions to calculus (see for example [Stewart (1999)]) contain the mathematical model which states that the rate of change of a nonnegative quantity P, is proportional to P. Expressed as a differential equation this statement becomes

$$\frac{dP}{dt} = \mu P \tag{5.20}$$

where μ is the proportionality constant and t usually represents time. The proportionality constant is called a "growth rate" if $\mu > 0$ and a "decay rate" if $\mu < 0$. If the value of P is known at a specified value of t (usually at $t = 0$) then this differential equation has solution $P(t) = P(0)e^{\mu t}$, see exercise 7. This mathematical model is described as **deterministic** since there is no place for random events to express themselves in the model. Once the initial value of P and the proportionality constant are set, the future evolution of $P(t)$ is completely determined.

Equation (5.20) can be rewritten as

$$\frac{dP}{P} = \mu\, dt, \tag{5.21}$$

and if we make the change of variable $Z = \ln P$ then this equation becomes

$$dZ = \mu\, dt. \tag{5.22}$$

Now suppose we add a stochastic component to the mix by introducing on the right-hand side of Eq. (5.22) a Wiener process with mean zero and standard deviation $\sigma\sqrt{dt}$ (which of course means the variance is $\sigma^2\,dt$). We obtain the mathematical model governing the time evolution of Z below.

$$dZ = \mu\,dt + \sigma\,dW(t) \tag{5.23}$$

A stochastic differential equation of this form is called a **generalized Wiener process**. Notice it possesses a deterministic part and a stochastic part. The constant μ is called the **drift** and the constant σ is called the **volatility**. The integral form of Eq. (5.23) is

$$Z(t) = Z(0) + \mu t + \int_0^t \sigma\,dW(\tau).$$

Do not lose sight of the fact that Z is a random variable. Therefore the $Z(t)$ on the left-hand side of the equation is really an expected value. Assuming that it is valid to interchange of the order of the limit and the expected value we see that

$$\mathrm{E}\left[\Delta Z\right] = \mathrm{E}\left[Z(t) - Z(0)\right]$$

$$= \mu t + \mathrm{E}\left[\int_0^t \sigma\,dW(\tau)\right]$$

$$= \mu t + \mathrm{E}\left[\lim_{n\to\infty}\sum_{k=1}^n \sigma(W(t_k) - W(t_{k-1}))\right]$$

$$= \mu t + \lim_{n\to\infty}\sum_{k=1}^n \sigma\mathrm{E}\left[W(t_k) - W(t_{k-1})\right]$$

$$= \mu t. \tag{5.24}$$

Meanwhile the variance is calculated as

$$\text{Var}\,(\Delta Z) = \text{Var}\,(Z(t) - Z(0))$$

$$= \text{Var}\left(\mu t + \lim_{n \to \infty} \sum_{k=1}^{n} \sigma(W(t_k) - W(t_{k-1}))\right)$$

$$= \sigma^2 \lim_{n \to \infty} \sum_{k=1}^{n} \text{Var}\,(W(t_k) - W(t_{k-1}))$$

$$= \sigma^2 \lim_{n \to \infty} \sum_{k=1}^{n} \left(\frac{t}{n}\right)$$

$$= \sigma^2 t.$$

Numerically the value of $Z(t)$ can be approximated by choosing a small $\Delta t = t/n$ and replacing

$$\int_0^t dW(\tau) \quad \text{with} \quad \sum_{j=1}^{n} X_j$$

where X_j is a normal random variable with mean 0 and variance $\sqrt{t/n}$. This approximation is used in the following two examples.

Example 5.4 Suppose the drift parameter is $\mu = 1$ and the volatility is $\sigma = 1/4$, then the expected value of the Wiener process is t and the standard deviation is $\sqrt{t}/4$. Plots of the expected value, the expected value plus and minus a standard deviation, and one realization of the random walk are illustrated in Fig. 5.5.

Example 5.5 Suppose the drift parameter is $\mu = 1/4$ and the volatility is $\sigma = 1$, then the expected value of the Wiener process is $t/4$ and the standard deviation is $\sqrt{t}$. Plots of the expected value, the expected value plus and minus a standard deviation, and one realization of the random walk are illustrated in Fig. 5.6.

Indeed we can generalize still further by imagining the drift and volatility are functions of t. For this case the stochastic differential equation has the form

$$dZ = \mu(t)\,dt + \sigma(t)\,dW(t).$$

Using the ordinary Riemann integral on the deterministic portion of the right-hand side and the stochastic integral on the random portion we can

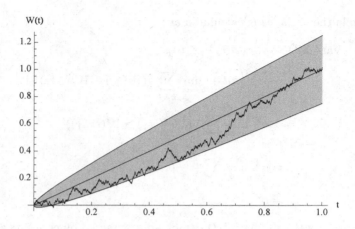

Fig. 5.5   The behavior of a Wiener process with drift $\mu = 1$ and volatility $\sigma = 1/4$.

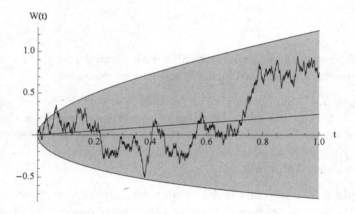

Fig. 5.6   The behavior of a Wiener process with drift $\mu = 1/4$ and volatility $\sigma = 1$.

find

$$Z(t) = Z(0) + \int_0^t \mu(\tau) \, d\tau + \int_0^t \sigma(\tau) \, dW(\tau).$$

If we wish to generalize still further by making μ and σ functions not only of t but of Z (and possibly other variables) then we can no longer integrate in closed form. However, a change of variable may enable us to convert these more general stochastic differential equations in Wiener form. In the next two sections we will develop a version of the chain rule for derivatives which is applicable to functions of stochastic variables and which can help overcome our current limitations in stochastic integration.

5.4 Itô Processes

One class of further generalization of Eq. (5.23) includes the **Itô processes** of the form

$$dZ = a(Z, t)\, dt + b(Z, t)\, dW(t) \tag{5.25}$$

where the expressions a and b are each now functions of time t and the random variable Z. In general the stochastic differential equation given in Eq. (5.25) cannot be solved explicitly through integration of both sides. Nevertheless it is an important type of equation for us to use because it will enable us to work with stochastic processes of quantities which depend in turn on other random variables having their own individual stochastic process descriptions. In this way we will develop the fundamentally important Black-Scholes partial differential equation. Before we can do this we must develop an analogue of the chain rule for differentiation found in elementary calculus.

In standard differential calculus if one makes the assignment $Z = \ln P$ then it is well understood that $dZ = dP/P$ according to the chain rule. Using this on a generalized Wiener process may lead us to believe that the following equations are equivalent.

$$dZ = \mu\, dt + \sigma\, dW(t)$$
$$dP = \mu P\, dt + \sigma P\, dW(t)$$

First we note that while the first equation is a generalized Wiener process the second is an Itô process. From the point of view of an applied mathematician the second equation has an advantage over the first. As P gets close to zero the "drift-like" quantity μP and the "volatility-like" quantity σP become small as well. This will prevent P from becoming negative (assuming that $P(0) > 0$). Non-negativity is certainly a property a mathematical model of the price of a financial instrument should possess. The first equation will allow Z to become negative even when the drift is positive. Secondly, if μ and σ represent the mean and standard deviation respectively of random variable dZ (as they did for the stock price data analyzed earlier), does the second equation imply that μ and σ are the mean and standard deviation of the underlying variable P? The answer is no.

What is needed is a valid procedure for changing variables in a stochastic process, in other words, a stochastic calculus version of the chain rule for differentiation. This change of variable differentiation result is known as Itô's Lemma and is covered in the next section.

5.5 Itô's Lemma

The proper procedure for changing variables in a stochastic differential equation can be derived using the multivariable version of Taylor's Theorem. A standard reference containing this result is [Taylor and Mann (1983)]. Here we will give a brief overview of a two-variable version of Taylor's formula with remainder. We start with the single-variable Taylor's formula. If $f(x)$ is an $(n+1)$-times differentiable function on an open interval containing x_0 then the function may be written as

$$f(x) = f(x_0) + f'(x_0)(x - x_0) + \frac{f''(x_0)}{2!}(x - x_0)^2$$
$$+ \cdots + \frac{f^{(n)}(x_0)}{n!}(x - x_0)^n + \frac{f^{(n+1)}(\theta)}{(n+1)!}(x - x_0)^{n+1} \qquad (5.26)$$

The last term above is usually called the Taylor remainder formula and is denoted by R_{n+1}. The quantity θ lies between x and x_0. The other terms form a polynomial in x of degree at most n and can be used as an approximation for $f(x)$ in a neighborhood of x_0. This form of Taylor's will be used to derive the two-variable form.

Suppose the function $F(y, z)$ has partial derivatives up to order three on an open disk containing the point with coordinates (y_0, z_0). Define the function $f(x) = F(y_0 + xh, z_0 + xk)$ where h and k are chosen small enough that $(y_0 + h, z_0 + k)$ lie within the disk surrounding (y_0, z_0). Since f is a function of a single variable then we can use the single-variable form of Taylor's formula in Eq. (5.26) with $x_0 = 0$ and $x = 1$ to write

$$f(1) = f(0) + f'(0)x + \frac{1}{2}f''(0)x^2 + R_3. \qquad (5.27)$$

Using the multivariable chain rule for derivatives we have, upon differentiating $f(x)$ and setting $x = 0$,

$$f'(0) = hF_y(y_0, z_0) + kF_z(y_0, z_0) \qquad (5.28)$$
$$f''(0) = h^2 F_{yy}(y_0, z_0) + 2hk F_{yz}(y_0, z_0) + k^2 F_{zz}(y_0, z_0). \qquad (5.29)$$

We have made use of the fact that $F_{yz} = F_{zy}$ for this function under the smoothness assumptions. The remainder term R_3 contains only third order partial derivatives of F evaluated somewhere on the line connecting the points (y_0, z_0) and $(y_0 + h, z_0 + k)$. Thus if we substitute Eqs. (5.28) and (5.29) into (5.27) we obtain

$$
\begin{aligned}
\Delta F &= f(1) - f(0) \\
&= F(y_0 + h, z_0 + k) - F(y_0, z_0) \\
&= hF_y(y_0, z_0) + kF_z(y_0, z_0) + \frac{1}{2}(h^2 F_{yy}(y_0, z_0) \\
&\quad + 2hkF_{yz}(y_0, z_0) + k^2 F_{zz}(y_0, z_0)) + R_3.
\end{aligned}
\tag{5.30}
$$

This last equation can be used to derive Itô's Lemma.

Let X be a random variable described by an Itô process of the form

$$
dX = a(X, t)\, dt + b(X, t)\, dW(t)
\tag{5.31}
$$

where $dW(t)$ is a normal random variable and a and b are functions of X and t. Let $Y = F(X, t)$ be another random variable defined as a function of X and t. Given the Itô process which describes X we will now determine the Itô process which describes Y.

Using a Taylor series expansion for Y detailed in (5.30) we find

$$
\begin{aligned}
\Delta Y &= F_X \Delta X + F_t \Delta t + \frac{1}{2}F_{XX}(\Delta X)^2 + F_{Xt}\Delta X \Delta t + \frac{1}{2}F_{tt}(\Delta t)^2 + R_3 \\
&= F_X(a\Delta t + b\, dW(t)) + F_t \Delta t + \frac{1}{2}F_{XX}(a\Delta t + b\, dW(t))^2 \\
&\quad + F_{Xt}(a\Delta t + b\, dW(t))\Delta t + \frac{1}{2}F_{tt}(\Delta t)^2 + R_3.
\end{aligned}
$$

Upon simplifying, the expression ΔX has been replaced by the discrete version of the Itô process. The dependence of a and b on X and t has been suppressed to simplify the notation. The reader is reminded that the Taylor remainder term R_3 contains terms of order $(\Delta t)^k$ where $k \geq 2$. Thus as Δt becomes small

$$
\Delta Y \approx F_X(a\,dt + b\,dW(t)) + F_t dt + \frac{1}{2!}F_{XX}b^2(dW(t))^2.
$$

Making use of Eq. (5.17) we can obtain the approximation

$$
\Delta Y \approx F_X(a\,dt + b\,dW(t)) + F_t dt + \frac{1}{2!}F_{XX}b^2 dt.
\tag{5.32}
$$

The reader should be aware that this is a merely an outline of a proof of Itô's Lemma. The interested reader should consult [Neftci (2000), Chap. 10] for a more rigorous proof. To summarize what we have done, the statement of the lemma is given below.

Lemma 5.4 *(Itô's Lemma) Suppose that the random variable X is described by the Itô process*

$$dX = a(X,t)\,dt + b(X,t)\,dW(t) \tag{5.33}$$

where $dW(t)$ is a normal random variable. Suppose the random variable $Y = F(X,t)$. Then Y is described by the following Itô process.

$$dY = \left(a(X,t)F_X + F_t + \frac{1}{2}(b(X,t))^2 F_{XX} \right) dt + b(X,t)F_X\,dW(t) \tag{5.34}$$

Now we can return to the question raised earlier regarding the stochastic process followed by a stock price P. If $Z = \ln P$ and

$$dZ = \mu\,dt + \sigma\,dW(t), \tag{5.35}$$

then we can let $F(Z,t) = e^Z = P$. According to Itô's Lemma then

$$dP = \left(\mu e^Z + 0 + \frac{1}{2}\sigma^2 e^Z \right) dt + \sigma e^Z\,dW(t)$$

$$= \left(\mu + \frac{1}{2}\sigma^2 \right) P\,dt + \sigma P\,dW(t)$$

Note that the coefficient of the deterministic portion of the equation contains the expression $\frac{1}{2}\sigma^2$ which would have been absent had the chain rule of deterministic calculus been used instead of Itô's Lemma. This lemma plays a role of central importance in the next chapter in which the Black-Scholes partial differential equations is derived.

5.6 Stock Market Example

In a previous section we explored a discrete approximation to a continuous model for generating a random walk which seemed to mimic the fluctuations of the price of a corporation's stock. The model depended on several parameters which were called μ, σ, and Δt. In this section we intend to analyze the data associated with the stock price of an actual corporation and determine how to assign values to these parameters which are appropriate to that corporation. The end-of-day closing price of a corporation's stock is generally available on the World Wide Web. One convenient source of such information is the website finance.yahoo.com. In this section we will use data for Sony Corporation stock. Appendix A contains the raw data analyzed here. The data is graphed in Fig. 5.7. The closing prices of the

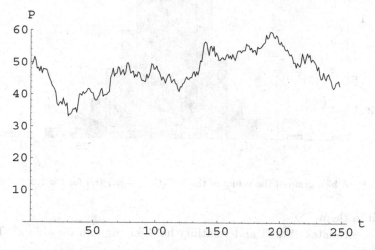

Fig. 5.7 The closing prices of Sony Corporation stock between August 13, 2001 (day 1) and August 12, 2002 (day 248).

stock for 248 consecutive trading days are represented in this figure.

Since the data is sampled once every trading day (and days on which no trading took place, such as weekends and holidays, are irrelevant) then the value of Δt appropriate for the model to be developed is $\Delta t = 1$. Simple descriptive statistics can be used to determine μ and σ. The raw stock prices are playing the role of the variable P in our model. To fit the discrete form of the model used in Eq. (5.23) we must take the natural logarithm of the the stock prices, since $Z = \ln P$. Then we must form ΔZ by subtracting consecutive days logarithmic prices. If the values of ΔZ just calculated fit the model then they will appear to be normally distributed. The mean of this data will be our estimate of μ and the standard deviation will be an estimate of σ which is also known as the volatility. A histogram of the values of ΔZ calculated from the data in Appendix A appears in Fig. 5.8. It is certainly plausible that the values of ΔZ summarized there are normally distributed. The mean of data is $\mu \approx -0.000555 \, \text{day}^{-1}$ and the standard deviation is $\sigma \approx 0.028139 \, \text{day}^{-1}$. Thus, despite the apparent randomness of the closing prices there is some structure and organization

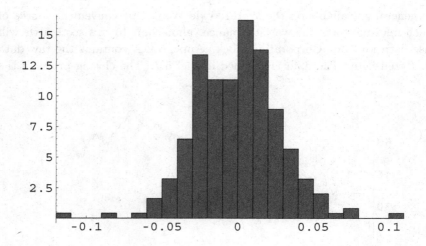

Fig. 5.8   A histogram of the values of the $\ln P(t_{i+1}) - \ln P(t_i)$ for $i = 1, 2, \ldots, 247$.

hidden in them.

The estimates of drift and volatility hold for random variable Z. Thus as before,

$$Z(t) = Z(0) + \mu t + \int_0^t \sigma \, dW(\tau)$$

which implies that the random motion of the stock prices obeys the formula

$$P(t) = P(0)e^{\mu t + \int_0^t \sigma \, dW(\tau)}. \tag{5.36}$$

The mean and variance of $P(t)$ are not μt and $\sigma^2 t$, but since P is a lognormal random variable, Lemma 3.1 implies its mean and variance are respectively

$$\mathrm{E}\left[P(t)\right] = e^{(\mu + \sigma^2/2)t}$$
$$\mathrm{Var}\left(P(t)\right) = e^{(2\mu + \sigma^2)t}\left(e^{\sigma^2 t} - 1\right)$$

The discrete approximation of the stochastic Eq. (5.23) evidently matches well the behavior of the stock price of an actual company. In the next chapters we will begin to make use of the brief information presented in this chapter on stochastic processes. A reader interested in a more rigorous introduction to the mathematics and statistics of stochastic processes and stochastic differential equations should consult one or more of the references such as [Durrett (1996)] or [Mikosch (1998)]. In this book we are merely taking an intuitive, non-rigorous approach to their use as tools for modeling prices of stocks and their derivatives.

5.7 Exercises

(1) Another, less formal, method for obtaining the probability that an unbiased random walk which begins in state $S(0) = i > 0$ and finishes in state $S = A$ while avoiding an absorbing boundary at 0 is to treat the Laplacian Eq. (5.5) as a continuous boundary value problem. Find all the functions $f(x)$ which satisfy $f''(x) = 0$ subject to the conditions that $f(0) = 0$ and $f(A) = 1$.

(2) Another, less formal, method for obtaining the stopping time for an unbiased random walk which begins in state $0 \leq S(0) = i \leq A$ and finishes in either state $S = 0$ or $S = A$ is to treat the Poisson Eq. (5.6) as a continuous boundary value problem. Find all the functions $g(x)$ which satisfy $g''(x) = -2$ subject to the conditions that $g(0) = 0$ and $g(A) = 0$.

(3) Suppose $f(x)$ is four times continuously differentiable at $x = x_0$. Use Taylor's Theorem [Stewart (1999)] to expand $f(x)$ about $x = x_0$. Then by using $f(x_0 + h)$ and $f(x_0 - h)$ show that

$$f''(x_0) \approx \frac{f(x_0 + h) - 2f(x_0) + f(x_0 - h)}{h^2}.$$

(4) If a stock price can change (up or down) by only one dollar per day and the probability of a unit increase is $1/2$ what is the probability that the stock price will increase by \$25 before decreasing from the present value by \$50?

(5) What is the expected number of days that will elapse before the stock described in exercise 4 increases in price by \$25 or decreases by \$50?

(6) What is the expected number of days that will elapse before the stock described in exercise 4 increases in price by \$25 while avoiding a decrease of \$50?

(7) Assuming that μ and $P(0)$ are constants, verify by differentiation and substitution that $P(0)e^{\mu t}$ solves Eq. (5.20).

(8) Using the growth constant and volatility of the Sony Corporation stock found earlier, use Eq. (5.36) to generate a realization of the random walk of stock prices for another year's duration.

(9) Investigate the random walk of another company's stock. Collect, if possible, a year's worth of closing prices $(P(t_1), P(t_2), \ldots, P(t_n))$

for the stock of your favorite company. Plot a histogram of $\Delta X = \ln P(t_{i+1}) - \ln P(t_i)$ and also calculate the mean and standard deviation of this random variable.

(10) Using the growth constant and volatility of the stock data you found in exercise 9, use Eq. (5.36) to generate a realization of the random walk of stock prices for another year's duration.

(11) Using the chain rule for derivatives from ordinary calculus (not the stochastic version governed by Itô's Lemma) verify the expressions found for $f'(0)$ and $f''(0)$ in Eqs. (5.28) and (5.29). Find an expression for the Taylor remainder R_3.

(12) Suppose that P is governed by the stochastic process

$$dP = \mu P\, dt + \sigma P\, dW(t)$$

and that $Y = P^n$. Find the process which governs Y.

(13) Suppose that P is governed by the stochastic process

$$dP = \mu P\, dt + \sigma P\, dW(t)$$

and that $Y = \ln P$. Find the process which governs Y.

(14) Suppose that X is a normally distributed random variable with mean $\mu = 0$ and variance σ^2. Show that $\mathrm{E}\left[X^4\right] = 3\sigma^4$.

Chapter 6

Forwards and Futures

This chapter introduces some of the concepts and terminology associated with the buying and selling of securities such as stocks. The main issue discussed will be the pricing of forward contracts and futures. The present chapter will give the reader the opportunity to apply the theory of interest, the arbitrage principle, and some elementary stochastic processes to the problem of pricing two commonly traded financial derivatives. The term "derivative" is used because the values of these financial instruments is "derived" from underlying securities or commodities. The material of this chapter can also be treated as a "warm-up" exercise for the later chapters on options and the development of the Black-Scholes option pricing formula.

6.1 Definition of a Forward Contract

At its essence, a **forward** is an agreement between two agents (which we will usually call the "party" and the "counterparty") to buy or sell a specified quantity of a commodity at a specified price on a specified date in the future. The forward is an obligation to buy or sell at the agreed upon quantity, price, and time. If the party or counterparty breaks the agreement, they may face legal and financial consequences. Consider the situation of a manufacturer producing portable MP3 players. Their product depends upon an adequate supply of solid state memory to match the manufacturing output. If the manufacturer is concerned that the readily available supply of memory may fall short of their needs three months from now, they may enter into a forward contract with a memory supplier to sell them say 100,000 units of memory for one million dollars in three months. Once the forward contract is established, the memory supplier must come up with the 100,000 units of memory in three months and must sell it for one million dollars even if another buyer would be willing to pay more for it. The MP3 manufacturer must buy the 100, 000 units of memory in three months for one million dollars even if another supplier is willing to sell it for less. While there are many reasons an institution may buy or sell a forward contract, the chief reason is that the forward reduces the risk of market prices moving against the party or counterparty. The MP3 manufacturer may fear the business disruption which a shortage or solid state memory at affordable prices may cause, while the memory supplier may worry that increased supply or improved technology may decrease the price of memory three months hence. While we will not usually concern ourselves with the issue, the party or counterparty of a forward contract can effectively "cancel" a forward contract by entering into an opposing forward contract. For example suppose the MP3 manufacturer after setting up the original forward contract decides they can will not need the solid state memory. They can enter into a forward contract with another memory buyer. The MP3 manufacturer becomes a seller of its unneeded memory.

A forward contract can be set up between two agreeing agents and can be customized to the precise needs of the buyer and seller. In contrast securities such as stocks (and later we will see futures) are often traded in a **market**. A stock market is created to increase the efficiency of the process of buying and selling stocks. Creating and maintaining this efficient environment for trading stocks is the business of **market makers**. Like any (for profit) business the market makers expect to earn money from the operation of the market. There are a number of ways this may happen. The market makers may charge buyers and sellers fees or commissions. Money can also be earned from the difference in the buying and selling price of a stock. In a stock market there may be several owners of a particular corporation's stock who are willing to sell and a number of buyers wishing to add that stock to their portfolio, and each investor, both buyers and sellers may have a different price for the stock. A price a buyer is willing to pay is called a **bid** price. A price a seller of a stock is willing to accept is known as a **ask** price. The ask price is known also as the **offer** price. Typically the ask prices exceed the bid prices. At a given moment in the market the difference between the lowest ask price and the highest bid price is called the **bid/ask spread** or **bid/offer spread**. The market maker earns the bid/ask spread on stock trades as reward for operating the market. Commissions, fees, and the bid/ask spread are examples of **transaction costs** associated with stock trading.

Example 6.1 Suppose the lowest ask price of a share of stock is $50.10 and the highest bid price for the stock is $50.00. The bid/ask spread is therefore $0.10 per share. A stock buyer who issues a buy order for 1000 shares will pay $50,100. The seller will receive $50,000 and the market maker will earn $100 on the trade (plus any other fees or commissions charged).

Without market makers and the markets for trading they create and operate, buyers and sellers of stocks and commodities would be responsible for locating each other. Anyone who has sold a used car knows that there are costs associated with finding potential buyers. When trading in a market buyers and sellers also benefit from knowing the last price at which a stock traded and from the price competition among traders.

6.2 Pricing a Forward Contract

Consumers are accustomed to an instantaneous transfer of ownership when purchasing most items. An MP3 player for instance becomes the buyer's property the instant the buyer transfers the appropriate amount of money to the seller. To describe the process of purchasing a share of stock we must generalize and will assume that events or actions take place at different times separated by a finite interval. To make the initial scenario simple we will assume that events may take place at time $t = 0$ and also later at time $t = T > 0$. There are three components or actions involved in the simplest description of a stock purchase: (1) fixing or agreeing on the price for the stock, (2) making payment for the stock, and (3) transferring ownership of the stock from the seller to the buyer. Logically the price is fixed before payment is made so it is assumed that the price of the stock is fixed at $t = 0$. However, the remaining two actions may occur at either $t = 0$ or $t = T$ depending on the arrangement made between buyer and seller.

The traditional buyer/seller mode of purchase in which all three events occur simultaneously is called an **outright purchase**. The situation in which the buyer receives ownership of the stock at time $t = 0$ but pays for the purchase at the later time $t = T$ is called a **fully leveraged purchase**. A fully leveraged purchase is equivalent to buying the stock on credit or with borrowed money. A buyer who pays for the stock at time $t = 0$ but does not receive ownership until time $t = T$ is said to have purchased the security using a **prepaid forward contract**. The prepaid forward contract allows the seller to retain certain rights associated with ownership of the stock until $t = T$. For example, the seller would retain voting rights inherent in ownership of the stock until $t = T$. Another important reserved right for the seller in the prepaid forward contract situation is the right to receive any dividend payments associated with the stock. Lastly if payment for the stock and transfer of ownership both take place at $t = T$, it is said that the stock is purchased using a **forward contract**. The timing and coordination of the payment for the purchase and the transfer of ownership have implications for the amount which should be paid for the contract and security.

The simplest case to analyze is the case of outright purchase of the stock. If the stock is worth $S(0)$ at time $t = 0$ and payment and transfer of ownership will take place at time $t = 0$ then the amount paid should be $S(0)$. To pay any other amount would create an arbitrage opportunity for either the buyer or the seller (see exercise 2). Since the fully leveraged purchase is equivalent to the buyer purchasing the stock with borrowed money, then if the continuously compounded interest rate for borrowing is r_b, the cost of the fully leveraged stock purchase is $S(0)e^{r_b T}$.

Determining the price of a prepaid forward or forward contract requires a more detailed (though still elementary) argument. To make the argument rigorous we must introduce some new terms and establish some assumptions. We will assume the stock pays no **dividends**. Dividends are periodic payments paid by a corporation to the stock holders. The funds for the dividends are paid out of a portion of the profit made by the corporation. We will also assume the stock obeys a Weiner process of the form

$$dS = \mu S \, dt + \sigma S \, dW(t).$$

Thus if the stock is worth $S(0)$ at time $t = 0$ then at time $t = T$, the expected value of the price of the stock will be $\mathrm{E}\,[S(T)] = S(0)e^{\mu T}$ according to Lemma 3.1. To develop the pricing argument we will make use of the terms long and short. While the implications of long and short positions can seem foreign (or perhaps even illegal) to novices, keep the following simple definitions in mind. An investor who owns (rather than merely has possession of) a stock, security, or other commodity is said to be in a **long position** relative to the item. If an investor outright purchases a stock, they have entered into a long position with regards to the stock. An investor who sells a stock, security, or commodity (perhaps by borrowing on short-term loan it from the owner) and must re-purchase it later is in a **short position** relative to the item. Typically an investor will short a stock if they believe the value of the stock will decrease in the near future.

They borrow the stock, sell it while the price is still relatively high, re-purchase it after the price has fallen, return the stock to the lender, and keep the difference as profit. A more detailed description of the practice of shorting stocks is contained in [McDonald (2006)]. The final assumption is that pricing is done within a "no-arbitrage" framework. This means that it is not possible for any party to make a risk-free positive profit. Now we may justify the price of a prepaid forward contract.

Theorem 6.1 *The price F of a prepaid forward contract on a non-dividend paying stock initially worth $S(0)$ at time $t = 0$ for which ownership of the stock will be transfered to the buyer at time $t = T > 0$ is $F = S(0)$.*

Proof. Suppose $F < S(0)$. The buyer can purchase the forward and sell the security (the buyer has entered into a long position on the forward and a short position on the security). Since $S(0) - F > 0$ the buyer has a positive cash flow at $t = 0$. At $t = T$, the buyer receives ownership of the security and immediately closes their short position in the security. The cash flow at $t = T$ is therefore zero. Thus the total cash flows at $t = 0$ and $t = T$ is $S(0) - F > 0$. There is no risk in obtaining this positive profit since the forward obligates the seller to deliver the security to the buyer so that the buyer's short position in the security can be closed out. Hence if $F < S(0)$ arbitrage is present.

Suppose $F > S(0)$. The buyer can purchase the security at time $t = 0$ and sell a prepaid forward (the buyer has entered into a long position on the security and a short position on the forward). Since $F - S(0) > 0$ the buyer has a positive cash flow at $t = 0$. At $t = T$, the buyer must transfer ownership of the security to the party who purchased the forward. The cash flow at $t = T$ is therefore zero. Thus the total cash flows at $t = 0$ and $t = T$ is $F - S(0) > 0$. There is no risk in this situation since the buyer owns the security at time $t = 0$ and thus will with certainty be able to transfer ownership at $t = T$. Hence if $F > S(0)$ arbitrage is present. Consequently $F = S(0)$. □

An alternative proof of the pricing formula for the prepaid forward contract makes use of a present value argument. Since the stock is worth $S(0)$ at time $t = 0$ and has an expected value of $E[S(T)] = S(0)e^{\mu T}$ at time $t = T$, then the continuously compounded rate of return for the stock is μ. Consequently the present value of the prepaid forward is

$$F = E[S(T)]e^{-\mu T} = S(0).$$

To summarize the prepaid forward contract, we now understand that for a non-dividend paying stock the value of the prepaid forward contract is the same as the present value of the stock. The reader may question the fairness of paying "full price" for a stock for which transfer of ownership will not be made until a future time. However, we have shown above this is the correct no-arbitrage price. To settle any misgivings a reader might have about this, conduct the following thought experiment. As long as the stock pays no dividends it can be thought of as inert during the interval $[0, T]$. Think of buying the stock through the mail. Even though the mail delivery will occur a few days after the purchase, the only behavior of the stock of interest is the market fluctuation of its value, which occurs whether or not the stock is physically in the buyer's possession.

It is more common that an investor will purchase a forward contract rather than a prepaid forward. Recall that a forward contract is similar to a prepaid forward except that the payment for the forward and the transfer of the ownership of the security take place simultaneously at $t = T > 0$ while the price F of the forward contract is determined at time $t = 0$. In this section we will assume that the risk-free interest rate (which may be the interest rate on US Treasury Bonds) is denoted by r and this interest is compounded continuously. The reader can easily modify the arguments given below to other interest compounding schedules. We will assume that parties can borrow and lend money at rate r. To determine the price of a forward contract the we can adopt a quick intuitive approach that since only the timing of the payment is different between the forward contract and the prepaid forward, the no-arbitrage price of a forward contract should be the future value of the prepaid forward, all other conditions being the same.

Theorem 6.2 *Suppose a share of a non-dividend paying stock is worth $S(0)$ at time $t = 0$ and that the continuously compounded risk-free interest rate is r, then the price of the forward contract is*

$$F = S(0)e^{rT}.$$

Proof. Suppose $F < S(0)e^{rT}$. The buyer can purchase the forward (which they will not have to pay for until $t = T$) and sell the security at time $t = 0$. The value of the security is $S(0)$ which is lent out at the risk-free rate compounded continuously. Thus the net cash flow at time $t = 0$ is $S(0) - S(0) = 0$. At $t = T$, when the borrower repays the loan, the buyer's cash balance is $S(0)e^{rT}$. The buyer pays F for the forward in order to receive the security which is then used to close out the forward position. The cash flow at $t = T$ is therefore $-F$. Thus the total cash flows at $t = 0$ and $t = T$ is $S(0)e^{rT} - F > 0$. There is no risk in obtaining this positive profit since the forward obligates the seller to deliver the security to the buyer so that the buyer's short position in the security can be closed out. Hence if $F < S(0)e^{rT}$ arbitrage is present.

Suppose $F > S(0)e^{rT}$. The buyer can sell a forward contract which will be paid for at time $t = T$ and borrow $S(0)$ to purchase the security at time $t = 0$. Thus the net cash flow at time $t = 0$ is $S(0) - S(0) = 0$. At $t = T$, the buyer must repay the loan of $S(0)e^{rT}$ and will sell the security for F. The cash flow at $t = T$ is therefore $F - S(0)e^{rT} > 0$. Thus the total cash flows at $t = 0$ and $t = T$ is $F - S(0)e^{rT} > 0$. There is no risk in this situation since the buyer owns the security at time $t = 0$ and thus will with certainty be able to transfer ownership at $t = T$. Hence if $F > S(0)e^{rT}$ arbitrage is present. Therefore the no-arbitrage price of the forward contract on the non-dividend paying stock is $F = S(0)e^{rT}$. ☐

Once a prepaid forward or forward contract has been priced and purchased, an investor will be interested in the profit from the transaction. The value of the stock at time $t = T$ is a random variable and may differ from $E[S(T)]$. The profit from the forward contract is defined as the difference between the price of the contract and the value of the stock when $t = T$. This is the amount of money the investor would make if they immediately sold the stock at time $t = T$, the time at which ownership is transferred to the investor. Mathematically we may express this as

$$\text{profit} = S(T) - S(0)e^{rT}.$$

Example 6.2 Suppose a share of stock is currently trading for \$25 and the risk-free interest rate is 4.65% per annum. The price of a two-month forward contract is

$$F = 25e^{0.0465(2/12)} \approx 25.1945.$$

A plot of the profit curve is shown in Fig. 6.1. A positive profit is made if at time $t = 2/12$ the stock is trading above F.

In the following section the pricing formulas developed above will be generalized to include the effects of transaction costs and stocks that pay dividends.

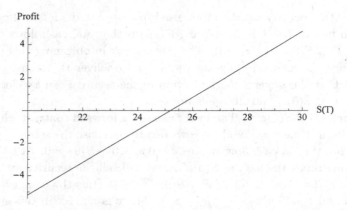

Fig. 6.1   The profit on a long forward contract is a linear function of the stock price at time $t = T$.

6.3   Dividends and Pricing

Dividends are paid to owners of stock shares from the profits earned by the corporation issuing the shares. Not all shares earn dividends. Dividends on shares from individual corporations may be paid annually or semi-annually. Occasionally it will be convenient to think of dividends on large and diverse collections of shares as paid continuously. Note that dividends are paid to the *shareholders*, not to the owners of prepaid forwards or forward contracts. When pricing a prepaid forward or forward contract we must carefully consider the effect of any dividends that will be paid to the shareholders between $t = 0$ and $t = T$, since the owner of the forward will not receive these disbursements. The value of the forward must be decreased by the present value of the dividends paid during the interval $[0, T]$. If the risk-free interest rate is r, then the present value of an amount D paid at time t where $0 \leq t \leq T$ is De^{-rt}. Consequently if n dividends in the amounts $\{D_1, D_2, \ldots, D_n\}$ are paid at times $\{t_1, t_2, \ldots, t_n\}$ in the interval $[0, T]$, the price of a prepaid forward on a stock currently valued at $S(0)$ becomes

$$F = S(0) - \sum_{i=1}^{n} D_i e^{-rt_i}.$$

Using the idea that the value of a forward contract is the future value of the prepaid forward we have the value of the forward contract for a stock paying dividends at discrete times expressed as

$$F = S(0)e^{rT} - \sum_{i=1}^{n} D_i e^{r(T-t_i)}.$$

Example 6.3 Suppose the risk-free interest rate is 4.75%. A share of stock whose current value is $121 per share will pay a dividend in six months of $3 and another in twelve months of $4. An investor is pricing a one-year forward contract and one-year prepaid forward on the stock assuming that transfer of ownership will take place immediately after the second dividend is paid. The value of the prepaid forward is

$$F = 121 - 3e^{-0.0475(6/12)} - 4e^{-0.0475(12/12)} \approx 114.256.$$

The value of a forward contract on the dividend paying stock is

$$F = 114.256e^{0.0475(12/12)} \approx 119.814.$$

If a forward is being purchased on a portfolio of stocks paying dividends at many times, it may be convenient to think of the investment as paying dividends continuously. In this case let the dividend rate be denoted by δ. Only simple modifications are needed in the pricing formulas above. The value of a prepaid forward on a stock currently worth $S(0)$ becomes

$$F = S(0)e^{-\delta T}.$$

The value of a forward contract on the same security is

$$F = S(0)e^{(r-\delta)T}.$$

Example 6.4 An investment valued at $117 pays dividends continuously at the annual rate of 2.55%. The risk-free interest rate is 3.95%. Therefore the price of a four-month prepaid forward on the investment is

$$F = 117e^{-0.0255(4/12)} \approx 116.01.$$

The value of a four-month forward contract on the investment is

$$F = 117e^{(0.0395-0.0255)(4/12)} \approx 117.547.$$

6.4 Incorporating Transaction Costs

In the previous sections we proved there is a single no-arbitrage price for a prepaid forward and a forward contract. This ignores the possibility of there being transaction costs associated with the buying and selling of the security and the forward. In this section we will generalize the specifications surrounding the forward contract and determine an interval of no-arbitrage prices for the forward rather than a single no-arbitrage price. In this section we will make use of the following definitions and notation for the components of the transactions.

S^a: the time $t = 0$ ask price at which the security can be bought.

S^b: the time $t = 0$ bid price at which the security can be sold. In general $S^b < S^a$.

r^b: the continuously compounded interest rate at which money may be borrowed.

r^l: the continuously compounded interest rate at which money may be lent. In general $r^l < r^b$.

k: the cost per transaction for executing a purchase or sale.

In this section we will derive an interval of forward contract prices of the form $[F^-, F^+]$ for which no arbitrage is possible when $F^- \leq F \leq F^+$. Outside of this interval arbitrage may be possible.

Suppose the forward contract has value F. We will show below that in the absence of arbitrage

$$F \leq (S^a + 2k)e^{r^b T} \equiv F^+.$$

For the sake of contradiction assume that $F > F^+$. At time $t = 0$ the buyer may borrow amount $S^a + 2k$ to purchase the security and sell the forward contract. Since a transaction cost of k is incurred for both the purchase of the security and the sale of the forward the amount $2k$ in addition to the ask price S^a, of the security at time $t = 0$ must be borrowed. As before the payment for the forward will be made at time $t = T > 0$. Thus the net cash flow at time $t = 0$ is zero. At time $t = T$ the loan must be repaid in the amount of $(S^a + 2k)e^{r^b T}$ and the buyer receives F for the forward. The total cash flow for times $t = 0$ and $t = T$ is therefore

$$F - (S^a + 2k)e^{r^b T} = F - F^+ > 0.$$

Hence when $F > F^+$ arbitrage results.

Now we will show that in the absence of arbitrage

$$F \geq (S^b - 2k)e^{r^l T} \equiv F^-.$$

For the sake of contradiction assume that $F < F^-$. At time $t = 0$ the buyer can purchase the forward contract (for which F will be paid at time $t = T > 0$) and sell short the security for S^b. A transaction cost of k is paid at time $t = 0$ for the forward contract and another transaction cost of k is incurred during the short sale the net proceeds from the sale are $S^b - 2k$. This amount is lent out at interest rate r^l until time $t = T$. At time $t = T$ the buyer's cash balance is $(S^b - 2k)e^{r^l T}$. Also at this time the buyer pays F for the forward contract and closes out the short position in the security. Thus the total cash flow at times $t = 0$ and $t = T$ is

$$(S^b - 2k)e^{r^l T} - F = F^- - F > 0.$$

Hence when $F < F^-$ arbitrage is possible. Therefore in summary for the situation when transaction costs are included in the mathematical model, the arbitrage-free forward contract price must satisfy the inequality

$$(S^b - 2k)e^{r^l T} \le F \le (S^a + 2k)e^{r^b T}. \tag{6.1}$$

Example 6.5 Suppose the asking price for a certain stock is $55 per share, the bid price is $54.50 per share, the fee for buying or selling a share or a forward contract is $1.50 per transaction, the continuously compounded lending rate is 2.5% per year, and the continuously compounded borrowing rate is 5.5% per year. The price of a three-month forward contract on the stock would fall in the interval

$$(54.50 - 2(1.50))e^{0.025(3/12)} \le F \le (55 + 2(1.50))e^{0.055(3/12)}$$
$$51.7223 \le F \le 58.8030$$

Throughout the remainder of this book, unless specifically mentioned, we will ignore the complications introduced by transaction costs and dividends.

6.5 Futures

A forward contract is very "customizable" in that the terms of the contract can be arranged to the satisfaction of the parties involved. The date of maturity of the forward, the volume of stocks or of a commodity to be exchanged at the maturity of the contract, and any necessary collateral to be held to reduce the risk of default by one or more of the parties may all be decided by the parties engaged in the contract. The parties may even decide that at maturity they will only exchange the net amount of profit earned by the parties on the transaction instead of actually selling or buying the underlying security or commodity.

Futures are like this last type of "cash-settled" forward contract with some additional differences. Futures are generally traded in a more structured exchange market. Futures have standardized maturity dates (typically a few months in the future) and standardized volumes of the underlying security or quantity. There are other important differences between forward and futures contracts. There is generally less risk of a party involved in a futures contract defaulting since daily adjustments to futures contracts take place and are managed by the clearinghouse associated with a futures exchange. The clearinghouse will require a deposit from both the party and the counterparty to the futures contract. This deposit is called a margin. The margin protects each party to the futures contract against default by the other party. The clearinghouse will then, based on subsequent changes in the futures price, require additional deposits to the margin so as to protect both parties from default. The process of adjusting the financial amounts owed to the parties in the futures contract is called **marking-to-market**. In contrast, recall that forward contracts are settled on the date of maturity of the contract. A futures exchange will generally have rules governing the practice of trading depending on changes in the price of the contract traded. For example trading on a particular futures contract may be temporarily halted if the price suddenly moves downward by a specified threshold proportion. The last difference we will mention is that due to the standardization and trading infrastructure provided by a futures exchange, futures are easily traded. An investor wishing to rid themselves of a particular obligation implied by a contract may easily purchase an offsetting opposite contract with the same date of maturity as the original contract.

If the risk-free interest rate is constant for the life of the contract, the price of a futures contract is the same as the price of a forward contract. This will be our assumption for the remainder of this chapter. We will also present an extended example of the process of marking-to-market. For the purpose of the example assume an investor is purchasing a 7-day futures contract whose initial price is $1000. The price may change daily until maturity. A volume of 5000 futures contracts will be purchased. The clearinghouse will require that a minimum margin of 10% of the current value of the futures contract be maintained until maturity. The margin will earn interest at the risk-free rate of 14% per annum compounded continuously.[1]

---

[1]The interest rate is set artificially high to magnify the daily changes in the margin balance.

Table 6.1 The daily values of the futures price and the margin balance for a hypothetical seven-day futures contract. The column headed "Margin Balance" shows the margin balance after interest is credited but before the amount in the "Margin Call" column is added to the margin balance.

Day	No. of Contracts	Futures Price	Price Change	Margin Balance	Margin Call
0	5000	1000.00	—	500, 000.00	—
1	5000	987.90	−12.10	439, 691.82	54, 258.18
2	5000	987.97	0.07	494, 489.50	0.00
3	5000	990.53	2.56	507, 479.20	0.00
4	5000	988.37	−2.16	496, 873.89	0.00
5	5000	973.89	−14.48	424, 664.51	62, 280.49
6	5000	968.70	−5.19	461, 181.81	23, 168.19
7	5000	980.82	12.12	545, 135.81	0.00

The ultimate profit to the investor will be the difference between the final margin balance and the future value of the initial margin balance. To start the investor deposits a margin of $(5000)(1000)(0.10) = 500,000$. Suppose that on the next day the price of the futures contract has fallen to \$987.90. The change in futures price is $\Delta F = -12.10$. Multiplied by the 5000 futures contracts the investor owns, the wealth of the investor has changed by $-60,500$. This loss will be taken from the margin balance. The initial margin balance has earned a day's interest, so its new post-loss balance is

$$500,000e^{0.14/365} - 60,500 \approx 439,691.82.$$

Minding the minimum 10% margin requirement, we see the margin should be at least $(5000)(987.90)(0.10) = 493,950$, therefore the owner of the futures contract must add

$$493,950 - 439,691.82 = 54,258.18$$

to the margin. The request of the additional margin deposit is called a **margin call**. After bringing the margin up to the minimum level (sometimes called the **maintenance margin**) the next day's adjustments can be made. This daily process of marking-to-market continues until the futures contract matures. The daily values of the futures price and the margin balance are shown in Table 6.1. The profit to the holder of the long position in the futures contract is the difference between the final margin balance and the future value of the initial margin. In this example

$$545,135.81 - 500,000e^{7(0.14)/365} = 43,791.50.$$

6.6　Exercises

(1) Suppose the bid/ask spread for a share of a particular stock is $0.25. An investor buys 1000 shares and then immediately sells them (before the bid and ask prices can change). What is the total transaction cost (the so-called **round trip cost**) to the investor?

(2) Show that in the absence of arbitrage, the price paid for outright purchase of a stock should be $S(0)$, the price of the stock at the time of outright purchase.

(3) Suppose the continuously compounded interest rate for borrowing is 5.05% per year. What is the cost to the buyer for a fully leveraged purchase of a stock worth now $17 and for which payment will be made in one month?

(4) Suppose the continuously compounded risk-free interest rate is 4.75% per year. What is the cost of a three-month forward contract on a non-dividend paying stock whose value currently is $23?

(5) Suppose the risk-free interest rate is 3.65%. A share of stock whose current value is $97 per share will pay a dividend in six months of $2.50 and another in twelve months of $2.75. Find the values of one-year forward contract and one-year prepaid forward on the stock assuming that transfer of ownership will take place immediately after the second dividend is paid.

(6) An investment valued at $195 pays dividends continuously at the annual rate of 1.95%. The risk-free interest rate is 4.55%. Find the prices of a three-month prepaid forward and a three-month forward contract on the investment.

(7) A security is currently priced at $1000. The continuously compounded risk-free interest rate is 5.05% annually. The price of a six-month forward contract for the security is $990. If the security pays dividends continuously at rate r, find r expressed as an annual percentage.

(8) Suppose the asking price for a certain stock is $75 per share, the bid price is $74 per share, the fee for buying or selling a share or a forward contract is $2 per transaction, the continuously compounded lending rate is 3% per year, and the continuously compounded borrowing rate is 4% per year. Find the interval of no-arbitrage prices for a six-month forward contract on the stock.

(9) A security is currently priced at $950 per share. An investor purchases a 6-month futures contract on 100 shares of the security. The continuously compounded risk-free interest rate is 6%. The initial margin on

the futures contract is 12.5% and the futures position will be marked to market monthly. The maintenance margin is likewise 12.5%. What is the highest security price for which a margin call will be made after one month?

(10) Suppose a stock pays no dividends and the risk-free rate of continuously compounded interest is r. An investor has the choice between the following investments:

(a) Purchasing one share of the stock at time $t = 0$ and selling it at time $t = T > 0$,

(b) Entering into a long forward contract on the stock at time $t = 0$ and lending the present value of the long forward contract until time $t = T$.

Show that the payoffs of the two investments are the same.

(11) An investor is purchasing a 10-day futures contract whose initial price is $850. The price of the contract changes daily following the path described in the table below.

Day	Price	Day	Price
1	774.67	6	735.64
2	779.39	7	741.59
3	778.42	8	759.88
4	749.56	9	766.25
5	742.87	10	805.36

A volume of 1500 futures contracts will be purchased. The clearinghouse will require that a minimum margin of 15% of the current value of the futures contract be maintained until maturity. The margin will earn interest at the risk-free rate of 10% per annum compounded continuously. Create a table for the daily accounting of marking to market for this futures contract similar to Table 6.1. What is the profit on the futures contract to the investor?

Chapter 7

Options

In the present world of finance there are many types of financial instruments which go by the name of **options**. At its simplest, an option is the right, but not the obligation, to buy or sell a security such as a stock for an agreed upon price at some time in the future. The agreed upon price for buying or selling the security is known as the **strike price**. Options come with a time limit at which (or prior to) they must be exercised or else they expire and become worthless. The deadline by which they must be exercised is known as the **exercise time, strike time**, or **expiry date**. Often the exercise time will simply be called **expiry**. In the remainder of the text the three terms will be used interchangeably. An option to buy a security in the future is called a **call option**. An option to sell is known as a **put option**.

Types of options can also be distinguished by their handling of the expiry date. The **European option** can only be exercised at maturity, while an **American option** can be exercised at or before expiry. Of the two types, the European option is simpler to treat mathematically, and will be the focus of much of the rest of this book. However, in practice, American options are more commonly traded. There is a mathematical price to be paid in terms of complexity for the added flexibility of the American-style option.

Suppose stock in a certain company is selling today for $100 per share. An investor may not want to buy this stock today, but may want to own it in the future. To reduce the risk of financial loss due to a potential large increase in the price of the stock during the next three-month period, they buy a European call option on the stock with a three-month strike time and a strike price of $110. At the expiry date, if the price of the stock is above $110 and the investor now wishes to buy the stock, they have the right as holder of the call option to purchase it for $110, even if the market value of the stock is higher. Otherwise if the value of the stock is below $110 and the investor still wishes to buy the stock, they will let the call option expire without exercising it, and purchase the stock at its market price. Call options allow investors to protect themselves against paying an unexpectedly high price in the future for a stock which they are considering purchasing.

We have used the language of "buying" an option. Thus options themselves have some value, just as the securities which underlie the options have a value. An important issue to consider is how are values assigned to options? In light of the Arbitrage Theorem of Chapter 4, if the option is mis-priced relative to the security, arbitrage opportunities may be created. In this chapter we will explore some of the relationships between option and security prices. We will derive the **Black-Scholes** partial differential equation, together with boundary and final conditions which govern the prices of European style options.

During the explanations of many of the concepts in this chapter, we will refer to buying or selling a financial instrument. In the financial markets it is often possible to sell something which we do not yet own by borrowing the object from a true owner. For example investor A may borrow 100 shares of stock from investor B and sell them with the understanding that by some time in the future investor A will purchase 100 shares of the same stock and return them to investor B. Normally one buys an object first and sells it later, but in the financial arena one can often sell first and then buy later. Borrowing and selling an object with the agreement to purchase it later is called adopting a **short position** in the object, or sometimes **shorting** the object. Adopting a short position can be a profitable transaction if the investor believes the price of the object is going to decrease. Purchasing an object first and then selling it in the future is called adopting a **long position**.

7.1   Properties of Options

There are many relationships between the values of options and their underlying securities. The maintenance of most of these relationships is necessary to eliminate the possibility of arbitrage. We will cover some of these relationships here and give the reader a taste of the way that one can prove these properties. Throughout this section we will use the following definitions.

C_a: value of an American-style call option
C_e: value of a European-style call option
K: strike price of an option
P_a: value of an American-style put option
P_e: value of a European-style put option
r: continuously compounded, risk-free interest rate
S: price of a share of a security
T: exercise time or expiry of an option

We have used the term "risk-free" to describe the interest rate r. It is assumed that this is the rate of return of an investment which carries no risk. While the reader may wish to debate whether a complete absence of risk can ever be achieved, there are some investments which carry very little risk, for example, U.S. Treasury Bonds.

Consider the cost of an American option compared with a European option. In the absence of arbitrage an American-style option must be worth at least as much as its European counterpart, *i.e.* $C_a \geq C_e$ and $P_a \geq P_e$. Naturally we are assuming that the strike prices, strike times, and underlying securities are the same. On the contrary, suppose that $C_a < C_e$. Taking the position of an informed consumer, knowing that the American-style option has all the characteristics of the European option and in addition has the increased flexibility that it may be exercised early, no investor would purchase a European-style call option if it cost more than the American-style option. To formulate a more mathematical proof, suppose an investor sells the European-style call option and purchases the American-style option. Since $C_e - C_a > 0$, the investor may purchase a risk-free bond paying interest at rate r compounded continuously. At the expiry date the bond will have value $(C_e - C_a)e^{rT}$. If the owner of the European option wishes to exercise the option, the investor insures this is possible by exercising their own American option. If the owner of the European option lets it expire unused, then the investor can do the same with the American option. Thus in both cases the investor makes a risk-free profit of $(C_e - C_a)e^{rT}$ which is an example of arbitrage.

It is also the case that $C_e \geq S - Ke^{-rT}$ or arbitrage is present. Assuming an absence of arbitrage and that $C_e < S - Ke^{-rT}$, an investor can purchase the call option and sell the security at time $t = 0$. Since $0 < Ke^{-rT} < S - C_e$, the investor can invest $S - C_e$ in a risk-free bond paying interest rate r. At the strike time the bond is worth $(S - C_e)e^{rT}$ which is greater than K. Thus if the investor chooses to exercise the option (because S at time

T is greater than K), the bond can be cashed out, the security purchased for K, and there is still capital left over. In other words there is risk-free profit. On the other hand if S at time T is less than K, the short position on the security is eliminated using the bond and purchasing the stock for S (which is less than K). Again there will still be capital left over. Once again there is a risk-free profit.

There is strict arbitrage-free relationship between the value of a European call and put with the same strike price and expiry date on the same underlying security. This is the important **Put-Call Parity Formula** of Eq. (7.1).

$$P_e + S = C_e + Ke^{-rT} \tag{7.1}$$

To see why this relationship must be true, imagine portfolio A represents the left-hand side of (7.1) while the right-hand side is represented by portfolio B. The Put-Call Parity Formula implies that in an arbitrage-free setting these portfolios must have the same value.

Suppose portfolio A is worth less than portfolio B, *i.e.*

$$P_e + S < C_e + Ke^{-rT}. \tag{7.2}$$

An investor can borrow at interest rate r an amount equal to $P_e + S - C_e$. This would allow the investor to purchase the security, the European put option, and to sell the European call option. At the strike time of the two options, the investor must repay principal and interest in the amount of $(P_e + S - C_e)e^{rT}$. If the security is worth more than K at time T, the put expires worthless and the call will be exercised by its owner. The investor must sell the security for K. Thus the net proceeds of this transaction are

$$K - (P_e + S - C_e)e^{rT} > 0 \tag{7.3}$$

since inequality (7.3) is equivalent to the one in (7.2). If the security is worth less than K at time T, the call expires worthless and the put will be exercised by the investor. Again the investor will sell the security for K. The net proceeds of this transaction are the same as in the previous case. Thus there is a risk-free profit to be realized if portfolio A is worth less than portfolio B.

Now suppose portfolio A is worth more than portfolio B, *i.e.*

$$P_e + S > C_e + Ke^{-rT}. \tag{7.4}$$

An investor can sell the security and the European put option and buy the call option. This generates an initial positive flow of capital in the amount of $S + P_e - C_e$. This amount will be invested in a risk-free bond earning interest at rate r until the expiry date arrives. At that time the investor will have $(S + P_e - C_e)e^{rT}$. If the security is worth more than K at time T, then the put option is worthless and investor will exercise the call option. The investor purchases the security for K (thus canceling their short position). This leaves the investor with a net gain of

$$(P_e + S - C_e)e^{rT} - K > 0 \tag{7.5}$$

since inequality (7.5) is equivalent to the one in (7.4). If the security is worth less than K at time T, the call option expires unused and the owner of the put option will exercise it. Thus the investor will clear their short position by buying the security for K and their net gain is as before. Consequently there exists an arbitrage opportunity if portfolio B is worth less than portfolio A. Therefore the two portfolios must have the same value, *i.e.* the Put-Call Parity Formula must be true.

Assuming that the prices of European call and put options are non-negative (a mild assumption), we can derive two corollary inequalities to the Put-Call Parity Formula.

$$C_e \geq S - Ke^{-rT} \tag{7.6}$$
$$P_e \geq -S + Ke^{-rT} \tag{7.7}$$

While none of the formulas or inequalities developed so far enable us to price options directly, they are useful checks for arbitrage opportunities.

7.2 Pricing an Option Using a Binary Model

In this section we will examine a very simple option/security situation and determine the arbitrage-free price for the security. The analysis done here is too simple to be applied directly to any real-world example of a security; however, it provides an example of an application of the Arbitrage Theorem (Theorem 4.5). This type of analysis can be generalized and extended to more relevant real-world examples and thus provides a good starting point for the goal of developing option pricing formulae. The analysis performed here is based on a binary model, so named since the security is assumed to take on one of two values in the future.

Suppose the price of a share of stock is \$100 currently. After a single, indivisible unit of time T, the price of the share will either be \$200 or \$50. The stock will be worth \$200 after time T with probability p or will be worth \$50 with probability $1 - p$. Currently an investor can purchase a European call option whose value is C. The exercise time and strike price of the option are respectively T and \$150. In the absence of arbitrage what is the value of C?

The investor could buy either the option or the stock at the beginning of the time interval. In the absence of arbitrage, the expected value of the investor's profit from either course of action should be zero regardless of which direction the price of the stock moves. Since the option or stock may be purchased at the beginning of the time interval but the profit (if any) arrives after time T has elapsed, we must calculate the present value of any potential profit. We will assume the interest rate per T unit of time is r. Suppose the investor purchases the stock initially. At time $t = 0$ their net gain is $-100 + 200(1 + r)^{-1}$ with probability p or $-100 + 50(1 + r)^{-1}$ with probability $1 - p$. In an arbitrage-free setting the expected value of this gain is zero.

$$0 = \left(-100 + 200(1 + r)^{-1}\right) p + \left(-100 + 50(1 + r)^{-1}\right) (1 - p)$$
$$0 = -100(1 + r) + 150p + 50$$
$$p = \frac{1 + 2r}{3} \tag{7.8}$$

On the other hand, if the investor purchases the option initially, at time $t = 0$ their net gain is $-C + (200 - 150)(1 + r)^{-1}$ with probability p or $-C$ with probability $1 - p$. Again the expected value of this gain will be zero in the absence of arbitrage.

$$0 = \left(-C + 50(1 + r)^{-1}\right) p + (-C)(1 - p)$$
$$0 = 50p(1 + r)^{-1} - C$$
$$C = \frac{50(1 + 2r)}{3(1 + r)} \tag{7.9}$$

Equation (7.9) was found using Eq. (7.8). The arbitrage-free probability and the option cost both depend on the interest rate. Figure 7.1 shows a parametric plot of the points $(p(r), C(r))$. If the probability p and the call option price deviate from the curve shown, then a risk-free profit can be made.

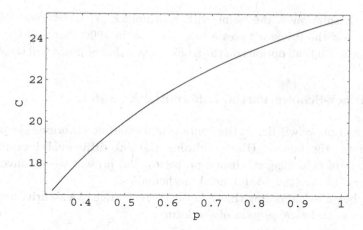

Fig. 7.1   A parametric plot of probability and call option cost as a function of interest rate.

Example 7.1   Suppose the current value of the stock is $S(0) = \$100$ and at time $T = 1$

$$S(1) = \begin{cases} \$200 \text{ with probability } p = 11/30, \\ \$50 \ \text{ with probability } 1 - p = 19/30. \end{cases}$$

Suppose further that the risk-free interest rate is $r = 5\%$ and that a European call option for the stock can be purchased for $C = \$17.50$. Note that by Eq. (7.9) an arbitrage-free option should be priced at $17.46. The strike price of the option is $K = \$150$. A risk-free profit can be generated in this situation since the option has been mis-priced according to Eqs. (7.9) and (7.8).

Suppose an investor borrows sufficient cash to purchase x shares of stock and y call options. The portfolio is originally worth $100x + 17.50y$ and they must pay back $e^{0.05}(100x + 17.50y)$ at time $T = 1$. If $S(1) = \$200$ then the call option will be exercised. Liquidating the portfolio yields $200x + (200 - 150)y$ in revenue. If $S(1) = \$50$ the call option is not be exercised. Cashing out the portfolio produces $50x$ in revenue. Thus if x and y can be found so that the following set of linear inequalities are simultaneously satisfied then a risk free profit can be realized.

$$200x + 50y > e^{0.05}(100x + 17.50y)$$
$$50x > e^{0.05}(100x + 17.50y)$$

One such solution is the point with coordinates $(x, y) = (1000, -3000)$. Therefore if the investor takes a long position in 1000 shares of the stock and shorts 3000 call options on the stock, their risk-free profit will be \$64.62.

7.3   Black-Scholes Partial Differential Equation

In this section we will derive the fundamental equation governing the pricing of options, the famous **Black-Scholes partial differential equation**. The ideas of arbitrage, stochastic processes, and present value converge at this point in the study of financial mathematics.

We begin by supposing that S is the current value of a security and that it obeys a stochastic process of the form

$$dS = \mu S\, dt + \sigma S\, dW(t) \tag{7.10}$$

If $F(S, t)$ is the value of any type of option (more generally called a financial derivative), then according to Itô's Lemma (Lemma 5.4), F obeys the following stochastic process.

$$dF = \left(\mu S F_S + \frac{1}{2}\sigma^2 S^2 F_{SS} + F_t\right) dt + \sigma S F_S\, dW(t) \tag{7.11}$$

Suppose a portfolio of value P is created by selling the option and buying Δ units of the security. Thus the value of the portfolio is $P = F - \Delta S$. The notation Δ for the number of units of the security purchased is standard in the derivation and analysis of the Black-Scholes equation. The reader should not let the notation ΔS suggest "change in S", it merely means Δ multiplied by S.

Since the portfolio is a linear combination of the the option and the security then the stochastic process governing the portfolio is

$$
\begin{aligned}
dP &= d(F - \Delta S) \\
&= dF - \Delta dS \\
&= \left(\mu S F_S + \frac{1}{2}\sigma^2 S^2 F_{SS} + F_t\right) dt + \sigma S F_S\, dW(t) \\
&\quad - \Delta\left(\mu S\, dt + \sigma S\, dW(t)\right) \\
&= \left(\mu S\left[F_S - \Delta\right] + \frac{1}{2}\sigma^2 S^2 F_{SS} + F_t\right) dt + \\
&\quad + \sigma S\left(F_S - \Delta\right) dW(t)
\end{aligned}
\tag{7.12}
$$

Notice the coefficient of the normal random variable, $dW(t)$, contains the factor $F_S - \Delta$. Equation (7.12) can be simplified if we assume $\Delta = F_S$. This has the beneficial effect of reducing the number of stochastic terms in Eq. (7.12). Randomness is not completely eliminated since the value of the security S remains and is stochastic. Under the assumption that $\Delta = F_S$, Eq. (7.12) becomes

$$dP = \left(\frac{1}{2}\sigma^2 S^2 F_{SS} + F_t\right) dt. \tag{7.13}$$

In an arbitrage-free setting the difference in the returns from investing in the portfolio described above or investing an equal amount of capital in a risk-free bond paying interest at rate r should be zero. Thus the following equations are true.

$$0 = rP\,dt - dP$$
$$= rP\,dt - \left(\frac{1}{2}\sigma^2 S^2 F_{SS} + F_t\right) dt$$
$$0 = \frac{1}{2}\sigma^2 S^2 F_{SS} + F_t - rP$$
$$0 = \frac{1}{2}\sigma^2 S^2 F_{SS} + F_t - r(F - \Delta S)$$
$$0 = F_t + rSF_S + \frac{1}{2}\sigma^2 S^2 F_{SS} - rF \tag{7.14}$$

Notice that Δ was replaced by F_S to obtain the final form of the equation. This is the well-known Black-Scholes equation for pricing financial derivatives. Equation (7.14) is an example of a **partial differential equation** or **PDE**, for short. While the general theory of solving PDEs is beyond the scope of this book, we will briefly mention a few concepts related to the Black-Scholes PDE. PDEs are often described by their order, type, and linearity properties. The Black-Scholes PDE is a **second order** equation since the highest order derivative of the unknown function F present in the equation is the second derivative. This PDE is of **parabolic type**. The best known example of the parabolic PDE is the **heat equation** which is used to describe the distribution of temperature along an object, for example a metal rod. For an introductory treatment of the heat equation see [Boyce and DiPrima (2001)]. Since the coefficients of F_t and F_{SS} have the same algebraic signs, the Black-Scholes PDE is sometimes referred to as a **backwards** parabolic equation. In [Wilmott *et al.* (1995)] the Black-Scholes PDE is solved by appropriate changes of variables until it becomes the heat equation. The heat equation can be solved by several independent techniques which all naturally, ultimately yield the same solution. One elementary solution technique is the method known as **separation of variables**. Lastly, the Black-Scholes PDE is an example of a **linear** partial differential equation since if F_1 and F_2 are two solutions to the equation then $c_1 F_1 + c_2 F_2$ is also a solution where c_1 and c_2 are any constants (see exercise 9).

Financial derivative products of many types obey the Black-Scholes equation. Different solutions correspond to different initial/final and boundary side conditions imposed while solving the equation. In the next section we will describe in more detail the initial and boundary conditions relevant to the Black-Scholes PDE. In Sec. 8.4 we will derive a solution to the Black-Scholes equation for the price of a European style call option.

7.4   Boundary and Initial Conditions

Without the imposition of side conditions in the form of initial or final conditions and boundary conditions, a partial differential equation can have many different solutions. Mathematicians refer to a differential equation with many possible solutions as "ill-posed." In this section we will discuss the appropriate conditions to impose in order to obtain a solution to Eq. (7.14) which describes the value of a European call option.

The domain of the unknown function F of Eq. (7.14) is a region in (S, t)-space. We will refer to this region as $\Omega \subset (S, t)$-space. For a European call option we are only interested in the value of the option during the time interval $[0, T]$, where T is the exercise time of the option. During this time interval the price of the security underlying the option may be any non-negative value. Thus the domain of the solution to Eq. (7.14) is

$$\Omega = \{(S, t) \,|\, 0 \leq S < \infty \text{ and } 0 \leq t \leq T\}.$$

At time $t = T$ the value of the security will either exceed the strike price (in which case the call option will be exercised generating an income flow of $S(T) - K > 0$) or the security will have a value less than or equal to the strike price (in which case the call option expires unused and has value 0). When $S(T) > K$ the call option is said to be **in the money**. When $S(T) \leq K$ the call option is said to be **out of the money**. Thus the terminal value of a European call option is

$$(S(T) - K)^+ = \max(S(T) - K, 0),$$

where $S(T)$ is the value of the underlying security at the exercise time and K is the strike price of the option. Graphically the payoff of the portfolio is said to resemble a "hockey stick". See Fig. 7.2. Hence we can use the

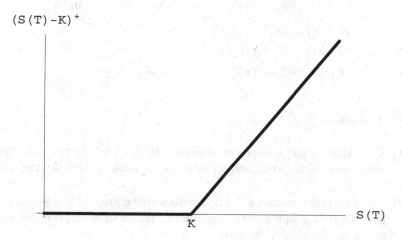

Fig. 7.2   The piecewise linear curve representing the payoff of a European call option.

following equation as the **final condition** for the Black-Scholes PDE.

$$F(S,T) = \max(S(T) - K, 0) \tag{7.15}$$

We see from Eq. (7.10) that if $S = 0$, then $dS = 0$, *i.e.* S never changes and hence remains zero. The boundary at $S = 0$ is said to be invariant. Thus on the portion of the boundary of Ω where $S = 0$, the call option would never be exercised and hence must have zero value. Thus we have derived one boundary condition, namely

$$F(0,t) = 0. \tag{7.16}$$

Now if we suppose that S is approaching infinity, it becomes increasingly likely that the call option will be exercised, since as $S \to \infty$, S will exceed any finite value of K. Likewise as $S \to \infty$ a put option would never be exercised. Thus according to the Put-Call Parity Formula in Eq. (7.1), as $S \to \infty$, $C \to S - Ke^{-r(T-t)}$. Also as $S \to \infty$, the difference $S - K \approx S$ and hence we have the second boundary condition

$$F(S,t) = S - Ke^{-r(T-t)}, \quad \text{as } S \to \infty. \tag{7.17}$$

Thus to summarize the Black-Scholes equation and its final and boundary conditions for a European call option we have the following set of equations.

$$rF = F_t + rSF_S + \frac{1}{2}\sigma^2 S^2 F_{SS} \quad \text{for } (S,t) \text{ in } \Omega,$$
$$F(S,T) = (S(T) - K)^+ \quad \text{for } S > 0,$$
$$F(0,t) = 0 \quad \text{for } 0 \le t < T,$$
$$F(S,t) = S - Ke^{-r(T-t)} \quad \text{as } S \to \infty.$$

7.5   Exercises

(1) Show that in the absence of arbitrage $P_a \ge P_e$ where the underlying security, exercise time, and strike price for both options are the same.

(2) Consider a European call option with a strike price of $60 which costs $10. Draw a graph illustrating the net payoff of the option for stock prices in the interval $[0, 100]$.

(3) Show that in the absence of arbitrage the value of a call option, either European or American, never exceeds the value of the underlying security.

(4) What is the minimum price of a European style call option with an exercise time of three months and a strike price of $26 for a security whose current value is $29 while the continuously compounded interest rate is 6%?

(5) If a share of a security is currently selling for $31, a three-month European call option is $3 with a strike price of $31, and the risk-free interest rate is 10%, what is the arbitrage-free European put option price for the security?

(6) If a share of a security is currently selling for $31, a three-month European call option is $3, a three-month European put option is $2.25, and the risk-free interest rate is 10%, what is the arbitrage-free strike price for the security?

(7) What is the minimum price of a European put option with an exercise time of two months and a strike price of $14 for a stock whose value is $11 while the continuously compounded interest rate is 7%?

(8) If a share of a security is currently selling for $31, a three-month European call option is $3, a three-month European put option is $1, the strike price for both options is $30, and the risk-free interest rate

is 10%, what arbitrage opportunities are open to investors?

(9) Suppose the functions $f_1(S, t)$ and $f_2(S, t)$ each solve the Black-Scholes Eq. (7.14). Show that if c_1 and c_2 are constants then the function $f(S, t) = c_1 f_1(S, t) + c_2 f_2(S, t)$ also solves the Black-Scholes equation.

(10) What final and boundary conditions are appropriate for the Black-Scholes Eq. (7.14) when F represents a European-style put option?

Chapter 8

Solution of the Black-Scholes Equation

In Chapter 7 the Black-Scholes partial differential equation was derived and summarized in Eq. (7.14). Every European style option satisfies this PDE. The differences in the options are due to different boundary and final conditions. In the present chapter we will solve the Black-Scholes equation with boundary and final conditions appropriate for a European style call option. Several methods are available to solve this equation. Some take the limit of a discrete time binomial model of security prices (see [Ross (1999)]). In [Wilmott *et al.* (1995)] the Black-Scholes PDE is transformed through a sequence of changes of variables to an ordinary differential equation which is solved by elementary means. Their change of variable approach will be mimicked here, but the **Fourier Transform** will be used to solve the equation. For readers unfamiliar with the Fourier Transform, a brief introduction to the relevant operations are included in this chapter. The Fourier Transform is a standard mathematical device used to convert certain types of partial differential equations into ordinary differential equations. A more complete introduction to Fourier Transforms can be found in either [Greenberg (1998)] or [Jeffrey (2002)].

If the boundary and final conditions of the Black-Scholes PDE are changed from those of the European calls and puts (examples of so-called vanilla options) to more esoteric conditions (of the exotic options), more sophisticated solution techniques or even numerical approximations of the solution may be required.

8.1 Fourier Transforms

Suppose the function $f(x)$ is defined for all real numbers and that the Fourier Transform of f is denoted by $\mathcal{F}\{f(x)\}$ or often as $\hat{f}(w)$. By definition the Fourier Transform of f is

$$\mathcal{F}\{f(x)\} = \hat{f}(w) = \int_{-\infty}^{\infty} f(x)e^{-iwx}\,dx, \qquad (8.1)$$

where $i = \sqrt{-1}$ and w is a parameter. The Fourier Transform maps the original function f which is a function of x to a new function $\hat{f}$ which depends on w which is often thought of as a frequency. The Fourier Transform of f is meaningful if and only if the improper integral converges. There are many theorems in the literature on the Fourier Transform which list conditions under which the Fourier Transform will exist. It is possible to state very general but technical conditions in such an existence theorem. For our purposes we will adopt a less technical theorem which states that if the domain of f is all real numbers and if

- f and f' are piecewise continuous on every interval of the form $[-M, M]$ for arbitrary $M > 0$, and
- $\int_{-\infty}^{\infty} |f(x)|\,dx$ converges,

then the Fourier Transform of f exists according to Theorem 17.9.1 of [Greenberg (1998)].

Example 8.1   Consider the piecewise-defined function

$$f(x) = \begin{cases} 1 & \text{if } |x| \leq 1 \\ 0 & \text{otherwise.} \end{cases}$$

Its Fourier Transform is readily found through the following calculation.

$$\begin{aligned}
\hat{f}(w) &= \int_{-\infty}^{\infty} f(x)e^{-iwx}\,dx \\
&= \int_{-1}^{1} e^{-iwx}\,dx \\
&= -\frac{1}{iw}\left(e^{-iw} - e^{iw}\right) \\
&= \frac{2}{w}\sin w
\end{aligned}$$

In the previous example we carried out integration of a complex exponential as if it followed the same rules of integration as real-valued functions. This is in fact true. The **Euler Identity**, $e^{i\theta} = \cos\theta + i\sin\theta$ was also used to simplify the result [Churchill *et al.* (1976)].

The Fourier Transform also has an interesting effect on derivatives of functions. Suppose the Fourier Transform of f exists and that f is differentiable for all real numbers.

$$
\begin{aligned}
\mathcal{F}\{f'(x)\} &= \int_{-\infty}^{\infty} f'(x)e^{-iwx}\,dx \\
&= f(x)e^{-iwx}\Big|_{-\infty}^{\infty} - \int_{-\infty}^{\infty} f(x)(-iw)e^{-iwx}\,dx \\
&= iw\int_{-\infty}^{\infty} f(x)e^{-iwx}\,dx \\
&= iw\hat{f}(w) \qquad\qquad\qquad\qquad\qquad\qquad\qquad (8.2)
\end{aligned}
$$

The technique of integration by parts was used to shift the derivative off of f and onto the exponential function. Since f vanishes as $|x| \to \infty$ the leading integral is zero. This property of the Fourier Transform can be extended to higher order derivatives.

Theorem 8.1 *If $f(x)$, $f'(x)$, ..., $f^{(n-1)}(x)$ are all Fourier transformable and if $f^{(n)}(x)$ exists (where $n \in \mathbb{N}$) then $\mathcal{F}\{f^{(n)}(x)\} = (iw)^n \hat{f}(w)$.*

Proof. The result has already been shown in Eq. (8.2) for the case where $n = 1$. The technique of proof by induction is used to establish this result for all higher order derivatives. Suppose the result has been proved for $n = k$.

$$
\begin{aligned}
\mathcal{F}\{f^{(k+1)}(x)\} &= \int_{-\infty}^{\infty} f^{(k+1)}(x)e^{-iwx}\,dx \\
&= f^{(k)}(x)e^{-iwx}\Big|_{-\infty}^{\infty} - \int_{-\infty}^{\infty} f^{(k)}(x)(-iw)e^{-iwx}\,dx \\
&= iw\int_{-\infty}^{\infty} f^{(k)}(x)e^{-iwx}\,dx \\
&= iw(iw)^k \hat{f}(w) \quad \text{(induction hypothesis)} \\
&= (iw)^{k+1} \hat{f}(w)
\end{aligned}
$$

Thus the result is true for $n = k + 1$ and by the principle of mathematical induction, hence for all $n \in \mathbb{N}$. $\qquad\qquad\qquad\qquad\qquad\qquad\square$

Another Fourier Transform result which we will have need for later concerns the transform of the convolution of two functions. By definition the **Fourier Convolution** of two functions f and g is

$$(f * g)(x) = \int_{-\infty}^{\infty} f(x - z)g(z)\,dz. \tag{8.3}$$

The Fourier Convolution is not the same as the Fourier Transform. The convolution involves no complex integration, merely the integration of the product of two real-valued functions. It is possible to calculate the Fourier Transform of this convolution.

Theorem 8.2 $\mathcal{F}\{(f * g)(x)\} = \hat{f}(w)\hat{g}(w)$, *in other words the Fourier Transform of the Fourier Convolution of f and g is the product of the Fourier Transforms of f and g.*

Proof.   We begin by applying the definitions of Fourier Transform and Convolution.

$$\mathcal{F}\{(f * g)(x)\} = \int_{-\infty}^{\infty} \left[\int_{-\infty}^{\infty} f(x - z)g(z)\,dz \right] e^{-iwx}\,dx$$

$$= \int_{-\infty}^{\infty} g(z) \left[\int_{-\infty}^{\infty} f(x - z)e^{-iwx}\,dx \right] dz$$

Since the order of integration in a multiple integral is irrelevant (only the region over which the integration takes place is meaningful), equality is preserved above. Now the change of variable $x = u + z$ is used to rewrite the interior integral above.

$$\mathcal{F}\{(f * g)(x)\} = \int_{-\infty}^{\infty} g(z) \left[\int_{-\infty}^{\infty} f(u)e^{-iw(u+z)}\,du \right] dz$$

$$= \int_{-\infty}^{\infty} g(z)e^{-iwz} \left[\int_{-\infty}^{\infty} f(u)e^{-iwu}\,du \right] dz$$

$$= \hat{f}(w)\hat{g}(w)$$

$\square$

8.2   Inverse Fourier Transforms

So far we have seen that the Fourier Transform can be used to map functions of x, their derivatives, and their convolutions to other functions which depend on a frequency-like variable w. We intend to apply this transformation to the Black-Scholes partial differential equation and its associated side conditions. However, any useful solution to the Black-Scholes equation should depend on the variables and constants in the original equation and not on the frequency variable w. It is necessary to have a method for transforming a function of w back into a function of x. This is the purpose of the **Inverse Fourier Transform**. By definition the inverse Fourier Transform of $\hat{f}(w)$ given by

$$\mathcal{F}^{-1}\{\hat{f}(w)\} = \frac{1}{2\pi} \int_{-\infty}^{\infty} \hat{f}(w)e^{iwx}\, dw. \tag{8.4}$$

Example 8.2 Suppose a is a positive constant, then the inverse Fourier Transform of $e^{-a|w|}$ is

$$\begin{aligned}
\mathcal{F}^{-1}\{e^{-a|w|}\} &= \frac{1}{2\pi} \int_{-\infty}^{\infty} e^{-a|w|}e^{iwx}\, dw \\
&= \frac{1}{2\pi} \int_{-\infty}^{0} e^{(a+ix)w}\, dw + \frac{1}{2\pi} \int_{0}^{\infty} e^{(-a+ix)w}\, dw \\
&= \frac{1}{2\pi(a+ix)} + \frac{1}{2\pi(a-ix)} \\
&= \frac{a}{\pi(a^2+x^2)}
\end{aligned}$$

Readers with some prior exposure to the Fourier Transform may prefer more symmetric definitions of the transform and the inverse transform. An infinite number of variations on the formulas can be stated, though they are all equivalent and compatible with the versions given above in Eqs. (8.1) and (8.4). Some people prefer each of the forward and inverse transforms to be scaled by a factor involving $\sqrt{2\pi}$ as in the following.

$$\hat{f}(w) = \frac{1}{\sqrt{2\pi}} \int_{-\infty}^{\infty} f(x)e^{-iwx}\, dx$$

$$f(x) = \frac{1}{\sqrt{2\pi}} \int_{-\infty}^{\infty} \hat{f}(w)e^{iwx}\, dw$$

8.3 Changing Variables in the Black-Scholes PDE

To recap the partial differential equation and its side conditions for a European style call option, we have the following three equations

$$rF = F_t + \frac{1}{2}\sigma^2 S^2 F_{SS} + rSF_S \quad \text{for } t \in (0,T),\ S \in (0,\infty) \quad (8.5)$$

$$F(S,T) = (S-K)^+ \quad \text{for } S \in (0,\infty) \tag{8.6}$$

$$F(0,t) = 0 \tag{8.7}$$

$$F(S,t) \to S - Ke^{-r(T-t)} \quad \text{as } S \to \infty \text{ for } t \in [0,T). \tag{8.8}$$

The final condition given in Eq. (8.6) states that when the expiry date for the option arrives, the call option will be worth nothing if the value of the stock is less than the strike price. Otherwise the call option is worth the excess of the stock's value over the strike price. The boundary conditions are specified in Eqs. (8.7) and (8.8). The first condition implies that if the stock itself becomes worthless before maturity, the call option is also worthless. An investor could buy the stock for nothing and then let the call option expire unused. The second part of this condition follows from the European call option property discussed in Chapter 7, namely $C_e \geq S - Ke^{-rT}$. By the Put/Call Parity Formula in Eq. (7.1), as $S \to \infty$ a put option becomes worthless and the value of a call option approaches becomes $S - Ke^{-r(T-t)}$ asymptotically.

A judicious change of variables for the Black-Scholes equation can simplify it. In fact, in only a few steps we can convert Eq. (8.5) into a more well known partial differential equation, namely the **heat equation**. Suppose F, S, and t are defined in terms of the new variables v, x, and τ as in the following equations.

$$S = Ke^x \Leftrightarrow x = \ln \frac{S}{K} \tag{8.9}$$

$$t = T - \frac{2\tau}{\sigma^2} \Leftrightarrow \tau = \frac{\sigma^2}{2}(T-t) \tag{8.10}$$

$$F(S,t) = Kv(x,\tau) \tag{8.11}$$

How are the Black-Scholes equation and its side conditions altered by this change of variables? The multivariable form of the chain rule can be used to determine new expressions for the derivatives present in Eq. (8.5).

$$F_S = (Kv(x,\tau))_S = K\left(v_x x_S + v_t t_S\right) = K\left(v_x \frac{1}{S} + 0\right) = e^{-x}v_x \tag{8.12}$$

The reader will be asked to verify that

$$F_t = -\frac{K\sigma^2}{2}v_\tau \quad \text{and} \tag{8.13}$$

$$F_{SS} = \frac{e^{-2x}}{K}(v_{xx} - v_x) \tag{8.14}$$

in exercises 4 and 5. Substituting the results in Eqs. (8.11)–(8.14) in the Black-Scholes Eq. (8.5) and simplifying produces the equation

$$v_\tau = v_{xx} + (k - 1)v_x - kv \tag{8.15}$$

where $k = 2r/\sigma^2$ (see exercise 6).

The final condition for F is converted by this change of variables into an initial condition since when $t = T$, $\tau = 0$. The initial condition $v(x, 0)$ is then found to be

$$\begin{aligned} Kv(x, 0) &= F(S, T) \\ &= (S - K)^+ \\ &= K(e^x - 1)^+ \\ v(x, 0) &= (e^x - 1)^+ \end{aligned} \tag{8.16}$$

Since $\lim_{S\to 0^+} x = -\infty$, then

$$0 = \lim_{S\to 0^+} F(S, t) = \lim_{x\to -\infty} Kv(x, \tau) \implies \lim_{x\to -\infty} v(x, \tau) = 0.$$

Likewise since as $\lim_{S\to\infty} x = \infty$,

$$\lim_{S\to\infty} F(S, t) = S - Ke^{-r(T-t)} = \lim_{x\to\infty} Kv(x, \tau)$$

$$e^x - e^{-k\tau} = \lim_{x\to\infty} v(x, \tau)$$

Thus we have derived a pair of boundary conditions for the partial differential equation in (8.15). So the original Black-Scholes initial, boundary value problem can be recast in the form of the following.

$$v_\tau = v_{xx} + (k - 1)v_x - kv \quad \text{for } x \in (-\infty, \infty), \tau \in (0, \tfrac{T\sigma^2}{2}) \tag{8.17}$$

$$v(x, 0) = (e^x - 1)^+ \quad \text{for } x \in (-\infty, \infty) \tag{8.18}$$

$$v(x, \tau) \to 0 \quad \text{as } x \to -\infty \text{ and} \tag{8.19}$$

$$v(x, \tau) \to e^x - e^{-k\tau} \quad \text{as } x \to \infty, \tau \in (0, \tfrac{T\sigma^2}{2}) \tag{8.20}$$

The independent variable x can be thought of as corresponding to a spatial variable. The reader should note that where previously the price of the security S was assumed to take on only non-negative values, now x can be any real number.

Now another change of variables is needed. If α and β are constants then we can introduce the new dependent variable u.

$$v(x, \tau) = e^{\alpha x + \beta \tau} u(x, \tau) \tag{8.21}$$

$$v_x = e^{\alpha x + \beta \tau} \left(\alpha u(x, \tau) + u_x \right) \tag{8.22}$$

$$v_{xx} = e^{\alpha x + \beta \tau} \left(\alpha^2 u(x, \tau) + 2\alpha u_x + u_{xx} \right) \tag{8.23}$$

$$v_\tau = e^{\alpha x + \beta \tau} \left(\beta u(x, \tau) + u_\tau \right) \tag{8.24}$$

Substituting the expressions found in (8.21)–(8.24) into Eq. (8.17) produces

$$u_\tau = (\alpha^2 + (k-1)\alpha - k - \beta)u + (2\alpha + k - 1)u_x + u_{xx} \tag{8.25}$$

Since α and β are arbitrary constants they can now be chosen appropriately to simplify Eq. (8.25). Ideally the coefficients of u_x and u would be zero. Solving the two equations:

$$0 = \alpha^2 + (k-1)\alpha - k - \beta$$
$$0 = 2\alpha + k - 1$$

yields $\alpha = (1-k)/2$ and $\beta = -(k+1)^2/4$. The initial condition for u can be derived from the initial condition given in (8.18).

$$v(x, 0) = (e^x - 1)^+$$
$$u(x, 0) = e^{(k-1)x/2}(e^x - 1)^+$$
$$= e^{(k-1)x/2} \begin{cases} e^x - 1 & \text{if } x > 0, \\ 0 & \text{if } x \le 0. \end{cases}$$
$$= \begin{cases} e^{(k+1)x/2} - e^{(k-1)x/2} & \text{if } x > 0, \\ 0 & \text{if } x \le 0. \end{cases}$$
$$= (e^{(k+1)x/2} - e^{(k-1)x/2})^+$$

Likewise we can derive boundary conditions at $x = \pm\infty$ for u from Eqs. (8.19) and (8.20). To summarize then the original Black-Scholes partial differential equation, initial, and boundary conditions have been converted to the following system of equations.

$$u_\tau = u_{xx} \quad \text{for } x \in (-\infty, \infty) \text{ and } \tau \in (0, T\sigma^2/2) \quad (8.26)$$

$$u(x, 0) = (e^{(k+1)x/2} - e^{(k-1)x/2})^+ \quad \text{for } x \in (-\infty, \infty) \quad (8.27)$$

$$u(x, \tau) \to 0 \quad \text{as } x \to -\infty \text{ for } \tau \in (0, T\sigma^2/2) \quad (8.28)$$

$$u(x, \tau) \to e^{\frac{(k+1)}{2}[x+(k+1)\tau/2]} - e^{\frac{(k-1)}{2}[x+(k-1)\tau/2]} \quad (8.29)$$

as $x \to \infty$ for $\tau \in (0, T\sigma^2/2)$. The PDE of Eq. (8.26) is the well-known **heat equation** of mathematical physics. The Fourier Transform technique introduced earlier will now be used to solve this initial boundary value problem.

8.4 Solving the Black-Scholes Equation

We begin by taking the Fourier transform of both sides of the heat equation (since the Black-Scholes equation has been through two changes of variables and has been converted into the heat equation on the real number line) in Eq. (8.26).

$$\mathcal{F}\{u_\tau\} = \mathcal{F}\{u_{xx}\}$$

$$\int_{-\infty}^{\infty} u_\tau e^{-iwx} \, dx = \int_{-\infty}^{\infty} u_{xx} e^{-iwx} \, dx$$

$$\frac{d}{d\tau} \int_{-\infty}^{\infty} u(x, \tau) e^{-iwx} \, dx = (iw)^2 \int_{-\infty}^{\infty} u(x, \tau) e^{-iwx} \, dx$$

$$\frac{d\hat{u}}{d\tau} = -w^2 \hat{u} \quad (8.30)$$

where $\hat{u}$ is the Fourier transform of $u(x, \tau)$. The reader should note that in deriving Eq. (8.30) Theorem 8.1 is used. This last equation is an ordinary differential equation of the type used to model exponential decay. Separating variables (see Sec. 5.3, especially Eq. (5.20) and the paragraphs which follow it) and solving this equation produces a solution of the form

$$\hat{u}(w, \tau) = De^{-w^2\tau}.$$

The expression represented by D is any quantity which is constant with respect to τ. Even though this equation was solved using ordinary differential equations techniques, the quantity $\hat{u}$ is a function of both w and τ. To evaluate D we can set $\tau = 0$ and determine that $\hat{u}(w, 0) = D$. Thus D is merely the Fourier transform of the initial condition in Eq. (8.27). For simplicity of notation we will write $D = \hat{f}(w)$. Thus the Fourier transformed solution to the heat equation is

$$\hat{u}(w,\tau) = \hat{f}(w)e^{-w^2\tau}. \tag{8.31}$$

Now this solution must be inverse Fourier transformed and then have its variables changed back to the original variables of the Black-Scholes equation.

$$\mathcal{F}^{-1}\{\hat{u}(w,\tau)\} = \mathcal{F}^{-1}\{\hat{f}(w)e^{-w^2\tau}\}$$

$$u(x,\tau) = (e^{(k+1)x/2} - e^{(k-1)x/2})^+ * \frac{1}{2\sqrt{\pi\tau}}e^{-x^2/(4\tau)}$$

$$= \frac{1}{2\sqrt{\pi\tau}}\int_{-\infty}^{\infty}(e^{(k+1)\frac{z}{2}} - e^{(k-1)\frac{z}{2}})^+ e^{-\frac{(x-z)^2}{4\tau}}\,dz \tag{8.32}$$

The reader should note that the Fourier convolution theorem was used above. In exercise 3 the reader will also provide the details of the essential result that

$$\mathcal{F}^{-1}\{e^{-w^2\tau}\} = \frac{1}{2\sqrt{\pi\tau}}e^{-x^2/(4\tau)}.$$

If the substitution $z = x + \sqrt{2\tau}y$ is made then Eq. (8.32) becomes

$$u(x,\tau)$$

$$= \frac{1}{\sqrt{2\pi}}\int_{-\infty}^{\infty}\left(e^{(k+1)(x+\sqrt{2\tau}y)/2} - e^{(k-1)(x+\sqrt{2\tau}y)/2}\right)^+ e^{-y^2/2}\,dy$$

$$= \frac{e^{(k+1)x/2}e^{(k+1)^2\tau/4}}{\sqrt{2\pi}}\int_{-x/\sqrt{2\tau}}^{\infty} e^{-(y-\frac{1}{2}(k+1)\sqrt{2\tau})^2/2}\,dy$$

$$- \frac{e^{(k-1)x/2}e^{(k-1)^2\tau/4}}{\sqrt{2\pi}}\int_{-x/\sqrt{2\tau}}^{\infty} e^{-(y-\frac{1}{2}(k-1)\sqrt{2\tau})^2/2}\,dy \tag{8.33}$$

The reader will be asked to fill in the details of this derivation in exercises 8 and 9. The two improper integrals of Eq. (8.33) can be expressed in terms of the cumulative distribution function for a normal random variable. We will make the substitution $w = y - \frac{1}{2}(k+1)\sqrt{2\tau}$ in the first integral and $w' = y - \frac{1}{2}(k-1)\sqrt{2\tau}$ in the second. The first integral of Eq. (8.33) equals the following expression

$$\int_{-\infty}^{x/\sqrt{2\tau}+\frac{1}{2}(k+1)\sqrt{2\tau}} e^{-w^2/2}\,dw = \sqrt{2\pi}\phi\left(\frac{x}{\sqrt{2\tau}} + \frac{1}{2}(k+1)\sqrt{2\tau}\right)$$

where ϕ is the cumulative distribution function for a normal random variable with mean zero and standard deviation one. The reader should verify this calculation (exercise 10) and a similar calculation for the second improper integral of Eq. (8.33).

Thus far the solution to the initial boundary value problem in Eqs. (8.26)–(8.29) is given by

$$u(x, \tau) = e^{(k+1)x/2+(k+1)^2\tau/4} \phi\left(\frac{x}{\sqrt{2\tau}} + \frac{1}{2}(k+1)\sqrt{2\tau}\right)$$
$$- e^{(k-1)x/2+(k-1)^2\tau/4} \phi\left(\frac{x}{\sqrt{2\tau}} + \frac{1}{2}(k-1)\sqrt{2\tau}\right). \qquad (8.34)$$

The reader should take a moment to check that the expression for $u(x, \tau)$ given in Eq. (8.34) satisfies the boundary conditions as $x \to \pm\infty$ specified in Eqs. (8.28) and (8.29).

Now we begin the task of reconverting variables to those of the Black-Scholes initial boundary value problem as stated in Eqs. (8.5)–(8.8). Using the change of variables in Eq. (8.21) this solution can be re-written in terms of the function $v(x, \tau)$.

$$v(x, \tau) = e^{-(k-1)x/2-(k+1)^2\tau/4} u(x, \tau)$$
$$= e^x \phi\left(\frac{x}{\sqrt{2\tau}} + \frac{1}{2}(k+1)\sqrt{2\tau}\right)$$
$$- e^{-k\tau} \phi\left(\frac{x}{\sqrt{2\tau}} + \frac{1}{2}(k-1)\sqrt{2\tau}\right) \qquad (8.35)$$

Now using the change of variables described in Eqs. (8.9) and (8.10) the reader can show in exercise 13 that

$$w = \frac{x}{\sqrt{2\tau}} + \frac{1}{2}(k+1)\sqrt{2\tau} = \frac{\ln(S/K) + (r + \sigma^2/2)(T-t)}{\sigma\sqrt{T-t}} \qquad (8.36)$$

$$w' = \frac{x}{\sqrt{2\tau}} + \frac{1}{2}(k-1)\sqrt{2\tau} = w - \sigma\sqrt{T-t}. \qquad (8.37)$$

Using these in Eq. (8.35) yields

$$v(x, \tau) = \frac{S}{K}\phi(w) - e^{-r(T-t)}\phi(w - \sigma\sqrt{T-t})$$

which upon using Eq. (8.11) produces the sought after **Black-Scholes European call option pricing formula**. For the sake of simplicity and meaningful notation the value of the European call will be denoted C.

$$C(S, t) = S\phi(w) - Ke^{-r(T-t)}\phi(w - \sigma\sqrt{T-t}) \qquad (8.38)$$

A surface plot of the value of the European call option as a function of S and t is shown in Fig. 8.1.

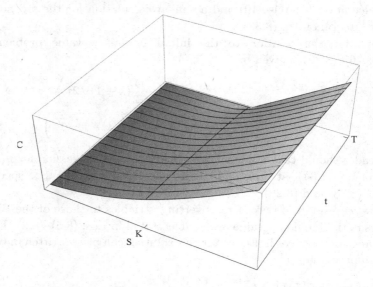

Fig. 8.1   A surface plot of C_e as a function of S and t.

The value of a European put option could be found via a similar set of calculations or the already determined value of the European call could be used along with the Put-Call Parity Formula (7.1) to derive

$$P(S, t) = Ke^{-r(T-t)}\phi(\sigma\sqrt{T-t} - w) - S\phi(-w). \qquad (8.39)$$

A surface plot of the value of the European put option as a function of S and t is shown in Fig. 8.2.

Example 8.3   Suppose the current price of a security is $62 per share. The continuously compounded interest rate is 10% per year. The volatility of the price of the security is $\sigma = 20\%$ per year. The cost of a five-month European call option with a strike price of $60 per share can be found using Eqs. (8.36) and (8.38). If we summarize the quantities we know, in the notation of this section, we have:

$$T = 5/12, \quad t = 0, \quad r = 0.10, \quad \sigma = 0.20, \quad S = 62, \quad \text{and} \quad K = 60.$$

Thus we have $w \approx 0.641287$ and therefore the price of the European call option is $C = \$5.80$.

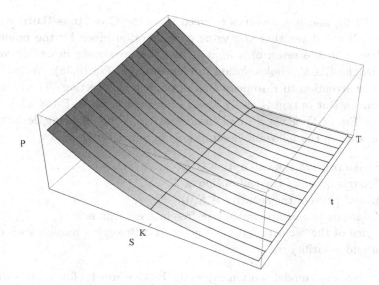

Fig. 8.2   A surface plot of P_e as a function of S and t.

Example 8.4   Suppose the current price of a security is $97 per share. The continuously compounded interest rate is 8% per year. The volatility of the price of the security is $\sigma = 45\%$ per year. The cost of a three-month European put option with a strike price of $95 per share can be found using Eqs. (8.36), and (8.39). If we summarize the quantities we know, in the notation of this section, we have:

$$T = 1/4, \quad t = 0, \quad r = 0.08, \quad \sigma = 0.45, \quad S = 97, \quad \text{and} \quad K = 95.$$

Thus we have $w \approx 0.293985$ and consequently the put option price is $P = \$6.71$.

8.5   Binomial Model (Optional)

As an alternative to the partial differential equation solution to the Black-Scholes equation, we may derive the same formulas from a simple, discrete model. Thus for readers uncomfortable with the formality and technicalities of the previous approach or those readers wanting to understand an alternative model which forms the basis of numerical methods for evaluating options, this section describes what is known as the **binomial model**. This derivation was initially developed by Cox, Ross, and Rubinstein [Cox *et al.* (1979)] and is sometimes referred to as the **Cox-Ross-Rubinstein model**. We will see that the value of the option given by the binomial model will, in the sense of a limit, equal the previously described value given by the Black-Scholes option pricing formula, Eq. (8.38). We will restrict our attention to European Call options in this section. The value of a European Put option is easily found either through the Put-Call Parity formula, Eq. (7.1) or via a simple modification to the following derivation.

We will use the following assumptions in our derivation:

- Strike price of the call option is K.
- Exercise time of the call option is T.
- Initial price of the security is $S(0)$.
- Continuously compounded risk-free interest rate is r.
- Price of the security follows a geometric Brownian motion with drift μ and volatility σ.

The binomial model is often called the **lattice model** for reasons which will soon become apparent. Suppose the time interval $[0, T]$ is partitioned into n equal subintervals of length $\Delta t = T/n$. At the start of the first subinterval $[0, \Delta t]$ the value of the security is $S(0)$. The binomial model assumes that the value of the security will evolve to $uS(0)$ where $u > 1$ with probability p or to $dS(0)$ where $0 < d < 1$ with probability $1 - p$. Since value of the security may only increase by the factor u or decrease by the factor d, the name "binomial model" is appropriate. A graph of the lattice model for a single time step is shown in Fig. 8.3.

The assumption that the value of the security follows a geometric Brownian motion implies that

$$dS = \mu S \, dt + \sigma S \, dW(t)$$
$$S(t) = S(0)e^{(\mu - \sigma^2/2)t + \sigma W(t)} \qquad \text{(exercise 19)}$$
$$\mathrm{E}\left[S(t)\right] = S(0)e^{\mu t}$$
$$\mathrm{Var}\left(S(t)\right) = (S(0))^2 \, e^{2\mu t} \left(e^{\sigma^2 t} - 1\right).$$

The expressions for the expected value and variance are a result of Lemma 3.1. The parameters of the binomial lattice model must be selected so that the discrete model of the security has the same expected value and variance as the continuous model.

Hence if the security can take on only values $uS(0)$ and $dS(0)$ with probabilities p and $1 - p$ respectively then after a time step of length Δt

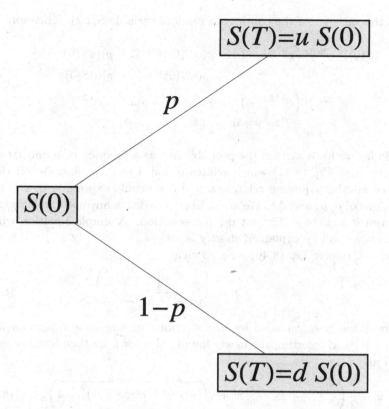

Fig. 8.3   A single time step of size Δt is shown for the binomial model of security values.

we have

$$puS(0) + (1 - p)dS(0) = S(0)e^{\mu\Delta t}$$

$$(8.40)$$

$$pu + (1 - p)d = e^{\mu\Delta t}.$$

Another assumption of the binomial model is that the variance in the value of the security at time Δt generated by the discrete model must agree with the variance of the continuous random variable $S(\Delta t)$. Therefore

$$
\begin{aligned}
(S(0))^2\, e^{2\mu\Delta t} \left(e^{\sigma^2\Delta t} - 1\right) &= p(uS(0))^2 + (1-p)(dS(0))^2 \\
&\quad - [p(uS(0)) + (1-p)(dS(0))]^2 \\
e^{2\mu\Delta t}\left(e^{\sigma^2\Delta t} - 1\right) &= pu^2 + (1-p)d^2 - (e^{\mu\Delta t})^2 \\
e^{(2\mu+\sigma^2)\Delta t} &= pu^2 + (1-p)d^2
\end{aligned}
\tag{8.41}
$$

So far we have written the probability p as a function of μ and Δt and now we have Eq. (8.41) which relates u and d to μ, σ, and Δt. If there existed another equation relating u and d we could express p, u, and d as functions of μ, σ, and Δt. We are at liberty to define any useful relationship between u and d for this last desired equation. A simple equation whose significance will be explained shortly is $ud = 1$.

We may solve Eq. (8.40) for p to yield

$$
p = \frac{ue^{\mu\Delta t} - 1}{u^2 - 1}
\tag{8.42}
$$

where d has been replaced by $1/u$. Performing the same replacement in Eq. (8.41) and inserting the newly found value for p we then find the value for u as

$$
u = \frac{1}{2}\left(e^{-\mu\Delta t} + e^{(\mu+\sigma^2)\Delta t} + \sqrt{\left(e^{-\mu\Delta t} + e^{(\mu+\sigma^2)\Delta t}\right)^2 - 4}\right).
\tag{8.43}
$$

For the sake of completeness, an expression for d is

$$
d = \frac{1}{2}\left(e^{-\mu\Delta t} + e^{(\mu+\sigma^2)\Delta t} - \sqrt{\left(e^{-\mu\Delta t} + e^{(\mu+\sigma^2)\Delta t}\right)^2 - 4}\right).
\tag{8.44}
$$

Just as we did for the continuous model we must impose a "no-arbitrage" condition on the binomial model. Π will denote the initial value of a portfolio consisting of a short position in the call option and a long position in the underlying security S. For every option sold, Δ units of the underlying are bought. As usual we will assume the underlying security is infinitely divisible. The value of Δ must be chosen so as to make the portfolio Delta neutral, in other words whether the price of the security rises or falls, the portfolio value must remain the same. We will introduce the symbols C_u and C_d to represent the value of the option for the cases in which the security is worth at time Δt, $uS(0)$ and $dS(0)$ respectively. The Delta neutrality of the portfolio can be expressed as

$$C_u - \Delta u S(0) = C_d - \Delta d S(0)$$

which implies

$$\Delta = \frac{C_u - C_d}{S(0)(u - d)}. \tag{8.45}$$

Now at time Δt the portfolio is worth $\Pi(\Delta t) = C(\Delta t) - \Delta S(\Delta t)$. At time Δt the underlying security will be worth $uS(0)$ (with probability p) or $dS(0)$ (with probability $1 - p$). If no arbitrage is permitted the value of the portfolio at time Δt should be equal to the future value of Π earning interest at the risk-free rate r. Thus

$$e^{r\Delta t}\Pi = \Pi(\Delta t)$$
$$e^{r\Delta t}(C - \Delta S(0)) = C_u - \Delta u S(0)$$
$$e^{r\Delta t}\left(C - \frac{C_u - C_d}{u - d}\right) = C_u - \frac{u(C_u - C_d)}{u - d}$$

which can be solved for C to produce

$$C = \frac{1 - de^{-r\Delta t}}{u - d}C_u + \frac{ue^{-r\Delta t} - 1}{u - d}C_d. \tag{8.46}$$

Another way of interpreting eq. (8.46) is as giving the future value of the option as a function of its two potential binomial values at the next time step.

$$e^{r\Delta t}C = \frac{e^{r\Delta t} - d}{u - d}C_u + \frac{u - e^{r\Delta t}}{u - d}C_d$$

Note that the coefficients of C_u and C_d sum to one. Thus setting

$$p' = \frac{e^{r\Delta t} - d}{u - d} = \frac{ue^{r\Delta t} - 1}{u^2 - 1} \tag{8.47}$$

we may write

$$e^{r\Delta t}C = p'C_u + (1 - p')C_d. \tag{8.48}$$

In other words the value of the option at time t is the present value of the expected value of the option at time $t + \Delta t$. The reader should note that the probability of an upward movement in the price of the security (refer to Eq. (8.42)) and the probability of an upward movement in the value of the call option (see Eq. (8.47)) are closely related. The main difference is that the drift parameter of the security has been replaced with the risk-free, continuously compounded interest rate. The no-arbitrage condition then imposes the restriction that the binomial probability and the increase and descrease factors (u and d respectively) must be calculated using r in place of σ. When calculating the value of options we will use the symbols p, u, and d but the reader should keep in mind that these quantities are computed using r in place fo μ.

So far a great deal of work has gone into constructing a discrete model of a single time step in the evolution of the value of the security. Once the model parameters p, u, and d are known a lattice of $n = T/\Delta t$ steps can be constructed. At each vertex of the lattice the value of the security may proceed up by a multiplicative factor u with probability p or down by a multiplicative factor d with probability $1 - p$. The parameters are assumed to remain constant across the time steps. For the sake of illustration a lattice of three time steps is shown in Fig. 8.4. Since we chose $ud = 1$ then we see that $udS(0) = duS(0) = S(0)$ and thus bifurcations in the lattice come together after two time steps during which one upward movement and one downward movement occurred. For this reason the graphs in this discrete model are sometimes called **recombining trees**. Now we may use the lattice model to approximate the value of a European call option.

Recall that the payoff value of a European call option is $(S(T) - K)^+$. Thus $C(T) = (S(T) - K)^+$. In an arbitrage-free setting the value of the call at any time must be the same as the present value of the expected payoff at expiry. Working backward in time from expiry, at $t = T - \Delta t$ the call option is worth

$$C(T - \Delta t) = e^{-r\Delta t} \mathrm{E}\left[(u^{2Y-1} - K)^+\right]$$

where $Y \in \{0, 1\}$. The assignment $Y = 0$ implies the security decreases in price over the next time step, while $Y = 1$ indicates it increases in price.

To generalize the discussion to the situation of an n-step binomial lattice, define Y to be the random variable denoting the number of upward movements in the value of the security since $t = 0$ (at which time the value of the security was $S(0)$). Note that Y is a binomial random variable in the sample space $\{0, 1, \ldots, n\}$. Thus if C represents the value of the call at

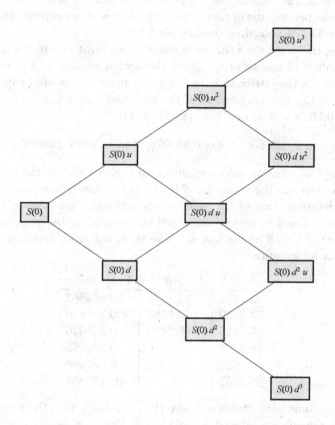

Fig. 8.4 A binomial model of three time steps.

time $t = 0$, then

$$C = e^{-rT}\mathrm{E}\left[(u^Y d^{n-Y} S(0) - K)^+\right]$$
$$= e^{-rT}\mathrm{E}\left[(u^{2Y-n} S(0) - K)^+\right]$$
$$= e^{-rT}\mathrm{E}\left[(u^{2Y-T/\Delta t} S(0) - K)^+\right]. \qquad (8.49)$$

Example 8.5 Now we may approximate the value of a European call option and compare the values generated by the Black-Scholes option pricing formula in Eq. (8.38) and the lattice method. Assume the current price of a security is \$62 per share, the continuously compounded interest rate is

10% per year, the volatility of the price of the security is $\sigma = 20\%$ per year. If the strike price of the option is $60 per share with an expiry of 5 months, then $C = \$5.789$ according to example 8.3.

To approximate the value of the option we must create a recombining tree of values of the security. Since the option expires in 5 months, it is convenient for the lattice to have 5 steps. The time step will be one month, so $\Delta t = 1/12$. The parameters of the binomial lattice can be found using Δt, $r = 0.10$, $\sigma = 0.20$ and Eqs. (8.42)–(8.44).

$$u = 1.06036, \quad d = 0.943073, \quad \text{and} \quad p = 0.556697.$$

The lattice of security values is shown in Fig. 8.5. Since the option is of European type on the values for S in the right-most column are important for the determination of the value of the option. The value of the European call option can be approximated by calculating the present value of the expected payoff of the option. The table below summarizes the most important information.

S	$(S - K)^+$	Probability
83.1122	23.1122	0.0534682
73.9189	13.9189	0.212886
65.7425	5.7425	0.339046
58.4705	0	0.269985
52.0029	0	0.107496
46.2507	0	0.0171199

The last column of the table contains the probability that the security price shown in the column labeled S is achieved. These values are calculated using the binomial probability density formula of Eq. (2.3). If we think of the columns labeled $(S - K)^+$ and probability as vectors then the approximate value of the option is the present value of their dot product. Therefore we find that

$$C \approx \frac{(23.1122)(0.0534682) + (13.9189)(0.212886) + (5.7425)(0.339046)}{e^{(0.10)(5/12)}}$$

$$= \frac{6.14588}{1.04255} = 5.89506.$$

This compares favorably to the value given by the Black-Scholes option pricing formula for European call options.

If the reader has access to spreadsheet software or a computer algebra system they may verify that if Δt for the discrete model is decreased from

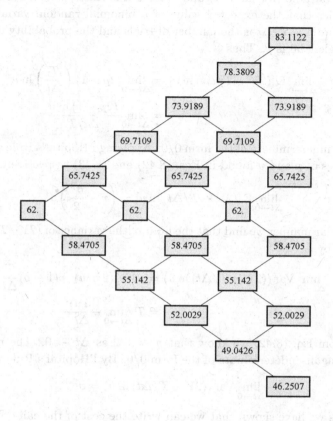

Fig. 8.5 A binomial lattice of security prices for example 8.5.

one month to one day, the binomial model yields a call option value of $C \approx 5.80305$ which provides numerical evidence that the discrete price is converging to the continuous price.

The reader should observe that

$$u^{2Y - T/\Delta t} = e^{(2Y - T/\Delta t)\ln u}$$

and since Y is a binomial random variable, as n becomes large (or equivalently as Δt becomes small) Y can be approximated by a normal random variable. Consequently $(2Y - T/\Delta t)\ln u$ is a linear function of a normal random variable (for large n) and is itself a normal random variable. We saw earlier that the expected value of a binomial random variable is np where here $n = T/\Delta t$ is the number of trials and the probability of success on a single trial is p. Thus

$$\lim_{\Delta t \to 0} \mathrm{E}\left[(2Y - T/\Delta t)\ln u\right] = \lim_{\Delta t \to 0}(2p-1)\left(\frac{T}{\Delta t}\right)\ln u$$

$$= \lim_{\Delta t \to 0}\frac{(2p-1)\ln u}{\Delta t/T}$$

which is indeterminate of the form $0/0$. Applying l'Hôpital's Rule and using the values of p and u found in Eqs. (8.42) and (8.43) respectively, we have

$$\lim_{\Delta t \to 0} \mathrm{E}\left[(2Y - T/\Delta t)\ln u\right] = \left(r - \frac{\sigma^2}{2}\right)T. \tag{8.50}$$

In a similar manner we find that the limit of the variance of $(2Y - T/\Delta t)\ln u$ is given by

$$\lim_{\Delta t \to 0} \mathrm{Var}\left((2Y - T/\Delta t)\ln u\right) = \lim_{\Delta t \to 0}(2\ln u)^2 p(1-p)\frac{T}{\Delta t}$$

$$= T\lim_{\Delta t \to 0}\frac{(\ln u)^2}{\Delta t}$$

since from Eq. (8.42) we know that $p \to 1/2$ as $\Delta t \to 0$. The remaining limit is again indeterminate of the form $0/0$. By l'Hôpital's Rule we obtain

$$\lim_{\Delta t \to 0} \mathrm{Var}\left((2Y - T/\Delta t)\ln u\right) = \sigma^2 T. \tag{8.51}$$

Thus we have shown that we can write the cost of the call option as

$$C = e^{-rT}\mathrm{E}\left[(S(0)e^X - K)^+\right], \tag{8.52}$$

where X is a normal random variable with

$$\mathrm{E}\left[X\right] = \left(r - \frac{\sigma^2}{2}\right)T \quad \text{and} \quad \mathrm{Var}\left(X\right) = \sigma^2 T,$$

and probability density function

$$g(x) = \frac{1}{\sigma\sqrt{T}\sqrt{2\pi}}e^{-(x - T(r - \sigma^2/2))^2/(2\sigma^2 T)}.$$

Define a new variable Z to be

$$Z = \frac{X - (r - \sigma^2/2)T}{\sigma\sqrt{T}}.$$

Since Z is a linear function of a normal random variable X, then Z is also a normal random variable. We note that

$$\mathrm{E}\,[Z] = \mathrm{E}\left[\frac{X - (r - \sigma^2/2)T}{\sigma\sqrt{T}}\right]$$

$$= \frac{\mathrm{E}\,[X]}{\sigma\sqrt{T}} - \frac{(r - \sigma^2/2)T}{\sigma\sqrt{T}}$$

$$= \frac{(r - \sigma^2/2)T}{\sigma\sqrt{T}} - \frac{(r - \sigma^2/2)T}{\sigma\sqrt{T}}$$

$$= 0,$$

and furthermore that

$$\mathrm{Var}\,(Z) = \mathrm{Var}\left(\frac{X - T(r - \sigma^2/2)}{\sigma\sqrt{T}}\right)$$

$$= \frac{\mathrm{Var}\,(X)}{\sigma^2 T}$$

$$= \frac{\sigma^2 T}{\sigma^2 T}$$

$$= 1.$$

Thus Z is a standard normal random variable. We know its probability distribution is given by the function,

$$h(z) = \frac{1}{\sqrt{2\pi}} e^{-z^2/2}.$$

Returning to the expected value to be calculated in Eq. (8.52), we have

$$\mathrm{E}\left[(S(0)e^X - K)^+\right]$$

$$= \int_{-\infty}^{\infty} (S(0)e^x - K)^+ g(x)\,dx$$

$$= \int_{\ln(K/S(0))}^{\infty} (S(0)e^x - K)g(x)\,dx$$

$$= \int_{w^*}^{\infty} (S(0)e^{w\sigma\sqrt{T}+(r-\frac{\sigma^2}{2})T} - K)g(w\sigma\sqrt{T} + (r - \frac{\sigma^2}{2})T)\sigma\sqrt{T}\,dw$$

$$= \int_{w^*}^{\infty} (S(0)e^{w\sigma\sqrt{T}+(r-\frac{\sigma^2}{2})T} - K)h(w)\,dw$$

$$= \int_{w^*}^{\infty} S(0)e^{w\sigma\sqrt{T}+(r-\frac{\sigma^2}{2})T}h(w)\,dw - \int_{w^*}^{\infty} Kh(w)\,dw \qquad (8.53)$$

where

$$w^* = \frac{\ln(K/S(0)) - T(r - \sigma^2/2)}{\sigma\sqrt{T}}.$$

Consider the expression contained in the first integrand above.

$$
\begin{aligned}
e^{w\sigma\sqrt{T}+T(r-\sigma^2/2)}h(w) &= \frac{1}{\sqrt{2\pi}}e^{w\sigma\sqrt{T}+T(r-\sigma^2/2)}e^{-w^2/2} \\
&= \frac{1}{\sqrt{2\pi}}e^{-(w^2-2\sigma w\sqrt{T}+\sigma^2 T)/2+rT} \\
&= \frac{1}{\sqrt{2\pi}}e^{(-(w-\sigma\sqrt{T})^2)/2+rT} \\
&= \frac{1}{\sqrt{2\pi}}e^{rT}e^{(-(w-\sigma\sqrt{T})^2)/2} \\
&= e^{rT}h(w-\sigma\sqrt{T})
\end{aligned}
$$

Now we will substitute this expression in Eq. (8.53) and find that

$$
\begin{aligned}
\mathrm{E}&\left[(S(0)e^X - K)^+\right] \\
&= S(0)e^{rT}\int_{w^*}^{\infty} h(w-\sigma\sqrt{T})\,dw - K\int_{w^*}^{\infty} h(w)\,dw \\
&= S(0)e^{rT}\left(1 - \int_{-\infty}^{w^*} h(w-\sigma\sqrt{T})\,dw\right) \\
&\quad - K\left(1 - \int_{-\infty}^{w^*} h(w)\,dw\right) \\
&= S(0)e^{rT}\left(1 - \int_{-\infty}^{w^*-\sigma\sqrt{T}} h(x)\,dx\right) \\
&\quad - K\left(1 - \int_{-\infty}^{w^*} h(w)\,dw\right) \\
&= S(0)e^{rT}\left(1 - \phi(w^* - \sigma\sqrt{T})\right) - K\left(1 - \phi(w^*)\right), \qquad (8.54)
\end{aligned}
$$

where $\phi(x)$ is the probability that a standard normal random variable is less than x. For convenience sake we will set $w = -(w^* - \sigma\sqrt{T})$. Since $\phi(-x) = 1 - \phi(x)$, then Eq. (8.54) can be rewritten as

$$\mathrm{E}\left[(S(0)e^X - K)^+\right] = S(0)e^{rT}\phi(w) - K\phi(w - \sigma\sqrt{T}). \qquad (8.55)$$

Now we can substitute this expression into Eq. (8.52) and finally we can write the **Black-Scholes Option Pricing Formula** as

$$C = e^{-rT}\left(S(0)e^{rT}\phi(w) - K\phi(w - \sigma\sqrt{T})\right)$$
$$= S(0)\phi(w) - Ke^{-rT}\phi(w - \sigma\sqrt{T}). \tag{8.56}$$

If we would like to know that value of the call option at any time $0 \le t \le T$ then a simple shift in the time variable can be used in the formula above in Eq. (8.56). This agrees with the formula derived in Eq. (8.38) and repeated below for completeness.

$$C(S,t) = S\phi(w) - Ke^{-r(T-t)}\phi(w - \sigma\sqrt{T-t})$$
$$\text{where} \quad w = \frac{\ln(S(0)/K) + (r + \sigma^2/2)(T-t)}{\sigma\sqrt{T-t}}$$

In conclusion, whether we derive the value of the European call option from the continuous model via the solution of the Black-Scholes partial differential equation or by developing a multi-time step binomial model and taking its limit as the time step becomes infinitesimally small, the formula remains the same.

8.6 Exercises

(1) Let the function $f(x)$ be defined as

$$f(x) = \begin{cases} e^{-ax} & \text{if } x \ge 0 \\ 0 & \text{otherwise} \end{cases}$$

where a is a positive constant. Show that the Fourier Transform of f is $\hat{f}(w) = \frac{1}{a+iw}$.

(2) Let the function $\hat{f}(w)$ be defined as

$$\hat{f}(w) = \begin{cases} e^{-aw} & \text{if } w \ge 0 \\ 0 & \text{otherwise} \end{cases}$$

where a is a positive constant. Show that the inverse Fourier Transform of $\hat{f}$ is $f(x) = \frac{1}{2\pi(a-ix)}$.

(3) If a is a positive constant verify that the inverse Fourier Transform of e^{-aw^2} is $\frac{1}{2\sqrt{\pi a}}e^{-x^2/4a}$.

(4) Using the new variables described in Eqs. (8.9), (8.10), (8.11), and the multivariable chain rule, verify the identity shown in Eq. (8.13).

(5) Using the new variables described in Eqs. (8.9), (8.10), (8.11), the expression for F_S shown in (8.12), and the multivariable chain rule, verify the identity shown in Eq. (8.14).

(6) Verify that the Black-Scholes partial differential Eq. (8.5) can be simplified to the form shown in Eq. (8.15) by using the variables described in (8.9), (8.10), and (8.11) and the partial derivatives found in (8.12), (8.13), and (8.14).

(7) Fill in some of the details in the derivation of Eqs.(8.26)–(8.29) by showing that

$$u(x,\tau) \to e^{\frac{(k+1)}{2}[x+(k+1)\tau/2]} - e^{\frac{(k-1)}{2}[x+(k-1)\tau/2]}$$

as $x \to \infty$.

(8) Verify that the expression

$$\left(e^{(k+1)(x+\sqrt{2\tau}y)/2} - e^{(k-1)(x+\sqrt{2\tau}y)/2} \right)^{+}$$

is non-zero whenever $y > -x/\sqrt{2\tau}$.

(9) Use completion of the square in the exponents of the integrand in Eq. (8.33) to derive Eq. (8.33).

(10) Using the change of variable $w = -y + \frac{1}{2}(k + 1)\sqrt{2\tau}$ in the first improper integral of Eq. (8.33) verify that

$$\frac{1}{\sqrt{2\pi}} \int_{-x/\sqrt{2\tau}}^{\infty} e^{-(y-\frac{1}{2}(k+1)\sqrt{2\tau})^2/2} \, dy$$

$$= \frac{1}{\sqrt{2\pi}} \int_{-\infty}^{x/\sqrt{2\tau}+\frac{1}{2}(k+1)\sqrt{2\tau}} e^{-\frac{w^2}{2}} \, dw$$

$$= \phi \left(x/\sqrt{2\tau} + \frac{1}{2}(k + 1)\sqrt{2\tau} \right)$$

where ϕ is the cumulative distribution function for a normal random variable with mean zero and standard deviation one.

(11) Using the change of variable $w' = -y + \frac{1}{2}(k - 1)\sqrt{2\tau}$ in the second

improper integral of Eq. (8.33) verify that

$$\frac{1}{\sqrt{2\pi}} \int_{-x/\sqrt{2\tau}}^{\infty} e^{-(y-\frac{1}{2}(k-1)\sqrt{2\tau})^2/2} \, dy$$

$$= \frac{1}{\sqrt{2\pi}} \int_{-\infty}^{x/\sqrt{2\tau}+\frac{1}{2}(k-1)\sqrt{2\tau}} e^{-\frac{w'^2}{2}} \, dw'$$

$$= \phi\left(x/\sqrt{2\tau} + \frac{1}{2}(k-1)\sqrt{2\tau}\right)$$

where ϕ is the cumulative distribution function for a normal random variable with mean zero and standard deviation one.

(12) Verify that the expression for $u(x, \tau)$ given in Eq. (8.34) satisfies the initial condition as specified in Eq. (8.27).

(13) Verify using the change of variable formulas in Eqs. (8.9) and (8.10) and the fact that $k = 2r/\sigma^2$ that

$$x/\sqrt{2\tau} + \frac{1}{2}(k+1)\sqrt{2\tau} = \frac{\ln(S/K) + (r + \sigma^2/2)(T-t)}{\sigma\sqrt{T-t}}$$

$$x/\sqrt{2\tau} + \frac{1}{2}(k-1)\sqrt{2\tau} = \frac{\ln(S/K) + (r + \sigma^2/2)(T-t)}{\sigma\sqrt{T-t}} - \sigma\sqrt{T-t}.$$

(14) What is the price of a European call option on a stock when the stock price is $52, the strike price is $50, the interest rate is 12%, the stock's volatility is 30%, and the exercise time is three months?

(15) What is the price of a European put option on a stock when the stock price is $69, the strike price is $70, the interest rate is 5%, the stock's volatility is 35%, and the exercise time is six months?

(16) A European call option on a stock has a market value of $2.50. The stock price is $15, the strike price is $13, the interest rate is 5%, and the exercise time is three months. What is the volatility of the stock?

(17) Show by differentiation and substitution in Eq. (8.5) that the price of the security itself, *i.e.* $f(S, t) = S$ solves the Black-Scholes equation.

(18) Show by differentiation and substitution in Eq. (8.5) that money earning the risk-free interest rate r compounded continuously solves the Black-Scholes equation.

(19) (optional) Show that if S obeys a stochastic process of the form

$$dS = \mu S \, dt + \sigma S \, dW(t)$$

with initial value $S(0)$ then for $t \geq 0$

$$S(t) = S(0)e^{(\mu - \sigma^2/2)t + \sigma W(t)}$$

(20) (optional) What is the price of a European call option on a stock when the stock price is \$43, the strike price is \$42, the risk-free interest rate is 11%, the stock's volatility is 25%, and the exercise time is four months? Use the binomial lattice model with a time step of one month to evaluate the option.

(21) (optional) What is the price of a European put option on a stock when the stock price is \$96, the strike price is \$100, the risk-free interest rate is 6%, the stock's volatility is 33%, and the exercise time is three months? Use the binomial lattice model with a time step of one month to evaluate the option.

Chapter 9

Derivatives of Black-Scholes Option Prices

Now that the solution to the Black-Scholes equation is known we can investigate the sensitivity of the solution to changes in the independent variables. Knowledge of these sensitivities allows a portfolio manager to minimize changes in the value of a portfolio when underlying variables such as the risk-free interest rate change. The material of the present chapter will be used extensively in Chapter 10. An understanding of the sensitivity of an option's value to changes in its independent variables will also provide a new way of interpreting the Black-Scholes equation itself.

To mathematicians the word *derivative* means an instantaneous rate of change in a quantity. To a quantitative analyst a *derivative* is a financial entity whose value is derived from the value of some underlying asset. Hence a European call option is a derivative (in the quantitative analytical sense) whose value is a function of (among other things) the value of the security underlying the option. In this chapter we explore derivatives from the mathematician's perspective. In Chapter 10 we will use these derivatives in the manner of a quantitative analyst. Members of the quantitative financial profession refer to the subject matter of this chapter as the "Greeks" since a Greek letter is used to name each derivative (except for one which we will meet in due time). Due to the typically large volume of options written in a contract, very accurate calculations of the Greeks are necessary to avoid round-off errors. Motivation for accuracy and accurate numerical techniques may be found in [Chawla (2006)] and [Chawla and Evans (2005)].

9.1 Theta

The quantity Θ is defined to be the rate of change of the value of an option with respect to t. We will calculate Θ for the European call and put options derived in Chapter 8. Since the valuation of European-style options depends heavily on the cumulative distribution function of the standard normal random variable and a composite variable w, we will begin by investigating their partial derivatives. The derivative with respect to t of quantity w defined in Eq. (8.36) is given by

$$\frac{\partial w}{\partial t} = \frac{1}{2\sigma\sqrt{T-t}} \left(\frac{\ln(S/K)}{T-t} - r - \sigma^2/2 \right). \tag{9.1}$$

By the Fundamental Theorem of Calculus [Smith and Minton (2002), Chap. 4] the cumulative distribution function for the standard normal distribution has derivative

$$\phi'(x) = \frac{1}{\sqrt{2\pi}} e^{-x^2/2}. \tag{9.2}$$

Hence using the Chain Rule [Stewart (1999), Chap. 3] for derivatives we can see that the time rate of change in the value of a European call option (defined in Eq. (8.38)) is

$$\begin{aligned}
\frac{\partial C}{\partial t} &= S\phi'(w)\frac{\partial w}{\partial t} - Ke^{-r(T-t)}\left(r\phi(w - \sigma\sqrt{T-t}) \right. \\
&\quad \left. + \phi'(w - \sigma\sqrt{T-t})\left[\frac{\partial w}{\partial t} + \frac{\sigma}{2\sqrt{T-t}} \right] \right) \\
&= \left(S\phi'(w) - Ke^{-r(T-t)}\phi'(w - \sigma\sqrt{T-t}) \right)\frac{\partial w}{\partial t} \\
&\quad - Ke^{-r(T-t)}\left(r\phi(w - \sigma\sqrt{T-t}) + \frac{\sigma}{2\sqrt{T-t}}\phi'(w - \sigma\sqrt{T-t}) \right) \\
\Theta_C &= \frac{Se^{-w^2/2} - Ke^{-r(T-t)-(w-\sigma\sqrt{T-t})^2/2}}{2\sigma\sqrt{2\pi(T-t)}}\left(\frac{\ln(S/K)}{T-t} - r - \sigma^2/2 \right) \\
&\quad - Ke^{-r(T-t)}\left(r\phi(w - \sigma\sqrt{T-t}) + \frac{\sigma e^{-(w-\sigma\sqrt{T-t})^2/2}}{2\sqrt{2\pi(T-t)}} \right). \tag{9.3}
\end{aligned}$$

To take an example, consider a European call option on a stock whose current price is \$250. The strike price is \$245, the annual risk-free interest rate is 2.5%, the volatility of the stock price is 20%, and the strike time is four months. Thus substituting $S = 250$, $K = 245$, $r = 0.025$, $\sigma = 0.20$, $T = 1/3$, and $t = 0$ into Eq. (9.3) produces

$$\Theta = \frac{\partial C}{\partial t} = -19.9836.$$

The case that Θ is negative for an option is typical since options lose value as the expiry date approaches.

The calculation of Θ for a European put option is carried out in a similar fashion to that for the call option. The reader will be asked to provide the details of the following derivation.

$$\Theta_P = \frac{\partial P}{\partial t} \tag{9.4}$$

$$= \frac{Se^{-w^2/2}}{2\sigma\sqrt{2\pi(T-t)}}\left(\frac{\ln(S/K)}{T-t} - r - \sigma^2/2\right)$$
$$+ Kre^{-r(T-t)}\phi(\sigma\sqrt{T-t}-w)$$
$$- \frac{Ke^{-r(T-t)-(w-\sigma\sqrt{T-t})^2/2}}{2\sigma\sqrt{2\pi(T-t)}}\left(\frac{\ln(S/K)}{T-t} - r + \sigma^2/2\right) \tag{9.5}$$

Equation (9.5) will be put to use in the following example. Consider a European put option on a stock whose current price is \$325. The strike price is \$330, the annual risk-free interest rate is 3.5%, the volatility of the stock price is 27%, and the strike time is three months. Thus substituting $S = 325$, $K = 330$, $r = 0.035$, $\sigma = 0.27$, $T = 1/4$, and $t = 0$ into Eq. (9.5) produces

$$\Theta = \frac{\partial P}{\partial t} = -28.7484.$$

The derivative Θ is unique among the derivatives covered in this chapter. The passage of time t is the only non-stochastic variable among the list of independent variables upon which C or P depend.

9.2   Delta

The value of an option is also sensitive to changes in the price of the underlying stock. This rate of change is called Delta, Δ. The astute reader may recall that Δ was introduced in Sec. 7.3 during the derivation of the Black-Scholes partial differential equation. Delta was used to eliminate a stochastic term from Eq. (7.12). The clever choice of Δ left us with a deterministic partial differential equation to solve rather than a stochastic differential equation. The starting point for deriving Δ is again Eq. (8.38).

$$\frac{\partial C}{\partial S} = \phi(w) + \left(S\phi'(w) - Ke^{-r(T-t)}\phi'(w - \sigma\sqrt{T-t})\right)\frac{\partial w}{\partial S} \tag{9.6}$$

We can make use of Eq. (9.2) and the following result.

$$\frac{\partial w}{\partial S} = \frac{1}{\sigma S \sqrt{T-t}} \tag{9.7}$$

Substituting Eqs. (9.2) and (9.7) into Eq. (9.6) produces

$$\frac{\partial C}{\partial S} = \phi(w) + \left(\frac{S e^{-w^2/2}}{\sqrt{2\pi}} - \frac{K e^{-r(T-t)} e^{-(w-\sigma\sqrt{T-t})^2/2}}{\sqrt{2\pi}} \right) \frac{1}{\sigma S \sqrt{T-t}}$$

$$= \phi(w) + \frac{e^{-w^2/2}}{\sigma\sqrt{2\pi(T-t)}} \left(1 - \frac{K e^{-r(T-t)}}{S} e^{-(\sigma^2(T-t) - 2\sigma w\sqrt{T-t})/2} \right)$$

$$= \phi(w) + \frac{e^{-w^2/2}}{\sigma\sqrt{2\pi(T-t)}} \times$$

$$\left(1 - \frac{K e^{-r(T-t)}}{S} e^{-(\sigma^2(T-t) - 2(r+\sigma^2/2)(T-t) - 2\ln(K/S))/2} \right)$$

$$= \phi(w) + \frac{e^{-w^2/2}}{\sigma\sqrt{2\pi(T-t)}} \left(1 - \frac{K e^{-r(T-t)}}{S} e^{r(T-t) + \ln(S/K)} \right)$$

$$= \phi(w) + \frac{e^{-w^2/2}}{\sigma\sqrt{2\pi(T-t)}} \left(1 - \frac{K}{S} e^{\ln(S/K)} \right)$$

$$= \phi(w). \tag{9.8}$$

We can put this formula to use in the following example. Suppose the price of a security is \$100, the risk-free interest rate is 4% per annum, the annual volatility of the stock price is 23%, the strike price is \$110, and the strike time is 6 months. Under these conditions $w = -0.381747$ and

$$\Delta_C = \frac{\partial C}{\partial S} = 0.351325.$$

Delta for a European put option could be calculated directly from the definition for P; however, the Put-Call parity formula (Eq. (7.1)) provides a convenient shortcut.

$$\frac{\partial}{\partial S}(P + S) = \frac{\partial}{\partial S}(C + K e^{-r(T-t)})$$

$$\frac{\partial P}{\partial S} + 1 = \frac{\partial C}{\partial S}$$

$$\Delta_P = \frac{\partial P}{\partial S} = \phi(w) - 1 \tag{9.9}$$

Thus using the same parameter values as in the previous example,

$$\frac{\partial P}{\partial S} = 0.351325 - 1 = -0.648675.$$

So, as the European call option increases in value with an increase in the value of the underlying security, the put option decreases in value.

9.3   Gamma

The **gamma**, Γ, of an option or collection of options is defined to be the second derivative of the option with respect to S, the price of the underlying security. Hence Γ is the partial derivative of Δ with respect to S. Thus for a European call option Γ is

$$\frac{\partial^2 C}{\partial S^2} = \phi'(w)\frac{\partial w}{\partial S}$$

$$= \frac{e^{-w^2/2}}{\sigma S\sqrt{2\pi(T-t)}} \tag{9.10}$$

We can readily see from Eq. (9.9) that the Γ for a European put is the same as that for a European call

$$\Gamma = \frac{\partial^2 P}{\partial S^2} = \frac{\partial^2 C}{\partial S^2} = \frac{e^{-w^2/2}}{\sigma S\sqrt{2\pi(T-t)}}. \tag{9.11}$$

If all independent variables are held fixed except for S, then Γ measures the concavity of the value of the option as a function of S. Suppose we consider a European call option on a security whose value is \$295. The strike price is \$290, the annual risk-free interest rate is 4.25%, the volatility of the security is 25% per annum, and the expiry date is two months. We can calculate the price of the option, its Δ, and its Γ as follows.

$$C = 15.7173, \quad \Delta_C = 0.613297, \quad \Gamma = 0.0127122$$

Thus the value of the option will increase with the price of the underlying security at an increasing rate. This will have important implications for the practice known as hedging which will be explored in Chapter 10. In terms of elementary calculus, the graph of C as a function of S (holding all other variables fixed) is increasing and concave upward on an interval containing $S = 295$. See Fig. 9.1.

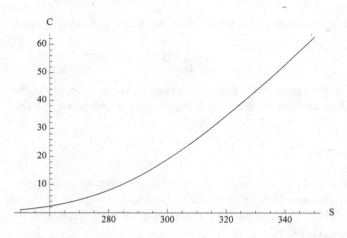

Fig. 9.1   Since the function $C(S)$ is increasing and concave upward on an interval containing $S = 295$ the price of the European call option increases at an increasing rate when the price of the underlying security is increased from \$295.

9.4   Vega

The **vega** of an option is the change in the value of the option as a function of the volatility of the underlying security. Among the partial derivatives of the value of an option vega distinguishes itself as the only partial derivative given a non-Greek letter name. Vega comes to us through medieval Latin from Arabic.

For a European call option defined as in Eq. (8.38),

$$\frac{\partial C}{\partial \sigma} = S\phi'(w)\frac{\partial w}{\partial \sigma} - Ke^{-r(T-t)}\phi'(w - \sigma\sqrt{T-t})\left(\frac{\partial w}{\partial \sigma} - \sqrt{T-t}\right), \quad (9.12)$$

where, according to exercise 8,

$$\frac{\partial w}{\partial \sigma} = \sqrt{T-t} - w/\sigma. \quad (9.13)$$

Substituting Eq. (9.13) into Eq. (9.12) produces

$$
\begin{aligned}
\frac{\partial C}{\partial \sigma} &- Ke^{-r(T-t)}\phi'(w - \sigma\sqrt{T-t})\left[\left(\sqrt{T-t} - \frac{w}{\sigma}\right) - \sqrt{T-t}\right] \\
&= S\phi'(w)\left(\sqrt{T-t} - \frac{w}{\sigma}\right) + Ke^{-r(T-t)}\phi'(w - \sigma\sqrt{T-t})\left(\frac{w}{\sigma}\right) \\
&= S\sqrt{T-t}\phi'(w) - \frac{w}{\sigma}\left(S\phi'(w) - Ke^{-r(T-t)}\phi'(w - \sigma\sqrt{T-t})\right) \\
&= S\sqrt{T-t}\phi'(w) - \frac{w}{\sigma\sqrt{2\pi}}\left(Se^{-w^2/2} - Ke^{-r(T-t)}e^{-(w-\sigma\sqrt{T-t})^2/2}\right) \\
&= S\sqrt{T-t}\phi'(w) \\
&\quad - \frac{w}{\sigma\sqrt{2\pi}}e^{-w^2/2}\left(S - Ke^{-r(T-t)+w\sigma\sqrt{T-t}-\sigma^2(T-t)/2}\right) \\
&= S\sqrt{T-t}\phi'(w) - \frac{w}{\sigma\sqrt{2\pi}}e^{-w^2/2}\left(S - Ke^{\ln(S/K)}\right) \\
&= \frac{S\sqrt{T-t}}{\sqrt{2\pi}}e^{-w^2/2}.
\end{aligned}
\tag{9.14}
$$

Note that the value of the European call option increases with increasing volatility in the underlying security. It is easily seen from the Put-Call Parity Formula (Eq. (7.1)) that

$$
\frac{\partial P}{\partial \sigma} = \frac{S\sqrt{T-t}}{\sqrt{2\pi}}e^{-w^2/2},
\tag{9.15}
$$

in other words the rate of change in the value of a European put option with respect to volatility is the same as the rate of change in the value of the corresponding European call option with respect to volatility. Thus we may unambiguously define vega for a European style option to be

$$
\mathcal{V} = \frac{S\sqrt{T-t}}{\sqrt{2\pi}}e^{-w^2/2}.
\tag{9.16}
$$

For example the vega for a European style option (either call or put) on a security whose current value is \$160, whose strike price is \$150, and whose expiry date is 5 months can be calculated using Eq. (9.16). Suppose the volatility of the security is 20% per year and the risk-free interest rate is 2.75% per year. Then $w = 0.653219$ and $\mathcal{V} = 33.2866$.

9.5 Rho

The **rho** of an option is the rate of change in the value of the option as a function of the risk-free interest rate r. Once again the European call option will be used to demonstrate the calculation of rho. Using Eq. (8.38) it is seen that

$$\frac{\partial C}{\partial r} = \left(S\phi'(w) - Ke^{-r(T-t)}\phi'(w - \sigma\sqrt{T-t}) \right) \frac{\partial w}{\partial r}$$
$$+ K(T-t)e^{-r(T-t)}\phi(w - \sigma\sqrt{T-t}) \tag{9.17}$$

Calling on the definition of w given in Eq. (8.36) we see that

$$\frac{\partial w}{\partial r} = \frac{\sqrt{T-t}}{\sigma}. \tag{9.18}$$

Substituting Eq. (9.18) into Eq. (9.17) yields

$$\frac{\partial C}{\partial r} = \left(S\phi'(w) - Ke^{-r(T-t)}\phi'(w - \sigma\sqrt{T-t}) \right) \frac{\sqrt{T-t}}{\sigma}$$
$$+ K(T-t)e^{-r(T-t)}\phi(w - \sigma\sqrt{T-t})$$
$$= \left(Se^{-w^2/2} - Ke^{-r(T-t)}e^{-(w-\sigma\sqrt{T-t})^2/2} \right) \frac{\sqrt{T-t}}{\sigma\sqrt{2\pi}}$$
$$+ K(T-t)e^{-r(T-t)}\phi(w - \sigma\sqrt{T-t})$$
$$= \left(S - Ke^{-r(T-t)+w\sigma\sqrt{T-t}-(T-t)\sigma^2/2} \right) \frac{e^{-w^2/2}\sqrt{T-t}}{\sigma\sqrt{2\pi}}$$
$$+ K(T-t)e^{-r(T-t)}\phi(w - \sigma\sqrt{T-t})$$
$$= \left(S - Ke^{\ln(S/K)} \right) \frac{e^{-w^2/2}\sqrt{T-t}}{\sigma\sqrt{2\pi}} + K(T-t)e^{-r(T-t)}\phi(w - \sigma\sqrt{T-t})$$
$$\rho_C = K(T-t)e^{-r(T-t)}\phi(w - \sigma\sqrt{T-t}). \tag{9.19}$$

Note that for $\rho_C > 0$ implying that the value of a European call option increases with increasing interest rate.

Consider the situation of a European call option on a security whose current value is \$225. The strike price is \$230, while the strike time is 4 months. The volatility of the price of the security is 27% annually and the risk-free interest rate is 3.25%. Under these conditions, $w = 0.0064434$ and

$$\frac{\partial C}{\partial r} = 33.4156.$$

The rho for a European put option can be determined from the Put-Call Parity Formula (Eq. (7.1)) and Eq. (9.19).

$$
\frac{\partial P}{\partial r} = \frac{\partial C}{\partial r} - K(T-t)e^{-r(T-t)}
$$
$$
= K(T-t)e^{-r(T-t)}\phi(w - \sigma\sqrt{T-t}) - K(T-t)e^{-r(T-t)}
$$
$$
= -K(T-t)e^{-r(T-t)}\left(1 - \phi(w - \sigma\sqrt{T-t})\right)
$$
$$
\rho_P = -K(T-t)e^{-r(T-t)}\phi(\sigma\sqrt{T-t} - w). \tag{9.20}
$$

In the last equation above we made use of the fact that $\phi(x) = 1 - \phi(-x)$ for the cumulative distribution function of a normally distributed random variable. Note that for $\rho_P < 0$ implying that the value of a European put option decreases with increasing interest rate.

To put Eq. (9.20) to work in an example, consider the situation of a European put option on a security whose underlying value is \$150. The strike price is \$152, the volatility is 15% per annum, the expiry date is three months, and the risk-free interest rate is 2.5%. Then $w = -0.0557697$ and

$$
\frac{\partial P}{\partial r} = -20.8461.
$$

9.6 Relationships Between Δ, Θ, and Γ

The value F of an option on a security S must satisfy the Black-Scholes partial differential Eq. (7.14). This equation is repeated below for convenience.

$$
rF = F_t + rSF_S + \frac{1}{2}\sigma^2 S^2 F_{SS}
$$

This equation assumes the risk-free interest rate is r and the volatility of the security is σ. In this chapter we have given names to the partial derivatives F_t, F_S, and F_{SS}, namely

$$
\Theta = F_t, \quad \Delta = F_S, \quad \Gamma = F_{SS}.
$$

Thus the Black-Scholes partial differential equation can be re-written as

$$
rF = \Theta + rS\Delta + \frac{1}{2}\sigma^2 S^2 \Gamma. \tag{9.21}
$$

The value of the option is a linear combination of Δ, Θ, and Γ.

Table 9.1   Listing of the derivatives of European option values and some of their properties.

Name	Symbol	Definition	Property
Theta	Θ_C	Eq. (9.3)	
	Θ_P	Eq. (9.5)	
Delta	Δ_C	$\phi(w)$	$\Delta_C > 0$
	Δ_P	$\phi(w) - 1$	$\Delta_P < 0$
Gamma	Γ	Eq. (9.10)	$\Gamma > 0$
Vega	$\mathcal{V}$	Eq. (9.16)	$\mathcal{V} > 0$
Rho	ρ_C	Eq. (9.19)	$\rho_C > 0$
	ρ_P	Eq. (9.20)	$\rho_P < 0$

When the value of the option is insensitive to the passage of time $\Theta = 0$ and Eq. (9.21) becomes

$$F = S\left(\Delta + \frac{\sigma^2 S}{2r}\Gamma\right).$$

If Δ should be of large magnitude then Γ must be large and of opposite sign. Returning to Eq. (9.21), if the rate of change in the value of the option with respect to the value of the underlying security is zero (or at least very small) then the Black-Scholes equation becomes

$$rF = \Theta + \frac{1}{2}\sigma^2 S^2 \Gamma.$$

Thus Θ and Γ tend to be of opposite algebraic signs. In practice, calculation of Θ can be used to approximate Γ.

Table 9.1 summarizes the option value derivatives discussed in this chapter and lists some of their relevant properties. This table is provided as a convenient place to find the definitions of these quantities in the future.

In the next chapter the practice of hedging will be described. Hedging is performed on a **portfolio** of securities, a collection which may include stocks, put and call options, cash in savings accounts, *etc.* The mathematical aspect of hedging is made possible by the linearity property of the Black-Scholes partial differential equation. An operator $L[\cdot]$ is linear if $L[aX + Y] = aL[X] + L[Y]$ for all scalars a and vectors X, Y in the domain of L. In the case of the Black-Scholes PDE the vectors are solution functions. If we define the operator $L[\cdot]$ to be

$$L[X] = \frac{\partial X}{\partial t} + rS\frac{\partial X}{\partial S} + \frac{1}{2}\sigma^2 S^2 \frac{\partial^2 X}{\partial S^2} - rX,$$

then any solution X, to the Black-Scholes partial differential Eq. (7.14), will satisfy $L[X] = 0$.

9.7 Exercises

(1) Find Θ for a European call option on a stock whose current value is \$300. The annual risk-free interest rate is 3%. The strike price is \$310 while the strike time is three months. The volatility of the stock is 25% annually.

(2) Carefully work through the details of the derivation of Θ for a European put option. Confirm that you obtain the result in Eq. (9.5).

(3) Find Θ for a European put option on a stock whose current value is \$275. The annual risk-free interest rate is 2%. The strike price is \$265 while the expiry date is four months. The volatility of the stock is 20% annually.

(4) Find the partial derivative of w (as defined in Eq. (8.36)) with respect to S.

(5) Find Δ for a European call option on a stock whose current value is \$150. The annual risk-free interest rate is 2.5%. The strike price is \$165 while the strike time is five months. The volatility of the stock is 22% annually.

(6) Find $\partial P / \partial S$ for a European put option on a stock whose current value is \$125. The annual risk-free interest rate is 5.5%. The strike price is \$140 while the expiry date is eight months. The volatility of the stock is 15% annually.

(7) Calculate Γ for a European put option for a security whose current price is \$180. The strike price is \$175, the annual risk-free interest rate is 3.75%, the volatility in the price of the security is 30% annually, and the strike time is four months.

(8) Using partial differentiation derive Eq. (9.13) from Eq. (8.36).

(9) Calculate the $\mathcal{V}$ of a European style option on a security whose current value is \$300, whose strike price is \$305, and whose expiry date is 6 months. Suppose the volatility of the security is 25% per year and the risk-free interest rate is 4.75% per year.

(10) Calculate the rho of a European style put option on a security whose current value is \$270, whose strike price is \$272, and whose strike time is 2 months. Suppose the volatility of the security is 15% per year and the risk-free interest rate is 3.75% per year.

(11) Calculate the rho of a European style call option on a security whose current value is $305, whose strike price is $325, and whose expiry date is 4 months. Suppose the volatility of the security is 35% per year and the risk-free interest rate is 2.55% per year.

Chapter 10

Hedging

Hedging is the practice of making a portfolio of investments less sensitive to changes in market variables such as the prices of securities and interest rates. If $F(S,t)$ is a solution to the Black-Scholes partial differential Eq. (7.14), the quantity $\Delta = F_S$ is central to hedging a portfolio. In fact this same quantity Δ was used to derive the Black-Scholes PDE. The term "Delta" may be used in two senses in this chapter. In some cases Δ will refer to the instantaneous rate of change of a financial derivative with respect to the underlying security. This is the mathematical Delta. The term may also be used as part of the phrase "Delta neutral" used to describe a portfolio consisting of the underlying and financial derivatives. The portfolio $\mathcal{P}$ made up of one option F and Δ shares of the underlying security S is delta neutral if $\mathcal{P}_S = 0$. This is the quantitative analytical sense of "Delta". The Black-Scholes equation can be thought of as the statement that when the Δ of a portfolio is zero, the rate of return from a portfolio consisting of ownership of the underlying security and sale of the option (or the opposite positions) should be the same as the rate of return from the risk-free interest rate on the same net amount of cash.

In Chapter 9 the partial derivatives of the values of European put and call options were calculated. In this chapter they will be used to hedge portfolios of investments.

10.1 General Principles

Before launching into a discussion of hedging, let us pause and recall what happens when one entity (say, a bank) sells a call option on a stock to another entity (say, an investor). The bank has promised the investor that they will be able to purchase the stock at a predetermined price (the strike price), at a predetermined time (the expiry date) in the future. The stock must be available to the investor at the strike price even if the market value of the stock exceeds the strike price. The bank is in a favorable position if, at the strike time, the market price of the stock is below the strike price set when the call option was sold to the investor. In this case the call option will not be exercised and the bank keeps the money it received when it sold the option. Conversely, if the market price of the stock exceeds the strike price of the option, the bank must ensure the investor can find a seller of the stock who will agree to accept the strike price. One way to accomplish this would be for the bank itself to be the seller to the investor.

Consider this strategy: the bank creates a European call option on a stock whose current price is $50 while the strike price is $52, the risk-free interest rate is 2.5% per annum, the expiry date is four months, and the volatility of the stock price is 22.5% per annum. According to the Black-Scholes option pricing formula, the value of this European call option is $1.91965. So that we deal with whole numbers, suppose an investor purchases one million of these call options. The bank accepts revenue equal to $1,919,650. The bank could wait until the expiry date to purchase the one million shares of the stock, or they could purchase one million shares at the time the investor bought the million call options. The former strategy is called a **naked** position since the bank's potential responsibility for providing the stock at the strike price is exposed to the movements in the market until expiry. The latter strategy is called a **covered** position since the bank is now protected against increases in the price of the security prior to the expiration of the option. Suppose the bank takes the covered position, they will expend $50M, considerably more than the revenue generated by the sale of the call options. If S represents the price per share of the stock at the strike time then the net revenue generated by the transaction obeys the formula

$$1919650 + (\min\{52, S\}e^{-0.025/3} - 50) \cdot 10^6.$$

Notice that the present value of the stock price must be incorporated rather than just S. The reader can check that when $S \approx 48.4827$ the net revenue is zero. Figure 10.1 shows if the stock is below that price the bank loses money (at $S = 46$ the loss comes to more than $2.4M), and if the stock price is higher the bank profits (at $S = 52$ the profit is nearly $3.5M). Since the strike price is $52, the maximum net revenue for the bank occurs at that price.

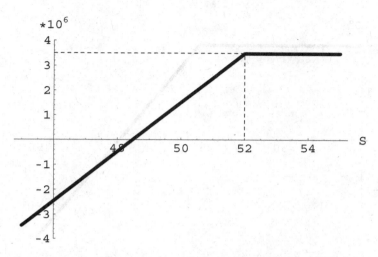

Fig. 10.1 The line shows the profit or loss to the seller of a European call option who adopts a covered position by purchasing the underlying security at the time they sell the option.

If the bank adopts a naked position it will purchase the stock at expiry and instantaneously sell it to the investor at the strike price. In this scenario the net revenue generated by this sequence of transactions is

$$1919650 + \min(0, 52 - S)e^{-0.025/3} \cdot 10^6.$$

The net revenue is zero when S is approximately \$53.9357. As long as the stock price remains below the strike price the bank keeps its revenue from the sale of the options. However the bank's losses could be dramatic when the price of the stock rises before the expiry date. At a price of \$56 per share the bank's net loss would be approximately \$2.0M. Figure 10.2 illustrates the profit or loss to the seller of the European call option.

Neither of these schemes are practical for hedging portfolios in the financial world due to their potentially large costs. The naked and the covered positions represent two extremes, one in which no stock is held and the other in which all the potentially needed shares are held from the moment the option is sold until expiry. A better strategy may be a compromise between these extremes. Perhaps some fraction of the total, potential number of necessary shares should be held by the bank. The optimal number would be the number which reduces the potential loss to the bank. In the next section a simple, practical method for hedging is explored.

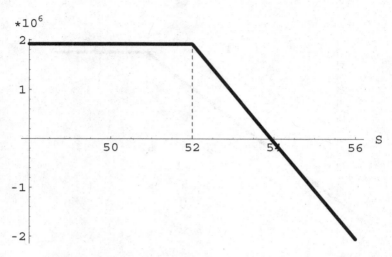

Fig. 10.2   The line shows the profit or loss to the seller of a European call option who adopts a naked position by purchasing the underlying security at the expiry date of the option.

10.2 Delta Hedging

The purpose of hedging is to eliminate or at least minimize the change in the value of a portion of an investor's or an institution's portfolio of investments as conditions change in the financial markets. The hedged portion of a portfolio may represent the value of some future financial obligation, for example a retirement pension. In general people and institutions invest in stocks to earn a greater rate of return per dollar invested than can be earned in highly stable government issued bonds. The higher rate of return can be thought of a payment for accepting greater risk. Hedging is a method of reducing the risk at the cost of some of the reward. The value of a portfolio containing stocks and options may change due to changes in the value of the stock underlying the option. The reader will recall from the previous chapter that Delta is the partial derivative of the value of an option with respect to the value of the underlying security. In Sec. 9.2 expressions for Delta for European call and put options were derived. One can think of Delta in the following way: for every unit change in the value of the underlying security, the value of the option changes by Δ. Delta is the marginal option value of the underlying security. A portfolio consisting of securities and options is called **Delta-neutral** if for every option sold, Δ units of the security are bought.

Example 10.1 Consider the case of a security whose price is $100, while the risk-free interest rate is 4% per annum, the annual volatility of the stock price is 23%, the strike price is set at $105, and the expiry date is 3 months. Under these conditions $w = -0.279806$, the value of a European call option is $C = 2.96155$ and Delta for the option is

$$\Delta = \frac{\partial C}{\partial S} = 0.389813.$$

Thus if a firm sold an investor European call options on ten thousand shares of the security[1], the firm would receive $29615.50 and purchase $389813 = (10000)(0.389813)(100)$ worth of the security, most likely with borrowed money.

The firm may choose to do nothing further until the strike time of the option. In that case, this is referred to as a "hedge and forget" scheme. On the other hand, since the price of the security is dynamic, the firm may choose to make periodic adjustments to the number of shares of the security it holds. This strategy is known as **rebalancing** the portfolio. The example begun above can be extended to include weekly rebalancing.

Assume that the value of the security follows the random walk illustrated in Fig. 10.3. In this case the European call option will be exercised since the price of the security at the expiry date exceeds the strike price of $105. At the end of the first week the value of the security has declined to $98.79. Upon re-computing, $\Delta = 0.339811$. Thus the investment firm would adjust its security holdings so that it now owned 3398.11 shares of the security[2] with a total value of $335699. Also at the end of the first week the firm will have incurred costs in the form of interest on the money initially borrowed to purchase shares of the security. This cost amounts to

$$(389813)(e^{0.04/52} - 1) = \$299.9717.$$

The interest for the current week is added to the cumulative cost entry of the next week. Table 10.1 summarizes the weekly rebalancing up to expiry. The entries in this table are computed iteratively. If the subscript i indicates a quantity evaluated at the beginning of the ith week, then the interest cost

---

[1]Typically an option is keyed to 100 shares of the underlying security, so in practice options on 10000 shares would amount to 100 options. To avoid confusion we will assume a one-to-one ratio of options to shares.

[2]The Black-Scholes equation treats the number of shares as a continuous, not discrete, quantity. To reduce round-off error we will calculate all quantities in this example to at least six significant digits.

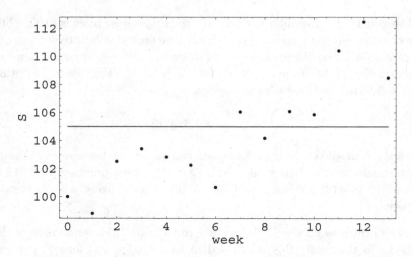

Fig. 10.3   A realization of a random walk taken by the value of a security. The horizontal line indicates the strike price of a European call on the security.

Table 10.1   Delta hedging using portfolio rebalancing at weekly intervals.

Week	S	Δ	Shares Held	Share Cost	Interest Cost	Cumulative Cost
0	100.00	0.389813	3898.13	389813	0	389813
1	98.79	0.339811	3398.11	335699	299.972	340716
2	102.52	0.462922	4629.23	474588	262.190	467192
3	103.41	0.490192	4901.92	506907	359.517	495751
4	102.82	0.460541	4605.42	473528	381.494	465646
5	102.25	0.428236	4282.37	437872	358.327	432972
6	100.67	0.347145	3471.46	349471	333.184	351671
7	106.05	0.589204	5892.05	624852	270.620	608646
8	104.17	0.491348	4913.49	511838	468.369	507177
9	106.08	0.595047	5950.47	631226	390.286	617571
10	105.86	0.585915	5859.15	620250	475.238	608379
11	110.40	0.878690	8786.90	970074	468.164	932071
12	112.46	0.985811	9858.11	1108643	717.253	1053256
13	108.47	1.000000	10000.00	1084700	810.509	1069458

initially is zero and the initial cumulative cost is $\Sigma_0 = 10000\Delta_0 S_0$. When $i \geq 1$, the interest cost is $\Sigma_{i-1}(e^{r/52} - 1)$ and the new cumulative cost is

$$\Sigma_i = (\Delta_i - \Delta_{i-1})S_i + \Sigma_{i-1}e^{r/52}.$$

Notice at the expiry date the investment firm has in its portfolio 10000 shares of the security for which the investor will pay a total of $1,050,000.

Thus the net proceeds to the investment firm of selling the call option and hedging this position are

$$1050000 + 29615.50e^{0.04(13/52)} - 1069458 = \$10455.10.$$

The reader should note that the value of the call options earns interest compounded continuously at the risk-free rate. In this case, the bank makes money by issuing the call option. If the bank had adopted a covered position from the moment the option contract was written, they would have made a profit of

$$1050000 + 29615.50e^{0.04(13/52)} - 1000000e^{0.04(13/52)} = \$69862.50.$$

If the bank had held a naked position until expiry, they would have had a loss of

$$1050000 + 29615.50e^{0.04(13/52)} - 1084700 = -\$4787.36.$$

Does this mean that adopting the covered position is preferable to hedging? We can only see it was preferable in hindsight. The potential losses from the covered position are large. Had the random walk of the stock moved higher by expiry, the losses from the naked position could have been much larger.

Alternatively the value of the security may evolve in such as fashion that the call option is not exercised. Such a scenario is depicted in Fig. 10.4. Since the value of the security finishes below the strike price, the call option expires unused. Table 10.2 summarizes the weekly rebalancing activity of the investment firm as it practices Delta hedging. Notice that at the expiry date the investment firm owns no shares of the security (none are needed since the call option will not be exercised). The net proceeds to the investment firm of selling the call option and hedging its position are

$$29615.50e^{0.04(13/52)} - 29667.7 = \$245.44.$$

In this case the firm earns a small amount of money on the collection of all these transactions.

Rebalancing of the portfolio can take place either more or less frequently than weekly as was done in the two previous extended examples. The quantity discussed in Sec. 9.3 known as Gamma (Γ) is the second partial derivative of the portfolio with respect to the value of the underlying security. In other words Gamma is the rate of change of Delta with respect to S. If $|\Gamma|$ is large then Δ changes rapidly with relatively small changes in S.

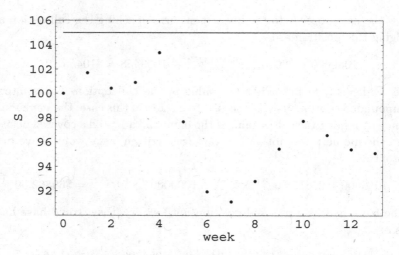

Fig. 10.4   Another realization of a random walk taken by the value of a security. The horizontal line indicates the strike price of a European call on the security.

Table 10.2   Delta hedging using portfolio rebalancing at weekly intervals for an option which will expire unused.

Week	S	Δ	Shares Held	Share Cost	Interest Cost	Cumulative Cost
0	100.00	0.389813	3898.13	389813	0.00000	389813
1	101.71	0.440643	4406.44	448179	299.972	441813
2	100.43	0.386757	3867.58	388421	339.987	388035
3	100.91	0.394649	3946.49	398240	298.603	396297
4	103.37	0.482725	4827.25	498993	304.961	487646
5	97.69	0.246176	2461.76	240489	375.257	256937
6	91.95	0.071229	712.290	65495.1	197.720	96270.5
7	91.12	0.043022	430.230	39202.5	74.0827	70643.2
8	92.81	0.050427	504.272	46801.4	54.3619	77569.4
9	95.45	0.078574	785.741	74999.0	80.4120	104495
10	97.75	0.110154	1101.54	107675	104.228	135445
11	96.58	0.036209	362.098	34971.4	49.3529	64134
12	95.40	0.001508	15.0811	1438.74	23.9154	31078.0
13	95.10	0.000000	0.00000	0.00000	0.00000	29667.7

In this case frequent rebalancing of the investment firm's position may be necessary. If $|\Gamma|$ is small then Δ is relatively insensitive to changes in S and hence rebalancing may be needed infrequently. Thus an investment firm can monitor Gamma in order to determine the frequency at which the firm's position should be rebalanced.

10.3   Delta Neutral Portfolios

In exercise 17 of Chapter 8 the reader was asked to verify that a stock or security itself satisfies the Black-Scholes PDE (7.14). Now suppose that a portfolio consists of a linear combination of options and shares of the underlying security. The portfolio contains a short position in a European call option and a long position in the security (hedged appropriately as described in the previous section). Thus the net value $\mathcal{P}$ of the portfolio is

$$\mathcal{P} = C - \Delta S = C - \left.\frac{\partial C}{\partial S}\right|_{S_0} S, \qquad (10.1)$$

where S_0 is the price of the security at the time the hedge is created. The quantity $\mathcal{P}$ satisfies the Black-Scholes equation since C and S separately do, and the equation is linear. Thus it is reasonable to contemplate Delta for the portfolio. The partial derivative of the entire portfolio with respect to S represents the sensitivity of the value of the portfolio to changes in S. Differentiating both sides of Eq. (10.1) with respect to S produces

$$\frac{\partial \mathcal{P}}{\partial S} = \frac{\partial C}{\partial S} - \left.\frac{\partial C}{\partial S}\right|_{S_0}$$

This quantity equals zero whenever $S = S_0$ (for example at the moment the hedge is set up) and remains very close to zero so long as S is near S_0. For this reason a portfolio which is hedged using Delta hedging is sometimes called **Delta neutral**.

Assuming that the risk-free interest rate remains constant and the volatility of the security does not change then a Taylor series expansion of the value of the portfolio in terms of t and S yields

$$\mathcal{P} = \mathcal{P}_0 + \frac{\partial \mathcal{P}}{\partial t}(t - t_0) + \frac{\partial \mathcal{P}}{\partial S}(S - S_0) + \frac{\partial^2 \mathcal{P}}{\partial S^2}\frac{(S - S_0)^2}{2} + \cdots$$
$$\delta\mathcal{P} = \Theta\delta t + \Delta\delta S + \frac{1}{2}\Gamma(\delta S)^2 + \cdots$$

All omitted terms in the Taylor series involve powers of δt greater than 1. The Gamma term is retained since the stochastic random variable S follows a stochastic process dependent on $\sqrt{\delta t}$, see Eq. (5.35). If the portfolio is hedged using Delta hedging then Δ for the portfolio is zero and thus

$$\delta\mathcal{P} \approx \Theta\delta t + \frac{1}{2}\Gamma(\delta S)^2, \qquad (10.2)$$

where the approximation omits terms involving powers of δt greater than 1. The term involving Θ is not stochastic and thus must be retained; however, the approximation can be further refined if the portfolio can be made **Gamma neutral**, *i.e.* if the makeup of the portfolio can be adjusted so that $\Gamma = 0$.

10.4 Gamma Neutral Portfolios

Gamma for a portfolio is the second partial derivative of the portfolio with respect to S. The portfolio cannot be made gamma neutral using a linear combination of just an option and its underlying security, since the second derivative of S with respect to itself is zero. As explained in the previous section, the portfolio can be made Delta neutral with the appropriate combination of the option and the underlying security. The portfolio can be made Gamma neutral by manipulating a component of the portfolio which depends non-linearly on S. One such component is the option. However, as mentioned earlier, the portfolio cannot contain only the option and its underlying security. One way to achieve the goal of a Gamma neutral portfolio is to include in the portfolio two (or perhaps more) different types of option dependent on the same underlying security. For example, suppose the portfolio contains options with two different strike times written on the same stock. An investment firm may sell a number of European call options with expiry three months hence and buy some other number of the European call options on the same stock with expiry arriving in six months. Let the number of the early option sold be w_e and the number of the later option be w_l. The Gamma of the portfolio would be

$$\Gamma_{\mathcal{P}} = w_e \Gamma_e - w_l \Gamma_l,$$

where Γ_e and Γ_l denote the Gammas of the earlier and later options respectively. The relative weights of the earlier and later options can be chosen so that $\Gamma_{\mathcal{P}} = 0$. Once the portfolio is made Gamma neutral, the underlying security can be added to the portfolio in such a way so as to make the portfolio Delta neutral. The reader should keep in mind that the underlying security will have a $\Gamma = 0$ and thus Gamma neutrality for the portfolio will be maintained while Delta neutrality is achieved. Thus Eq. (10.2) for a Gamma neutral portfolio reduces to

$$\delta \mathcal{P} \approx \Theta \delta t. \tag{10.3}$$

Equation (10.3) implies that the change in the value of a Gamma (and simultaneously Delta) neutral portfolio is proportional to the size of the time step.

Example 10.2 Suppose a stock's current value is \$100 while its volatility is $\sigma = 0.22$ and the risk-free interest rate is 2.5% per year. An investment firm sells European call options on this stock with a strike time of three months and a strike price of \$102. The firm buys European call options on the same stock with the same strike price but with a strike time of six months. According to Eq. (9.10) the Gamma of the three-month option is $\Gamma_3 = 0.03618$, while that of the six-month options is $\Gamma_6 = 0.02563$. The portfolio becomes Gamma neutral at any point in the first quadrant of $w_3 w_6$-space where the equation

$$0.03618 w_3 - 0.02563 w_6 = 0$$

is satisfied. Suppose $w_3 = 100000$ of the three-month option were sold, then the portfolio is Gamma neutral if $w_6 = 141163$ of the six-month options are purchased. Thus prior to the inclusion of the underlying stock in the portfolio, the Delta of the portfolio is

$$w_3 \Delta_3 - w_6 \Delta_6 = (100000)(0.4728) - (141163)(0.5123) = -25038.$$

Therefore the portfolio can be made Delta neutral if 25038 shares of the underlying stock are sold short. Figure 10.5 shows that over a fairly wide range of values of the underlying stock the value $\mathcal{P}$ of the Gamma and Delta neutral portfolio remains nearly the same.

This discussion of hedging is far from complete. The present chapter has focused mainly on making a portfolio's value resistant to changes in the value of a security. In reality the risk-free interest rate and the volatility of the stock also affect the value of a portfolio. The quantities Rho and Vega discussed in Chapter 9 can be used to set up portfolios hedged against changes in the interest rate and volatility. In the present chapter it was also assumed that the necessary options and securities could be bought so as to form the desired hedge. In practice this may not always be possible. For example an investment firm may not be able to purchase sufficient quantities of a stock to make their portfolio Delta or Gamma neutral. In this case they may have to substitute a different, but related security or other financial instrument in order to set up the hedge. This strategy will be discussed in the next chapter after the prerequisite statistical concepts are introduced.

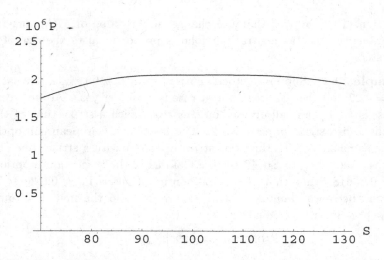

Fig. 10.5   The aggregate value of a Gamma and Delta neutral portfolio is insensitive to a range of changes in the value of the underlying securities.

10.5   Exercises

(1) A call option closes "in the money" if the market price of the underlying security exceeds the strike price for the call option at the time the option matures. Use Eq. (8.36) to show that $\Delta \to 1$ as $t \to T^-$ for an "in the money" European call option.

(2) A call option closes "out of the money" if the strike price for the call option exceeds the market price of the underlying security at the time the option matures.. Use Eq. (8.36) to show that $\Delta \to 0$ as $t \to T^-$ for an "out of the money" European call option.

(3) Suppose the price of a stock is $45 per share with volatility $\sigma = 0.20$ per annum and the risk-free interest rate is 4.5% per year. Create a table similar to Table 10.1 for European call options on 5000 shares of the stock with a strike price of $47 per share and a strike time of 15 weeks. Rebalance the portfolio weekly assuming the weekly prices of the stock are as follows.

Week	0	1	2	3	4	5	6	7
S	45.00	44.58	46.55	47.23	47.62	47.28	49.60	50.07
Week	8	9	10	11	12	13	14	15
S	47.79	48.33	48.81	51.36	52.06	51.98	54.22	54.31

(4) Repeat exercise 3 for the following scenario in which the call option

finishes out of the money.

Week	0	1	2	3	4	5	6	7
S	45.00	44.58	45.64	44.90	43.42	42.23	41.18	41.52
Week	8	9	10	11	12	13	14	15
S	41.94	42.72	44.83	44.93	44.19	41.77	39.56	40.62

(5) Repeat exercise 3 assuming this time that the financial institution has issued a European put option for the stock.

(6) Repeat exercise 4 assuming this time that the financial institution has issued a European put option for the stock.

(7) Set up a Gamma neutral portfolio consisting of European call options on a security whose current value is $85. Suppose the risk-free interest rate is 5.5% and the volatility of the security is 17% per year. The portfolio should contain a short position in four-month options with a strike price of $88 and a long position in six-month options with the same strike price.

(8) For the portfolio described in exercise 7 add a position in the underlying security so that the portfolio is also Delta neutral.

(9) Set up a Gamma neutral portfolio consisting of European call options on a security whose current value is $95. Suppose the risk-free interest rate is 4.5% and the volatility of the security is 23% per year. The portfolio should contain a short position in three-month options with a strike price of $97 and a long position in five-month options with a strike price of $98.

(10) For the portfolio described in exercise 9 add a position in the underlying security so that the portfolio is also Delta neutral.

Chapter 11

Optimizing Portfolios

There are several notions of the idea of optimizing a portfolio of securities, options, bonds, cash, *etc.* An optimal portfolio could be defined as one with a maximal rate of return or it could mean a portfolio for which the probability of a large fluctuation (usually in the downward direction) in the value of the portfolio is minimized. Alternatively an optimal portfolio for some investors could combine the two notions allowing the investor to specify for example an acceptable level of return and then designing the portfolio with the minimum chance of deviating from that return. The probability of deviating from a desired rate of return will serve as our definition of risk associated with a portfolio. Thus in this chapter we will explore portfolio optimality in the sense of maximizing the rate of return while minimizing the variance in the rate of return, and hence minimizing the risk to the investor. We will also introduce the Capital Assets Pricing Model which attempts to relate the rate of return of a specific investment to the rate of return for the entire market of investments. We will see in this chapter that the difference in the expected rates of return for a specific security and the risk-free interest rate is proportional to the difference in the expected rates of return for the market and the risk-free interest rate. In order to explain this proportionality relationship we will introduce two additional statistical measures, covariance and correlation. We will also make use of these concepts in developing additional hedging strategies.

11.1 Covariance and Correlation

The concept of **covariance** is related to the degree to which two random variables tend to change in the same or opposite direction relative to one another. If X and Y are the random variables then mathematically the

Table 11.1   A sample of heights and arm spans for children.

Child	Ht. (cm)	Span (cm)	Child	Ht. (cm)	Span (cm)
1	142	138	11	150	147
2	148	144	12	152	141
3	152	148	13	148	144
4	150	145	14	152	148
5	141	136	15	144	140
6	142	139	16	148	143
7	149	144	17	150	146
8	151	145	18	138	134
9	147	144	19	145	142
10	152	148	20	142	138

covariance, denoted $\text{Cov}(X, Y)$, is defined as

$$\text{Cov}(X, Y) = \text{E}\left[(X - \text{E}[X])(Y - \text{E}[Y])\right]. \tag{11.1}$$

Making use of the definition and properties of expected value one can see that

$$\begin{aligned}
\text{Cov}(X, Y) &= \text{E}[XY - Y\text{E}[X] - X\text{E}[Y] + \text{E}[X]\text{E}[Y]] \\
&= \text{E}[XY] - \text{E}[Y]\text{E}[X] - \text{E}[X]\text{E}[Y] + \text{E}[X]\text{E}[Y] \\
&= \text{E}[XY] - \text{E}[X]\text{E}[Y], \tag{11.2}
\end{aligned}$$

where Eq. (11.2) is generally more convenient to use than the expression in the right-hand side of Eq. (11.1).

Example 11.1   Table 11.1 lists the heights and arm spans of a sample of 20 children [Shodor (2007)]. This data will be used to illustrate the concept of covariance. Let X represent height and Y represent arm span for each child. The reader can readily calculate that $\text{E}[X] = 147.15$ cm and $\text{E}[Y] = 142.70$ cm. The expected value of the pairwise product of height and arm span is $\text{E}[XY] = 21013.8$. Thus the covariance of height and arm span for this sample is $\text{Cov}(X, Y) = 15.445$. Note that the covariance is positive, indicating that, in general, as height increases so does arm span.

From the definition of covariance several properties of this concept follow almost immediately. If X and Y are independent random variables then $\text{E}[XY] = \text{E}[X]\text{E}[Y]$ by Theorem 2.5 and thus the covariance of independent random variables is zero. The following set of relationships can also be established.

Theorem 11.1   *Suppose X, Y, and Z are random variables, then the following statements are true:*

(1) $\text{Cov}(X, X) = \text{Var}(X)$,
(2) $\text{Cov}(X, Y) = \text{Cov}(Y, X)$,
(3) $\text{Cov}(X + Y, Z) = \text{Cov}(X, Z) + \text{Cov}(Y, Z)$.

The proofs of the first two statements are left for the reader as exercises. Justification for statement 3 is given below.

Proof. Let X, Y, and Z be random variables, then

$$\begin{aligned}
\text{Cov}(X + Y, Z) &= \text{E}[(X + Y)Z] - \text{E}[X + Y]\text{E}[Z] \\
&= \text{E}[XZ] + \text{E}[YZ] - \text{E}[X]\text{E}[Z] - \text{E}[Y]\text{E}[Z] \\
&= \text{E}[XZ] - \text{E}[X]\text{E}[Z] + \text{E}[YZ] - \text{E}[Y]\text{E}[Z] \\
&= \text{Cov}(X, Z) + \text{Cov}(Y, Z)
\end{aligned}$$

$\square$

The third statement of Theorem 11.1 can be generalized as in the following corollary.

Corollary 11.1 *Suppose $\{X_1, X_2, \ldots, X_n\}$ and $\{Y_1, Y_2, \ldots, Y_m\}$ are sets of random variables where $n, m \geq 1$.*

$$\text{Cov}\left(\sum_{i=1}^{n} X_i, \sum_{i=1}^{m} Y_i\right) = \sum_{i=1}^{n}\sum_{j=1}^{m} \text{Cov}(X_i, Y_j) \tag{11.3}$$

Proof. We begin by demonstrating the result when $m = 1$. The corollary holds trivially when $n = 1$ and follows from the third statement of Theorem 11.1 when $n = 2$. Suppose the claim holds when $n \leq k$ where $k \in \mathbb{N}$. Let $\{X_1, X_2, \ldots, X_k, X_{k+1}\}$ be random variables.

$$\begin{aligned}
\text{Cov}\left(\sum_{i=1}^{k+1} X_i, Y_1\right) &= \text{Cov}\left(\sum_{i=1}^{k} X_i, Y_1\right) + \text{Cov}(X_{k+1}, Y_1) \\
&= \sum_{i=1}^{k} \text{Cov}(X_i, Y_1) + \text{Cov}(X_{k+1}, Y_1) \\
&= \sum_{i=1}^{k+1} \text{Cov}(X_i, Y_1)
\end{aligned}$$

Therefore by induction we may show that the result is true for any finite, integer value of n (at least when $m = 1$). When m is an integer larger than

1 we can argue that

$$\text{Cov}\left(\sum_{i=1}^{n} X_i, \sum_{j=1}^{m} Y_j\right) = \sum_{i=1}^{n} \text{Cov}\left(X_i, \sum_{j=1}^{m} Y_j\right) \quad \text{(shown above)}$$

$$= \sum_{i=1}^{n} \text{Cov}\left(\sum_{j=1}^{m} Y_j, X_i\right) \quad \text{(by Thm. 11.1 (2))}$$

$$= \sum_{i=1}^{n} \sum_{j=1}^{m} \text{Cov}\left(Y_j, X_i\right)$$

$$= \sum_{i=1}^{n} \sum_{j=1}^{m} \text{Cov}\left(X_i, Y_j\right).$$

Therefore Eq. (11.3) holds for all finite, positive integer values of m and n.
$\square$

Yet another corollary follows from Corollary 11.1. This one generalizes statement 1 of Theorem 11.1.

Corollary 11.2 *If $\{X_1, X_2, \ldots, X_n\}$ are random variables then*

$$\text{Var}\left(\sum_{i=1}^{n} X_i\right) = \sum_{i=1}^{n} \text{Var}\left(X_i\right) + \sum_{i=1}^{n} \sum_{j \neq i}^{n} \text{Cov}\left(X_i, X_j\right). \tag{11.4}$$

Proof. Let $Y = \sum_{i=1}^{n} X_i$, then according to the first statement of Theorem 11.1,

$$\text{Var}\left(Y\right) = \text{Cov}\left(Y, Y\right)$$

$$\text{Var}\left(\sum_{i=1}^{n} X_i\right) = \text{Cov}\left(\sum_{i=1}^{n} X_i, \sum_{j=1}^{n} X_j\right)$$

$$= \sum_{i=1}^{n} \sum_{j=1}^{n} \text{Cov}\left(X_i, X_j\right) \quad \text{(by Corollary 11.1)}$$

$$= \sum_{i=1}^{n} \text{Cov}\left(X_i, X_i\right) + \sum_{i=1}^{n} \sum_{j \neq i}^{n} \text{Cov}\left(X_i, X_j\right)$$

$$= \sum_{i=1}^{n} \text{Var}\left(X_i\right) + \sum_{i=1}^{n} \sum_{j \neq i}^{n} \text{Cov}\left(X_i, X_j\right)$$

Once more the first statement of Theorem 11.1 was used to reintroduce the variance in the last line of the derivation.
$\square$

Often a quantity related to covariance is used as a measure of the degree to which large values of a random variable X are associated with large values of another random variable Y. This quantity is known as **correlation** and is denoted $\rho(X, Y)$. The correlation of two random variables is defined as

$$\rho(X, Y) = \frac{\text{Cov}(X, Y)}{\sqrt{\text{Var}(X)\,\text{Var}(Y)}}. \tag{11.5}$$

The correlation of two random variables can be interpreted as a measure of the degree to which monotonic changes (increases or decreases) in one of the variables is reflected in similar changes (increases with increases and decreases with decreases) in the other variable. The correlation is more than just a simple re-scaling of the covariance. While the covariance of X and Y may numerically be positive, negative, or zero, the correlation always lies in the interval $[-1, 1]$. Once again we see that independent random variables have a correlation of zero and hence are described as **uncorrelated**.

Theorem 11.2 *Suppose X and Y are random variables such that $Y = aX + b$ where $a, b \in \mathbb{R}$ with $a \neq 0$. If $a > 0$ then $\rho(X, Y) = 1$, while if $a < 0$ then $\rho(X, Y) = -1$.*

Proof. We start by calculating the covariance of X and Y.

$$\begin{aligned}
\text{Cov}(X, Y) &= \text{Cov}(X, aX + b) \\
&= \text{E}[X(aX + b)] - \text{E}[X]\,\text{E}[aX + b] \\
&= \text{E}[aX^2 + bX]] - \text{E}[X](a\text{E}[X] + b) \\
&= a\text{E}[X^2] + b\text{E}[X] - a\text{E}[X]\,\text{E}[X] - b\text{E}[X] \\
&= a\left(\text{E}[X^2] - \text{E}[X]^2\right) \\
&= a\text{Var}(X)
\end{aligned}$$

The reader should make note of the use of Theorem 2.3. Therefore using the result of exercise 18 from Chapter 2

$$\rho(X, Y) = \frac{a\text{Var}(X)}{\sqrt{\text{Var}(X) \cdot a^2\text{Var}(X)}} = \frac{a}{|a|}$$

which is -1 when $a < 0$ and 1 when $a > 0$. □

The converse of Theorem 11.2 is false. A correlation close to unity does not indicate a linear relationship between the random variables. See exercise 6.

Before we can bound the correlation in the interval $[-1, 1]$ we will need to prove the following lemma.

Lemma 11.1 *(Schwarz Inequality) If X and Y are random variables then $(\mathrm{E}\,[XY])^2 \leq \mathrm{E}\,[X^2]\,\mathrm{E}\,[Y^2]$.*

Proof. The cases in which $\mathrm{E}\,[X^2] = 0$, $\mathrm{E}\,[X^2] = \infty$, $\mathrm{E}\,[Y^2] = 0$, or $\mathrm{E}\,[Y^2] = \infty$ are left as exercises. Suppose for the purposes of this proof that $0 < \mathrm{E}\,[X^2] < \infty$ and $0 < \mathrm{E}\,[Y^2] < \infty$. If a and b are real numbers then the following two inequalities hold:

$$0 \leq \mathrm{E}\,[(aX + bY)^2] = a^2\mathrm{E}\,[X^2] + 2ab\mathrm{E}\,[XY] + b^2\mathrm{E}\,[Y^2]$$
$$0 \leq \mathrm{E}\,[(aX - bY)^2] = a^2\mathrm{E}\,[X^2] - 2ab\mathrm{E}\,[XY] + b^2\mathrm{E}\,[Y^2]$$

If we let $a^2 = \mathrm{E}\,[Y^2]$ and $b^2 = \mathrm{E}\,[X^2]$ then the first inequality above yields

$$0 \leq \mathrm{E}\,[Y^2]\,\mathrm{E}\,[X^2] + 2\sqrt{\mathrm{E}\,[Y^2]\,\mathrm{E}\,[X^2]}\mathrm{E}\,[XY] + \mathrm{E}\,[X^2]\,\mathrm{E}\,[Y^2]$$
$$-2\mathrm{E}\,[X^2]\,\mathrm{E}\,[Y^2] \leq 2\sqrt{\mathrm{E}\,[Y^2]\,\mathrm{E}\,[X^2]}\mathrm{E}\,[XY]$$
$$-\sqrt{\mathrm{E}\,[X^2]\,\mathrm{E}\,[Y^2]} \leq \mathrm{E}\,[XY].$$

By a similar set of steps the second inequality produces

$$\mathrm{E}\,[XY] \leq \sqrt{\mathrm{E}\,[X^2]\,\mathrm{E}\,[Y^2]}.$$

Therefore, since

$$-\sqrt{\mathrm{E}\,[X^2]\,\mathrm{E}\,[Y^2]} \leq \mathrm{E}\,[XY] \leq \sqrt{\mathrm{E}\,[X^2]\,\mathrm{E}\,[Y^2]}$$
$$(\mathrm{E}\,[XY])^2 \leq \mathrm{E}\,[X^2]\,\mathrm{E}\,[Y^2].$$

Occasionally $(\mathrm{E}\,[XY])^2$ is written as $\mathrm{E}\,[XY]^2$ or even as $\mathrm{E}^2\,[XY]$. A careful reading of the expression will avoid any possible confusion. $\qquad\square$

Now we are in a position to prove the following theorem.

Theorem 11.3 *If X and Y are random variables then $-1 \leq \rho\,(X,Y) \leq 1$.*

Proof. Consider the covariance of X and Y.

$$(\mathrm{Cov}\,(X,Y))^2 = (\mathrm{E}\,[(X - \mathrm{E}\,[X])(Y - \mathrm{E}\,[Y])])^2$$
$$\leq \mathrm{E}\,[(X - \mathrm{E}\,[X])^2]\,\mathrm{E}\,[(Y - \mathrm{E}\,[Y])^2] \quad \text{(by Lemma 11.1)}$$
$$= \mathrm{Var}\,(X)\,\mathrm{Var}\,(Y)$$

Thus $|\text{Cov}(X,Y)| \le \sqrt{\text{Var}(X)\,\text{Var}(Y)}$, which is equivalent to the inequality,

$$-1 \le \frac{\text{Cov}(X,Y)}{\sqrt{\text{Var}(X)\,\text{Var}(Y)}} < 1$$
$$-1 \le \rho(X,Y) < 1,$$

which follows from the definition of correlation. $\qquad\square$

Example 11.2 Referring to the data in Table 11.1 the correlation between height and arm span is calculated as

$$\rho(X,Y) = \frac{\text{Cov}(X,Y)}{\sqrt{\text{Var}(X)\,\text{Var}(Y)}} = \frac{15.445}{\sqrt{(18.6605)(16.8526)}} \approx 0.870948.$$

For this data set there seems to be an especially strong linear relationship between height and weight. Figure 11.1 illustrates this as well.

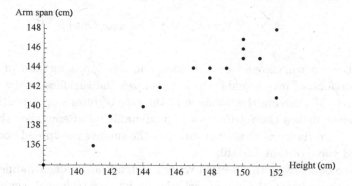

Fig. 11.1   A scatter plot of height *versus* arm span for a sample of twenty children.

We end this section with a more financial application of the concepts of covariance and correlation. A portfolio may consist of n different investments. Suppose that an investor may invest an amount w_i for one time period and at the end of the time period, the investment will return wealth in the amount of $w_i X_i$. In Sec. 1.5 the concept of rate of return was defined to be the interest rate equivalent to this growth in value. For the ith investment the rate of return will be determined by

$$w_i(1 + R_i) = w_i X_i \quad \text{which implies} \quad R_i = X_i - 1.$$

Thus the total accumulated wealth after one time period from all n investments is $W = \sum_{i=1}^{n} w_i(1 + R_i)$. If the rate of return for each of the n investments is treated as a random variable, then

$$E[W] = E\left[\sum_{i=1}^{n} w_i(1 + R_i)\right] = \sum_{i=1}^{n} w_i + \sum_{i=1}^{n} w_i E[R_i].$$

The variance in the accumulated wealth is, according to Eq. (11.4),

$$\begin{aligned}
\text{Var}(W) &= \text{Var}\left(\sum_{i=1}^{n} w_i(1 + R_i)\right) \\
&= \text{Var}\left(\sum_{i=1}^{n} w_i R_i\right) \\
&= \sum_{i=1}^{n} \text{Var}(w_i R_i) + \sum_{i=1}^{n}\sum_{j \neq i} \text{Cov}(w_i R_i, w_j R_j) \\
&= \sum_{i=1}^{n} w_i^2 \text{Var}(R_i) + \sum_{i=1}^{n}\sum_{j \neq i} w_i w_j \text{Cov}(R_i, R_j).
\end{aligned}$$

Later sections will expand on this idea and develop a method of selecting a portfolio of investments which minimizes the variance in the wealth generated. Minimizing the variance in the rate of return on a portfolio of investments makes the return more "predictable". Remember that the smaller the variance of a random variable, the smaller the spread about the expected value of that variable.

Before concluding this section we will extend our understanding of covariance and lognormal random variables to derive a technical result which will be needed in Sec. 11.6.

Lemma 11.2 *If X is a lognormal random variable with drift parameter μ and volatility σ^2 and $K > 0$ is a constant then*

$$\text{Cov}\left(X, (X - K)^+\right) = E\left[X(X - K)^+\right] - E[X]E\left[(X - K)^+\right], \quad (11.6)$$

where

$$E\left[X(X - K)^+\right] = e^{2(\mu + \sigma^2)}\phi(w + 2\sigma) - Ke^{\mu + \sigma^2/2}\phi(w + \sigma), \quad (11.7)$$

with $w = (\mu - \ln K)/\sigma$.

Proof. Equation (11.6) follows from the definition of covariance. The technical portion of the lemma is the derivation of Eq. (11.7).

$$E\left[X(X-K)^+\right] = \frac{1}{\sqrt{2\pi}\sigma} \int_0^\infty x(x-K)^+ \frac{1}{x} e^{-(\ln x - \mu)^2/2\sigma^2} \, dx$$

$$= \frac{1}{\sqrt{2\pi}\sigma} \int_K^\infty (x-K) e^{-(\ln x - \mu)^2/2\sigma^2} \, dx$$

$$= \frac{1}{\sqrt{2\pi}} \int_{(\ln K - \mu)/\sigma}^\infty (e^{\sigma z + \mu} - K) e^{\sigma z + \mu} e^{-z^2/2} \, dz$$

The last equation is derived by making the substitution $\sigma z = \ln x - \mu$. Therefore

$$E\left[X(X-K)^+\right] = \frac{e^{2(\mu+\sigma^2)}}{\sqrt{2\pi}} \int_{(\ln K - \mu)/\sigma}^\infty e^{-(z-2\sigma)^2/2} \, dz$$

$$- \frac{Ke^{\mu+\sigma^2/2}}{\sqrt{2\pi}} \int_{(\ln K - \mu)/\sigma}^\infty e^{-(z-\sigma)^2/2} \, dz$$

$$= e^{2(\mu+\sigma^2)} \phi\left(\frac{\mu - \ln K}{\sigma} + 2\sigma\right)$$

$$- Ke^{\mu+\sigma^2/2} \phi\left(\frac{\mu - \ln K}{\sigma} + \sigma\right).$$

The proof is complete if we let $w = (\mu - \ln K)/\sigma$. $\qquad\square$

11.2 Optimal Portfolios

In the previous chapter on hedging, all of the discussion and examples assumed that a hedged position with respect to a particular option could be set up by taking a position in the stock or security underlying the option. However, it may not always be possible to purchase (or sell) sufficient shares of the underlying security (or the option itself) to create the hedged position. In these cases the manager of a portfolio may have to find a surrogate for the security. Upon reflection, a particular financial instrument will be a better or worse surrogate for a security depending on how the values of the security and the other instrument change. If the changes in the two values are highly correlated then the portfolio manager will have more confidence in using the surrogate.

Suppose an investment firm sells a European call option for stock A and wishes to create a Delta-neutral portfolio. If sufficient shares of stock

A cannot be purchased to create the hedge then the firm can investigate hedging their short position in the option for stock A with a long position in n shares of stock B. If the hedge could be created with stock A in place of stock B then the Delta-neutral hedge is achieved when $n = \Delta$ as we saw in Sec. 10.2. The issue confronting the investment firm now is, what should n be when the hedge must be created with stock B? An **optimal portfolio** is one for which the variance in the value of the portfolio is minimized. The value of the portfolio consisting of a short position on a call option for stock A and a long position of n shares of stock B will be denoted

$$\mathcal{P} = C_A - nB.$$

Note that C_A denotes a European call option on the underlying security A (earlier the notation C_a was used to denote an American call option). The variance in the value of the portfolio is

$$\text{Var}\left(\mathcal{P}\right) = \text{E}\left[\left(C_A - nB\right)^2\right] - \text{E}\left[C_A - nB\right]^2$$
$$= n^2\text{Var}\left(B\right) - 2n\text{Cov}\left(C_A, B\right) + \text{Var}\left(C_A\right).$$

Thus it is seen that the variance in the value of the portfolio is a quadratic function of the hedging parameter n. Thus the minimum variance is achieved when

$$n = \frac{\text{Cov}\left(C_A, B\right)}{\text{Var}\left(B\right)} = \rho\left(C_A, B\right)\sqrt{\frac{\text{Var}\left(C_A\right)}{\text{Var}\left(B\right)}}. \tag{11.8}$$

Thus if the correlation between the values of the option on stock A and stock B were unity then the hedging parameter would be the ratio of the standard deviations in the value of the option on stock A and the value of stock B. A small correlation between C_A and B would indicate that stock B is a poor surrogate for stock A. Equation (11.8) also implies that n will decrease as $\text{Var}\left(B\right)$ increases. The variance in the portfolio will be minimized by decreasing the amount of large variance component in the portfolio.

The analysis leading up to Eq. (11.8) overlooks several important assumptions about the hedging situation. First the expiry date of the call option did not come into play. If expiry is some finite (as opposed to instantaneously short) time from the moment the hedge is created, then rightfully we should consider the stochastic behavior of B over the interval to expiry. The concept of hedging as described in Chapter 10 is a continuous dynamic process (even though in practice it may be carried out discretely). To illustrate the importance of considering the time horizon of the calculation consider a hedging problem for which we already know the correct answer. Namely consider the calculation of the value of n for the case where the long position in the European call option C_A is hedged with a short position in n shares of security A (rather than security B). This time the call option is hedged using the underlying security. We know from the derivation of the Black-Scholes PDE that the n which minimizes the variance of $\mathcal{P}$ is the Delta ($\Delta = \partial C_A/\partial A$) of the call option. Consider the value of n given in Eq. (11.8) with B replaced by A and considering only the change in the value of the call option and the security over an instantaneously short time interval ΔT. According to Eq. (11.8)

$$n = \frac{\text{Cov}\,(\Delta C, \Delta A)}{\text{Var}\,(\Delta A)}.$$

We have dropped the subscript of the call option since there is only one security relevant to the portfolio. Making use of differentials

$$\frac{\text{Cov}\,(\Delta C, \Delta A)}{\text{Var}\,(\Delta A)} = \frac{\text{Cov}\,\left(\frac{\partial C}{\partial A}\Delta A, \Delta A\right)}{\text{Var}\,(\Delta A)}$$

$$= \frac{\partial C}{\partial A}\frac{\text{Cov}\,(\Delta A, \Delta A)}{\text{Var}\,(\Delta A)}$$

$$n = \frac{\partial C}{\partial A}.$$

Thus the minimizing n is the familiar hedging ratio Δ for the call option. The quantity $\partial C/\partial A$ can be pulled out of the covariance since it is an instantaneously determined quantity and thus constant. The last sequence of equations rightly should have been a sequence of approximations which become equations by passing to the limit as $\Delta T \to 0$. The reader who is interested in more rigor and in reviewing the stochastic calculus introduced in Chapter 5 should consult exercise 8.

The idea of selecting a portfolio by minimizing the variance of its value will be extended in later sections of this chapter. Before returning to minimum variance analysis, we must introduce the concept known as utility and describe its properties.

11.3   Utility Functions

This section will focus on defining and understanding a class of functions which can be used as the basis of rational decision making. Suppose that an investor is faced with the choice of two different investment products. Suppose further that the set of outcomes resulting from investing in either of the products is $\{C_1, C_2, \ldots, C_n\}$. The reader should think of this set of outcomes as the union of the possible outcomes of investing in the first product and the possible outcomes resulting from investing in the second product. The probability of outcome C_i coming about as the result of investing in the first product will be denoted p_i for $i = 1, 2, \ldots, n$. If outcome C_i cannot result from investing in the first product then, of course, $p_i = 0$. For the second product the values of the probabilities will be denoted q_i.

The investor can rank the outcomes in order of desirability. Without loss of generality suppose the outcomes have been ranked from least to most desirable as

$$C_1 \leq C_2 \leq \cdots \leq C_n.$$

To each of the possible outcomes a **utility** can be assigned. The utility function $u(C_i)$ is defined as follows. To start, $u(C_1) = C_1$ and $u(C_n) = C_n$. The values of $u(C_i)$ for $1 < i < n$ will be defined by referring to C_1 and C_n. Suppose that for each i the investor is given the following choice: participate in a random experiment in which they receive outcome C_i with certainty, or participate in a random experiment where they will receive C_1 with probability ϕ_i or receive C_n with probability $1 - \phi_i$. The expected value of the outcome of the first experiment is C_i and the expected value of the outcome of the second experiment is $E_i = \phi_i C_1 + (1 - \phi_i) C_n$. If $\phi_i = 1$ then the investor will surely participate in the random experiment with the certain outcome of C_i, since they rank this as a more desirable outcome than C_1. If $\phi_i = 0$ then the investor will participate in the second random experiment since its expected outcome is C_n. They would not better their result by taking the sure outcome C_i. At some value of $\phi_i \in [0, 1]$ the investor will be indifferent to the choice. The utility of C_i, in other words $u(C_i)$ is defined to be the E_i for which the investor is indifferent to the choice of experiment.

Utility functions are specific to individual investors, like personality traits. In general a utility function is an increasing function. Most rational investors assign greater utility to more preferable outcomes. It is assumed that the value of ϕ_i at which the investor is indifferent to the choice of experiment is unique. The rational investor once having decided to participate in the second random experiment will not at a higher expected value decide to once again accept the certain result. Thus the utility function is well-defined.

The utility function provides a means by which two outcomes of unequal desirability may be compared. The utility of outcome C_i is equivalent to receiving C_1 with probability ϕ_i or receiving C_n with probability $1-\phi_i$. The reader should beware that there are multiple notions of probability at work here. The utility function was defined in terms of a person's preference for receiving outcome C_i with certainty or willingness to participate in a specific type of random experiment. However, the probability of the occurrence of outcome C_i is not $u(C_i)$. Recall we have assumed that $P(C_i) = p_i$ or q_i depending on which of two investment products the hypothetical investor chooses. Returning to the original task of deciding between two different investment products, suppose the investor decides to invest in the first product. For the first investment product the expected value of the utility function is then

$$E\left[u(\mathbf{p})\right] = \sum_{i=1}^{n} p_i u(C_i)$$

$$= C_1 \sum_{i=1}^{n} p_i \phi_i + C_n \sum_{i=1}^{n} p_i(1 - \phi_i).$$

Thus the expected utility of the first investment is equivalent to the expected value of a simple random experiment in which the investor will receive the least desirable outcome C_1 with probability $\sum_{i=1}^{n} p_i \phi_i$ and the most desirable outcome C_n with probability $\sum_{i=1}^{n} p_i(1 - \phi_i)$. Similarly the expected value of the utility function for the second investment product is found to be

$$E\left[u(\mathbf{q})\right] = \sum_{i=1}^{n} q_i u(C_i).$$

Therefore the investor will choose the first product whenever

$$\sum_{i=1}^{n} p_i u(C_i) > \sum_{i=1}^{n} q_i u(C_i), \tag{11.9}$$

and otherwise will choose the second product.

In this section we have spoken of outcomes or consequences C_i of making an investment decision. More concretely these outcomes can be thought of as receiving differing amounts of money for an investment (some of which may be negative). Thus in general the **utility function** denoted $u(x)$ is the investor's utility of receiving an amount x. In the remainder of this section categories and properties of utility functions will be explored.

A general property of utility functions is that the amount of extra utility that an investor experiences when x is increased to $x + \Delta x$ is non-increasing. In other words $u(x + \Delta x) - u(x)$ is a non-increasing function of x. This property is illustrated in Fig. 11.2. We will describe a function $f(t)$ as

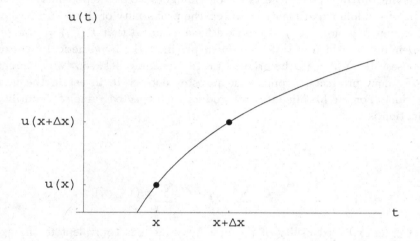

Fig. 11.2   The extra utility received, $u(x + \Delta x) - u(x)$, is a decreasing function of x.

concave on an open interval (a, b) if for every $x, y \in (a, b)$ and every $\lambda \in [0, 1]$ we have

$$\lambda f(x) + (1 - \lambda)f(y) \le f(\lambda x + (1 - \lambda)y). \tag{11.10}$$

Graphically this may be interpreted as meaning that all the secant lines lie below the graph of $f(t)$. See Fig. 11.3. A utility function $u(t)$ obeying inequality (11.10) will also be called concave. An often repeated statement about investing is that an investor may receive greater rewards by taking greater risks. However, for most rational investors there is a level of reward beyond which in order to reap greater reward, they are unwilling to accept the higher level of risk. While a casino gambler may be willing to roll the

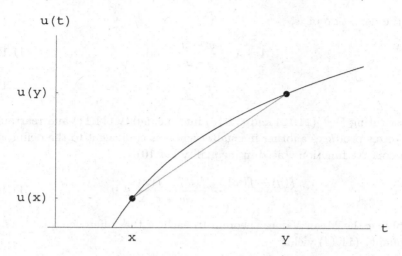

Fig. 11.3   For concave functions the secant line parameterized by $\lambda \in [0,1]$ will lie below the corresponding point on the graph of the function.

dice for a chance to double an investment of $20, it would be the rare gambler who would be willing to roll the dice in order to double $20,000,000. The utility of that extra reward is not as high as the initial reward. These are gamblers and investors who will avoid games or investments for which the risk is (in their belief) too high even if the potential rewards are commensurately high. An investor whose utility function is concave is said to be **risk-averse**. An investor with a linear utility function of the form $u(x) = ax + b$ with $a > 0$ is said to be **risk-neutral**. An investor whose utility function increases more rapidly as the reward increases is said to be **risk-loving**.

The following result can be demonstrated regarding concave functions.

Theorem 11.4   *If $f \in C^2(a,b)$ then f is concave on (a,b) if and only if $f''(t) \leq 0$ for $a < t < b$.*

Proof.   If f is concave on (a,b) then by definition f satisfies inequality (11.10). Let $x, y \in (a,b)$. Without loss of generality we may assume $x < y$. If $w = \lambda x + (1 - \lambda)y$ and if $0 < \lambda < 1$ then $a < x < w < y < b$. Inequality (11.10) is then equivalent to

$$(1 - \lambda)\left[f(y) - f(w)\right] \leq \lambda\left[f(w) - f(x)\right]. \tag{11.11}$$

By the definition of w,

$$1 - \lambda = \frac{w - x}{y - x} \quad \text{and} \tag{11.12}$$

$$\lambda = \frac{y - w}{y - x}. \tag{11.13}$$

Substituting Eqs. (11.12) and (11.13) into inequality (11.11) and rearranging terms produces another inequality which is equivalent to the definition of a concave function stated in inequality (11.10).

$$\frac{f(y) - f(w)}{y - w} - \frac{f(w) - f(x)}{w - x} \leq 0 \tag{11.14}$$

Applying the Mean Value Theorem to each of the difference quotients of inequality (11.14) yields

$$f'(\beta) - f'(\alpha) \leq 0 \quad \text{with } x < \alpha < w < \beta < y$$
$$f''(t)(\beta - \alpha) \quad \text{with } \alpha < t < \beta.$$

The last inequality made yet another use of the Mean Value Theorem. The reader should keep in mind that this last inequality is still equivalent to the inequality defining a concave function. Thus inequality (11.10) holds for a twice continuously differentiable function if and only if $f''(t) \leq 0$ for $a < t < b$. $\qquad\square$

In the next section we will make use of the following result known as Jensen's Inequality.

Theorem 11.5 *(Jensen's Inequality (Discrete Version))* *Let f be a concave function on the interval (a, b), suppose $x_i \in (a, b)$ for $i = 1, 2, \ldots, n$, and suppose $\lambda_i \in [0, 1]$ for $i = 1, 2, \ldots, n$ with $\sum_{i=1}^{n} \lambda_i = 1$, then*

$$\sum_{i=1}^{n} \lambda_i f(x_i) \leq f\left(\sum_{i=1}^{n} \lambda_i x_i\right). \tag{11.15}$$

Proof. Let $\mu = \sum_{i=1}^{n} \lambda_i x_i$ and note that since $\lambda_i \in [0, 1]$ for $i = 1, 2, \ldots, n$ and $\sum_{i=1}^{n} \lambda_i = 1$, then $a < \mu < b$. The equation of the line tangent to the graph of f at the point $(\mu, f(\mu))$ is $y = f'(\mu)(x - \mu) + f(\mu)$. Since f is concave on (a, b) then

$$f(x_i) \leq f'(\mu)(x_i - \mu) + f(\mu) \quad \text{for } i = 1, 2, \ldots, n.$$

Therefore

$$\sum_{i=1}^{n} \lambda_i f(x_i) \leq \sum_{i=1}^{n} \left(\lambda_i \left[f'(\mu)(x_i - \mu) + f(\mu) \right] \right)$$

$$= f'(\mu) \sum_{i=1}^{n} (\lambda_i x_i - \lambda_i \mu) + f(\mu) \sum_{i=1}^{n} \lambda_i$$

$$= f(\mu)$$

$$= f \left(\sum_{i=1}^{n} \lambda_i x_i \right).$$

$\square$

Jensen's inequality may be interpreted as saying that for a concave function, the mean of the values of the function applied to a set of values of a random variable is no greater than the function applied to the mean of the values of the random variable. If u is a concave utility function, then $E[u(X)] \leq u(E[X])$. Thus a risk-averse investor prefers the certain return of $C = E[X]$ to receiving a random return with this expected value.

Example 11.3 Suppose $f(x) = \tanh x$ and let

$$\{\lambda_1, \lambda_2, \lambda_3, \lambda_4, \lambda_5\} = \{0.30, 0.07, 0.23, 0.20, 0.20\}.$$

The reader can verify that $f(x)$ is concave on $(0, \infty)$. If $X_i = i^2$ for $i = 1, 2, \ldots, 5$ then

$$\sum_{i=1}^{5} \lambda_i f(X_i) = 0.928431 \leq 1 = f \left(\sum_{i=1}^{5} \lambda_i X_i \right)$$

There is also a continuous version of Jensen's Inequality.

Theorem 11.6 *(Jensen's Inequality (Continuous Version))* *Let $\phi(t)$ be an integrable function on $[0, 1]$ and let f be a concave function, then*

$$\int_0^1 f(\phi(t)) \, dt \leq f \left(\int_0^1 \phi(t) \, dt \right). \tag{11.16}$$

Proof. For the sake of compactness of notation let $\alpha = \int_0^1 \phi(t) \, dt$, and let $y = f'(\alpha)(x - \alpha) + f(\alpha)$, the equation of the tangent line passing through the point with coordinates $(\alpha, f(\alpha))$. Since f is concave then

$$f(\phi(t)) \leq f'(\alpha)(\phi(t) - \alpha) + f(\alpha),$$

which implies that

$$\int_0^1 f(\phi(t))\, dt \le \int_0^1 [f'(\alpha)(\phi(t) - \alpha) + f(\alpha)]\, dt$$

$$= f(\alpha) + f'(\alpha) \int_0^1 (\phi(t) - \alpha)\, dt$$

$$= f(\alpha)$$

$$= f\left(\int_0^1 \phi(t)\, dt\right).$$

$\square$

11.4 Expected Utility

As the name suggests, **expected utility** is the expected value of a utility function. Returning to the previous discussion of choosing between two possible investment instruments, we see that the investment with the greater expected utility is preferable. The inequality in (11.9) indicated that if the first investment choice results in outcomes $\{C_1, C_2, \ldots, C_n\}$ with respective probabilities p_i for $i = 1, 2, \ldots n$ and the second investment choice produces the same outcomes but with probabilities $\{q_i, q_2, \ldots, q_n\}$ then a rational investor would select the first investment whenever

$$\sum_{i=1}^n p_i u(C_i) > \sum_{i=1}^n q_i u(C_i).$$

Suppose the random variable X is thought of as the set of outcomes with probabilities $\{p_i, p_2, \ldots, p_n\}$ while the random variable Y represents the outcomes with probabilities $\{q_i, q_2, \ldots, q_n\}$. Then the first investment is preferable whenever

$$\mathrm{E}\left[u(X)\right] > \mathrm{E}\left[u(Y)\right].$$

Example 11.4 An investor must choose between the following two "investments":

A: Flip a fair coin, if the coin lands heads up the investor receives $10, otherwise they receive nothing.

B: Receive an amount $M with certainty.

The investor is risk-averse with a utility function $u(x) = x - x^2/25$. The rational investor will select the investment with the greater expected utility.

The expected utility for investment A is

$$\frac{1}{2}u(10) + \frac{1}{2}u(0) = \frac{1}{2}\left(10 - \frac{10^2}{25}\right) = 3.$$

The expected utility for B is $u(M) = M - M^2/25$. Thus the investor will choose the coin flip whenever

$$3 > M - \frac{M^2}{25}$$

$$M^2 - 25M + 75 > 0$$

$$\frac{25 - 5\sqrt{13}}{2} > M.$$

Thus investment A is preferable to B whenever $M < \$3.49$.

The astute reader will note that the quadratic inequality solved above is satisfied for M in the sets defined by the relationships

$$M < \frac{25 - 5\sqrt{13}}{2} \approx 3.49 \quad \text{or} \quad M > \frac{25 + 5\sqrt{13}}{2} \approx 21.51.$$

Mathematically this implies the investor would prefer investment A even if the certain amount of investment B is greater than \$21.51. A wise investor would never accept a chance at a maximum payoff of \$10 if they can receive a guaranteed amount of \$21.51 or more. This type of situation leads us to adopt the following logical convention.

If $M < 3.49$ the investor of the previous example will choose investment A, while for $M \geq \$3.49$ the investor will choose the investment B. This example illustrates the concept known as the **certainty equivalent** which is defined as the minimum value C of a random variable X at which $u(C) = E[u(X)]$.

Example 11.5 An investor wishes to find the certainty equivalent C for the following investment choice:

A: Flip a fair coin, if the coin lands heads up the investor receives $0 < X \leq 10$, otherwise they receive $0 < Y < X$.
B: Receive an amount C with certainty.

Assuming the investor's utility function is the same as in the previous example, the certainty equivalent and payoffs of investment A must satisfy the equation

$$C - \frac{C^2}{25} = \frac{1}{2}\left(X - \frac{X^2}{25} + Y - \frac{Y^2}{25}\right)$$

$$\left(C - \frac{25}{2}\right)^2 = \frac{1}{2}\left[\left(X - \frac{25}{2}\right)^2 + \left(Y - \frac{25}{2}\right)^2\right]$$

$$C = \frac{25}{2} - \frac{1}{\sqrt{2}}\sqrt{\left(X - \frac{25}{2}\right)^2 + \left(Y - \frac{25}{2}\right)^2}$$

Again note that we choose the certainty equivalent to be the smallest value of C satisfying the equation. The design of investment A specifies that $0 < Y < X < 10$. Thus the certainty equivalent can be thought of as a surface plotted over the triangular region bounded by $0 < X < 10$ with $0 < Y < X$. Figure 11.4 illustrates the certainty equivalent as a function of X and Y.

11.5 Portfolio Selection

In the previous sections we have treated the investor's choice between two different investments as a binary, all-or-nothing decision. In reality an investor may choose to split investment capital between two (or more) investments. In this section we will explore the **portfolio selection problem**, the task of allocating funds in an optimal fashion among several investment options.

We begin with a simple example. Suppose an investor has a total of x amount of capital to invest. Assuming that the investor may use any proportion of this capital, let $\alpha \in [0, 1]$ be the proportion they invest. The investment is structured such that an allocation of αx will earn αx with probability p and lose αx with probability $1 - p$. Thus for an investment of αx the random variable representing the investor's financial position after the conclusion of the investment is

$$X = \begin{cases} x(1 + \alpha) \text{ with probability } p, \\ x(1 - \alpha) \text{ with probability } 1 - p. \end{cases}$$

The allocation proportion α will be optimal when the expected value of the

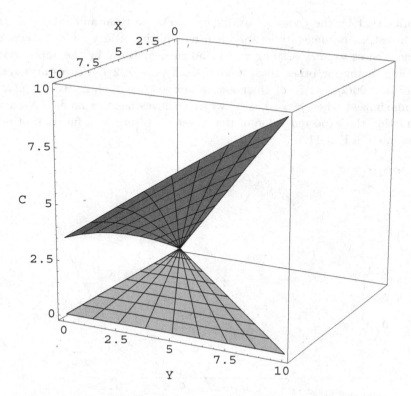

Fig. 11.4 The certainty equivalent in this example is a section of an ellipsoid. The domain of interest in the XY-plane is drawn as a shadow below the surface.

utility is maximized.

$$\mathrm{E}\left[u(X)\right] = pu(x(1+\alpha)) + (1-p)u(x(1-\alpha))$$

$$\frac{d}{d\alpha}\mathrm{E}\left[u(X)\right] = pxu'(x(1+\alpha)) - x(1-p)u'(x(1-\alpha))$$

$$0 = pu'(x(1+\alpha)) - (1-p)u'(x(1-\alpha))$$

The critical value of α which solves the last equation above will correspond to a maximum for the expected value of the utility as long as the utility function is concave.

Example 11.6 If $u(x) = \ln x$, then the expected value of the investor's utility is maximized when

$$\frac{p}{1+\alpha} - \frac{1-p}{1-\alpha} = 0, \quad \text{which implies} \quad \alpha = 2p - 1.$$

Notice that by the choice of utility function, the total amount of capital to invest, x, becomes irrelevant when maximizing utility. For the sake of concreteness we may assume $x = 1$ and then multiply by the appropriate scaling factor for other total amounts. If $p > 1/2$ the investor should allocate $100(2p - 1)\%$ of their capital. If $0 \le p \le 1/2$, then $\mathrm{E}\,[u(X)]$ is maximized when $\alpha = 0$, *i.e.* when no investment is made. A curve depicting the expected value of the investor's utility as a function of p is presented in Fig. 11.5.

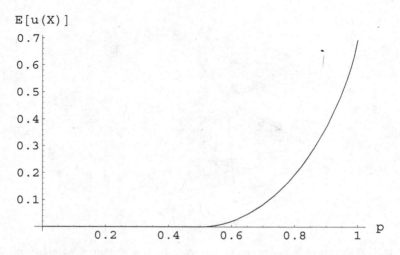

Fig. 11.5   The expected value of utility for a risk-averse investor allocating $100(2p-1)\%$ of their capital.

With this simple example understood, we are now ready to consider a more general situation. Suppose that an infinitely divisible unit amount of capital may be allocated among n different investments. A proportion x_i will be allocated to security i for $i = 1, 2, \ldots, n$. The vector notation $\langle x_1, x_2, \ldots, x_n \rangle$ will be used to represent the portfolio of proportions of investments. The return from investment i will be denoted W_i for $i = 1, 2, \ldots, n$. The quantity $W_i = 1 + R_i$ where R_i is the rate of return for investment i. The portfolio selection problem is then defined to be that of determining x_i for $i = 1, 2, \ldots, n$ such that

(1) $0 \le x_i \le 1$ for $i = 1, 2, \ldots, n$, and
(2) $\sum_{i=1}^{n} x_i = 1$,

and which maximizes the investor's total expected utility $E[u(W)]$, where

$$W = \sum_{i=1}^{n} x_i W_i.$$

Throughout this section it will be assumed that n is large and that the W_i's are not too highly correlated. Under these assumptions the Central Limit Theorem of Sec. 3.5 implies that W is a normally distributed random variable.

The allocation of investment funds can be driven by the objective of maximizing the expected value of the investor's utility function. Suppose an investor's utility function is given by $u(x) = 1 - e^{-bx}$ where $b > 0$. The reader is asked in exercise 15 to show that this utility function is concave and monotonically increasing. Assuming that W is a normally distributed random variable then $-bW$ is also normally distributed with

$$E[-bW] = -bE[W] \quad \text{and} \quad \text{Var}(-bW) = b^2\text{Var}(W).$$

Consequently,

$$E[u(W)] = E\left[1 - e^{-bW}\right] = 1 - E\left[e^{-bW}\right].$$

Making the assignment $Y = e^{-bW}$, then Y is a lognormal random variable (see Sec. 3.6). According to Lemma 3.1,

$$E[Y] = E\left[e^{-bW}\right] = e^{-bE[W]+b^2\text{Var}(W)/2}$$

which implies that

$$E[u(W)] = 1 - e^{-b(E[W]-b\text{Var}(W)/2)}.$$

Thus the expected utility is maximized when $E[W] - b\text{Var}(W)/2$ is maximized. If two different portfolios, $\langle x_1, x_2, \ldots, x_n \rangle$ and $\langle y_1, y_2, \ldots, y_n \rangle$, give rise respectively to returns X and Y then the portfolio represented by $\langle x_1, x_2, \ldots, x_n \rangle$ will be preferable if

$$E[X] \geq E[Y] \quad \text{and} \quad \text{Var}(X) \leq \text{Var}(Y).$$

To extend this discussion and make it more concrete numerically, suppose $b = 0.005$ and that an investor wishes to choose among an infinite number of different portfolios which can be created by investing in two securities denoted A and B. The investor has \$100 to invest and will allocate

y dollars to security A and $100 - y$ dollars to B. The following table summarizes what is known about the rates of return on the hypothetical securities A and B.

Security	A	B
E [rate of return]	0.16	0.18
$\sqrt{\text{Var}}$ (rate of return)	0.20	0.24

Assume that the correlation between the rates of return is $\rho = -0.35$. Once again W will represent the wealth returned.

$$E[W] = 100 + 0.16y + 0.18(100 - y)$$
$$= 118 - 0.02y$$
$$\text{Var}(W) = y^2(0.20)^2 + (100 - y)^2(0.24)^2 + 2y(100 - y)(0.20)(0.24)(-0.35)$$
$$= 0.04y^2 + 0.0576(100 - y)^2 - 0.0336y(100 - y)$$

Since the optimal portfolio is the one which maximizes

$$E[W] - \frac{b}{2}\text{Var}(W)$$
$$= 118 - 0.02y - 0.0025(0.04y^2 + 0.0576(100 - y)^2 - 0.0336y(100 - y))$$
$$= -0.000328y^2 + 0.0172y + 116.56.$$

This occurs when $y \approx 26.2195$. For this choice of y

$$E[W] \approx 117.476$$
$$\text{Var}(W) \approx 276.049$$
$$E[u(W)] \approx 0.442296.$$

In the next section we will generalize and extend these ideas to situations in which investment capital is partitioned in n ways.

11.6   Minimum Variance Analysis

Building on the ideas contained in the previous section we consider the case of an investor allocating a portion of their capital between two potential investments. Without loss of generality it is assumed that the investor has an infinitely divisible unit of capital to invest and will invest a fraction $\alpha \in [0, 1]$ in one security and $1 - \alpha$ in a second security. The rates of return of the two securities are respectively the random variables R_1 and R_2. The variances of these rates of return will be denoted σ_1^2 and σ_2^2. The covariance in the rates of return is assumed to be c. Thus the variance in the wealth W returned from the investments is

$$\text{Var}(W) = \alpha^2 \sigma_1^2 + (1-\alpha)^2 \sigma_2^2 + 2c\alpha(1-\alpha). \qquad (11.17)$$

One measure of an optimal portfolio of investments would be the one with the least variance in the returned wealth. Differentiating $\text{Var}(W)$ with respect to α and solving for the critical value of α produces

$$0 = 2\left(\alpha\sigma_1^2 - (1-\alpha)\sigma_2^2 + c(1-2\alpha)\right)$$

$$\alpha^* = \frac{\sigma_2^2 - c}{\sigma_1^2 + \sigma_2^2 - 2c}$$

The denominator of α^* is the covariance of $R_1 - R_2$ with itself and thus is non-negative. In order for the critical value of α to be defined, it is assumed that $\sigma_1^2 + \sigma_2^2 > 2c$. This condition is also sufficient to guarantee that the variance of the wealth in Eq. (11.17) will have a global minimum. The parameter α^* must also fall within the closed interval $[0,1]$. This occurs whenever

$$c \leq \min\{\sigma_1^2, \sigma_2^2\}.$$

If $c \leq 0$ then this condition is always satisfied. According to the Extreme Value Theorem [Stewart (1999)], the minimum variance will occur when $\alpha = 0$, $\alpha = 1$, or $\alpha = \alpha^*$.

$$\text{Var}(W)|_{\alpha=0} = \sigma_2^2$$

$$\text{Var}(W)|_{\alpha=1} = \sigma_1^2$$

$$\text{Var}(W)|_{\alpha=\alpha^*} = \frac{\sigma_1^2 \sigma_2^2 - c^2}{\sigma_1^2 + \sigma_2^2 - 2c}$$

Consider the special case when the rates of return of the two investments are uncorrelated. Under this assumption $c = 0$ and the critical value of α at which the variance in the returned wealth is minimized reduces to

$$\alpha^* = \frac{\sigma_2^2}{\sigma_1^2 + \sigma_2^2} = \frac{\frac{1}{\sigma_1^2}}{\frac{1}{\sigma_1^2} + \frac{1}{\sigma_2^2}}.$$

There is an appealing simplicity and symmetry to this critical value. This is not a coincidence, it is in fact a pattern seen in the more general case of allocating a unit of investment among n different potential investments.

Theorem 11.7 *Suppose that $0 \leq \alpha_i \leq 1$ will be invested in security i for $i = 1, 2, \ldots, n$ subject to the constraint that $\alpha_1 + \alpha_2 + \cdots + \alpha_n = 1$. Suppose the rate of return of security i is a random variable R_i and that all the rates of return are mutually uncorrelated. The optimal, minimum variance portfolio described by the allocation vector $\langle \alpha_1^*, \alpha_2^*, \ldots, \alpha_n^* \rangle$, is the one for which*

$$\alpha_i^* = \frac{\frac{1}{\sigma_i^2}}{\sum_{j=1}^n \frac{1}{\sigma_j^2}} \quad \text{for } i = 1, 2, \ldots, n,$$

where $\sigma_i^2 = \text{Var}(R_i)$.

Proof. Since the rates of return are uncorrelated then the variance in the returned wealth W is

$$\text{Var}(W) = \sum_{i=1}^n \alpha_i^2 \sigma_i^2,$$

and is subject to the constraint that $1 = \sum_{i=1}^n \alpha_i$. Minimizing the variance subject to the constraint is accomplished by use of Lagrange Multipliers [Stewart (1999)]. This technique states that $\text{Var}(W)$ will be optimized at one of the solutions $(\alpha_1, \alpha_2, \ldots, \alpha_n, \lambda)$ of the set of simultaneous equations:

$$\nabla \left(\sum_{i=1}^n \alpha_i^2 \sigma_i^2 \right) = \lambda \nabla \left(\sum_{i=1}^n \alpha_i \right)$$

$$\sum_{i=1}^n \alpha_i = 1$$

The symbol ∇ denotes the gradient operator. These equations are equivalent to respectively:

$$2\alpha_i \sigma_i^2 = \lambda \quad \text{for } i = 1, 2, \ldots, n, \text{ and}$$

$$\sum_{i=1}^n \alpha_i = 1.$$

Solving for α_i in the first equation and substituting into the second equation determines that

$$\lambda = \frac{2}{\sum_{j=1}^n \frac{1}{\sigma_j^2}}.$$

Substituting this expression for λ into the first equation yields

$$\alpha_i = \frac{\frac{1}{\sigma_i^2}}{\sum_{j=1}^n \frac{1}{\sigma_j^2}} \quad \text{for } i = 1, 2, \ldots, n.$$

The reader can confirm that for each i, $\alpha_i \in [0, 1]$. $\qquad \square$

The previous discussion can be generalized yet again to the situation in which the portfolio of securities is financed with borrowed capital. Let $\mathbf{w} = \langle w_1, w_2, \ldots, w_n \rangle$ represent the portfolio of investments. As before, R_i will represent the rate of return on investing w_i in the ith security. For the sake of simplicity we will assume this is a one-period model in which investments are purchased by borrowing money which must be paid back at simple interest rate r. The net wealth generated by the portfolio financed with borrowed capital after one time period is

$$R(\mathbf{w}) = \sum_{i=1}^n w_i(1 + R_i) - (1 + r)\sum_{i=1}^n w_i = \sum_{i=1}^n w_i(R_i - r). \qquad (11.18)$$

The expected value of the net wealth generated by the portfolio and the variance in the net wealth are functions of the vector $\mathbf{w}$ and are defined to be respectively,

$$r(\mathbf{w}) = \mathrm{E}\left[R(\mathbf{w})\right] \qquad (11.19)$$
$$\sigma^2(\mathbf{w}) = \mathrm{Var}\left(R(\mathbf{w})\right). \qquad (11.20)$$

In this situation, the borrowed amounts $w_1, w_2, \ldots, w_n$ do not have to sum to unity, or to any other prescribed value.

Lemma 11.3 *Assuming the rates of return on the securities are uncorrelated, the optimal portfolio generating an expected unit amount of net wealth with the minimum variance in the return net wealth is*

$$\mathbf{w}^* = \left\langle \frac{\frac{r_1 - r}{\sigma_1^2}}{\sum_{j=1}^n \frac{(r_j - r)^2}{\sigma_j^2}}, \frac{\frac{r_2 - r}{\sigma_2^2}}{\sum_{j=1}^n \frac{(r_j - r)^2}{\sigma_j^2}}, \ldots, \frac{\frac{r_n - r}{\sigma_n^2}}{\sum_{j=1}^n \frac{(r_j - r)^2}{\sigma_j^2}} \right\rangle \qquad (11.21)$$

where $r_i = \mathrm{E}\left[R_i\right]$ and $\sigma_i^2 = \mathrm{Var}\left(R_i\right)$ for $i = 1, 2, \ldots, n$.

The proof of this lemma is similar to the proof of Theorem 11.7 and is left as an exercise.

The existence of the optimal portfolio $\mathbf{w}^*$ provided by Lemma 11.3 will be used in the following result known as the **Portfolio Separation Theorem**. The name is suggestive of the result which indicates to an investor how portions of a portfolio should be invested so as to minimize the variance in the wealth generated.

Theorem 11.8   *(Portfolio Separation Theorem) If b is any positive scalar, the variance of all portfolios with expected wealth generated equal to b is minimized by portfolio $b\mathbf{w}^*$ where $\mathbf{w}^*$ is described in Eq. (11.21).*

Proof.   Suppose $\mathbf{x}$ is a portfolio for which $r(\mathbf{x}) = b$, then

$$\frac{1}{b}r(\mathbf{x}) = r\left(\frac{1}{b}\mathbf{x}\right) \quad \text{(by exercise 18)}$$
$$= 1.$$

Thus we see that $\frac{1}{b}\mathbf{x}$ is a portfolio with unit expected rate of return. For the portfolio $\mathbf{w}^*$,

$$\sigma^2(b\mathbf{w}^*) = b^2\sigma^2(\mathbf{w}^*)$$
$$\leq b^2\sigma^2\left(\frac{1}{b}\mathbf{x}\right) \quad \text{(by Lemma 11.3)}$$
$$= \sigma^2(\mathbf{x}) \quad \text{(by exercise 18).}$$
$$\square$$

As a consequence of the Portfolio Separation Theorem all expected wealths generated by portfolios can be normalized to unity when determining the optimal portfolio.

As a special case of the preceding discussion the investor may wish to divide their portfolio between a risk-free investment such as a savings account and a "risky" investment such as a security. The risk-free interest rate will be symbolized by r_f. By assumption the rate of return on the risk-free investment is guaranteed, *i.e.* has a variance of zero. If the rate of return of the security is R_S and $x \in [0, 1]$ is invested in the security while $1 - x$ is placed in the risk-free investment, the rate of return on the portfolio is

$$R = (1 - x)r_f + xR_S,$$

which implies the expected rate of return for this portfolio is

$$\mathrm{E}[R] = (1 - x)\mathrm{E}[r_f] + x\mathrm{E}[R_S]$$
$$= (1 - x)r_f + xr_S$$
$$r = r_f + (r_S - r_f)x \tag{11.22}$$

To achieve a more compact notation we have set $r = \mathrm{E}[R]$ and $r_S = \mathrm{E}[R_S]$. Since the risk-free return is a constant, its expected value is this same constant value. The expected rate of return on the portfolio is a linear function of x. The line has its r-intercept at r_f and slope $r_S - r_f$. Thus if the expected rate of return of the security exceeds the risk-free rate of return, the slope of the line is positive and the expected rate of return of the portfolio increases with x, otherwise it decreases. The variance in the rate of return on the portfolio is

$$\text{Var}\,(R) = (1-x)^2 \text{Var}\,(r_f) + x^2 \text{Var}\,(R_S) + 2x(1-x)\text{Cov}\,(r_f, R_S)$$
$$\sigma^2 = x^2 \sigma_S^2 \tag{11.23}$$

where $\text{Var}\,(R) = \sigma^2$ and $\text{Var}\,(R_S) = \sigma_S^2$. The reader will note that since the rate of return on the risk-free investment is constant, its variance is zero and the covariance with the rate of return on the security also vanishes. The standard deviation of the return on the portfolio is also an increasing linear function of x. Equation (11.23) can be used to eliminate the parameter x from Eq. (11.22) and to derive a formula relating the expected rate of return and the standard deviation of the rate of return.

$$r = r_f + \frac{r_S - r_f}{\sigma_S}\sigma$$

Once again, if the expected rate of return on the security exceeds the rate of return of the risk-free investment, the expected rate of return of the portfolio increases with the standard deviation in the return on the portfolio. Thus an investor willing to accept a wider variation in the return on funds invested can expect a higher rate of return.

If the idealized single security of the previous discussion is replaced with an investment spread evenly throughout the securities market then the same form of linear relationship is determined, but the symbols r_M and σ_M will be used in place of r_S and σ_S respectively. This is known as the equation of the **capital market line**.

$$r = r_f + \frac{r_M - r_f}{\sigma_M}\sigma \tag{11.24}$$

The capital market line is illustrated in Fig. 11.6. Equation (11.24) can also be written as

$$\frac{r - r_f}{\sigma} = \frac{r_M - r_f}{\sigma_M}$$

which can be interpreted as saying that the excess rate of return (above the risk-free rate) normalized by the standard deviation in the rate of return for an investment spread evenly throughout the market is (naturally) the same as the normalized excess rate of return on the entire market.

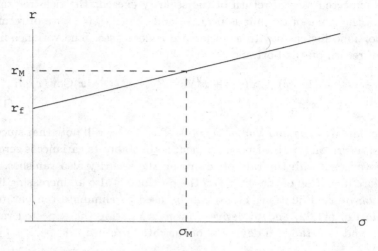

Fig. 11.6   The capital market line illustrates the linear relationship between the standard deviation in the return on a portfolio and the return on the portfolio.

Before moving on, the reader should reflect on the σr-plane. For a fixed level of expected return, risk-adverse investors will want to minimize the variance (or standard deviation) in the rate of return. A fixed level of return corresponds to a horizontal line in the σr-plane. Thus a risk-adverse investor would prefer a portfolio for which the standard deviation in the rate of return is as far to the left on the horizontal line as is feasible. The phrase "as is feasible" is used because it may not be possible to achieve a given rate of return with zero standard deviation. Referring to the illustration of the capital market line in Fig. 11.6, for a rate of return above r_f it is not feasible to have zero variance in the rate of return. Similarly, for a fixed standard deviation in the rate of return, an investor should prefer the highest expected return. Thus along a vertical line in the σr-plane an investor should prefer a portfolio as high on the line as is feasible. Again, for a fixed σ not all levels of r are achievable. These ideas could be extended a great deal and the interested reader is encouraged to consult [Luenberger (1998)].

Finally we arrive at a discussion of the main topic of this chapter, the **Capital Asset Pricing Model** (CAPM). This mathematical model will relate the return on the investment in an individual security to the return on the entire market. In this way an investor can weigh the return on the investment against the risk involved in the investment. The expression R_i will be the return on investment i and R_M will denote the return on the entire market. An investor will invest a portion x of his wealth in investment i and the remainder $1-x$ in the market. The following calculations will seem familiar to the reader. They are similar to those done when determining the optimal allocation of a portfolio between two securities. This time the second security is the entire market. The rate of return on the portfolio is given by

$$R = xR_i + (1-x)R_M,$$

which implies the expected rate of return on the portfolio and the variance in the expected rate of return are respectively,

$$E[R] = xE[R_i] + (1-x)E[R_M]$$
$$\text{Var}(R) = x^2\text{Var}(R_i) + (1-x)^2\text{Var}(R_M) + 2x(1-x)\text{Cov}(R_i, R_M).$$

Without further information we cannot assume that the rate of return for investment i is uncorrelated from the rate of return for the market. To simplify the notation we will make the following assignments: $r = E[R]$, $r_i = E[R_i]$, $r_M = E[R_M]$, $\sigma^2 = \text{Var}(R)$, $\sigma_i^2 = \text{Var}(R_i)$, and $\sigma_M^2 = \text{Var}(R_M)$. Furthermore we will replace $\text{Cov}(R_i, R_M)$ by its slightly more compact equivalent $\rho_{i,M}\sigma_i\sigma_M$.[1] Thus the equations above become

$$r = xr_i + (1-x)r_M \qquad (11.25)$$
$$\sigma^2 = x^2\sigma_i^2 + (1-x)^2\sigma_M^2 + 2x(1-x)\rho_{i,M}\sigma_i\sigma_M. \qquad (11.26)$$

Equations (11.25) and (11.26) can be thought of as the parametric form of a curve in the σr-plane. When $x = 0$, corresponding to the situation in which the entire portfolio is invested in the market, the parametric curve intersects the capital market line. In fact the parametric curve must be tangent to the capital market line at $x = 0$. If the intersection were transverse rather than tangential, then for some value of x near zero the parametric curve would lie above the capital market line and hence represent an infeasible expected return. See Fig. 11.7. Thus we can obtain a relationship between r_i, r_M, and r_f. Using Eq. (11.24) to obtain the slope of the capital market

---

[1]This is a small abuse of notation since we normally use $\rho(R_i, R_M)$ to denote the correlation of R_i and R_M. Here there is little chance of confusion.

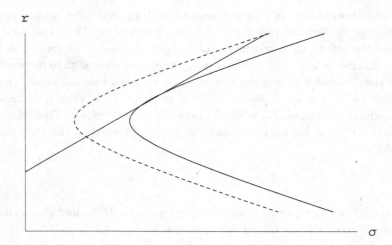

Fig. 11.7   The solid line shows once again the capital market line. The solid curve tangent to the capital market line represents feasible allocations of a portfolio. The dashed curve is infeasible since a portion of it lies above the capital market line, and hence would have a greater return with lower risk than the entire market.

line we have

$$\frac{r_M - r_f}{\sigma_M} = \left.\frac{dr}{d\sigma}\right|_{x=0}$$

$$= \frac{\left.\frac{dr}{dx}\right|_{x=0}}{\left.\frac{d\sigma}{dx}\right|_{x=0}}$$

$$= \frac{r_i - r_M}{\frac{\rho_{i,M}\sigma_i\sigma_M}{\sigma_M} - \sigma_M}.$$

Rewriting the last equation in the form

$$\frac{r_i - r_f}{\sigma_i} = \rho_{i,M}\frac{r_M - r_f}{\sigma_M}$$

allows us to interpret the equation as saying that the normalized excess rate or return on investment i is proportional to the normalized excess rate of return on the market, where the proportionality constant is the correlation in the rates of return on investment i and the market. Solving the last equation for $r_i - r_f$ yields

$$r_i - r_f = \frac{\rho_{i,M}\sigma_i}{\sigma_M}(r_M - r_f). \qquad (11.27)$$

The expression $\rho_{i,M}\sigma_i/\sigma_M$ is commonly denoted β_i and is referred to as the **beta** of security i. The Capital Asset Pricing Model, or CAPM for short, is the name given to Eq. (11.27). The CAPM can be interpreted as stating that the excess rate of return of security i above the risk-free interest rate is proportional to the excess rate of return of the market above the risk-free interest rate. The proportionality constant is β_i. For the case in which the rate of return on security i is uncorrelated with the rate of return on the market, the excess expected return will be zero. When the correlation is positive, excess positive expected return for the market implies excess positive expected return for security i. When the correlation is negative, excess positive expected return for the market will imply excess negative return for security i.

Example 11.7 Suppose the risk-free interest rate is 5% per year, the expected rate of return on the market is 9% per year, and the standard deviation in the return on the market is 15% per year. If the covariance in the expected returns on a particular stock and the market is 10% then

$$\beta = \frac{0.10}{(0.15)^2} = 4.44444.$$

Therefore the expected rate of return on the stock will be

$$r = 0.05 + 4.44444(0.09 - 0.05) = 0.22777,$$

or expressed as a percentage return, 22.78%.

The variance in the rate of return of a portfolio can be thought of as a measure of the risk in the portfolio. For a given level of variance in the rate of return, the investor will expect the highest expected value of the rate of return. Thus once again we turn to Lagrange Multipliers to solve this constrained optimization problem. The maximum and minimum value of expected rate of return will occur when

$$\frac{dr}{dx} = \lambda \frac{d(\sigma^2)}{dx} \tag{11.28}$$

subject to the constraint expressed by Eq. (11.26). The reader will be asked to verify in the exercises that the solution to this set of equations is

$$\lambda = \pm \frac{r_i - r_M}{2\sqrt{(\rho_{i,M}^2 - 1)\sigma_i^2 \sigma_M^2 + (\sigma_i^2 + \sigma_M^2 - 2\rho_{i,M}\sigma_i\sigma_M)\sigma^2}} \qquad (11.29)$$

$$x = \frac{\sigma_M^2 - \rho_{i,M}\sigma_i\sigma_M}{\sigma_i^2 + \sigma_M^2 - 2\rho_{i,M}\sigma_i\sigma_M}$$
$$\pm \frac{\sqrt{(\rho_{i,M}^2 - 1)\sigma_i^2\sigma_M^2 + (\sigma_i^2 + \sigma_M^2 - 2\rho_{i,M}\sigma_i\sigma_M)\sigma^2}}{\sigma_i^2 + \sigma_M^2 - 2\rho_{i,M}\sigma_i\sigma_M}. \qquad (11.30)$$

Under these conditions the maximum expected rate of return is

$$r = r_M - \frac{(r_i - r_M)\left(\rho_{i,M}\sigma_i\sigma_M - \sigma_M^2\right)}{\sigma_i^2 + \sigma_M^2 - 2\rho_{i,M}\sigma_i\sigma_M} \qquad (11.31)$$
$$\mp \frac{(r_i - r_M)\sqrt{(\rho_{i,M}^2 - 1)\sigma_i^2\sigma_M^2 + (\sigma_i^2 + \sigma_M^2 - 2\rho_{i,M}\sigma_i\sigma_M)\sigma^2}}{\sigma_i^2 + \sigma_M^2 - 2\rho_{i,M}\sigma_i\sigma_M}.$$

In most scenarios the variance in the rate of return for investment i will exceed the variance for the market. Thus we will assume that $\sigma_i > \sigma_M$. As two special cases, we will explore the return on the portfolio when the rates of return on the market and investment i are perfectly correlated, *i.e.* when $\rho_{i,M} = 1$. In this case

$$r = r_M - \frac{(r_i - r_M)(\sigma_M \pm \sigma)}{\sigma_i - \sigma_M}.$$

If the rate of return on investment i is anti-correlated with the rate of return on the market, then

$$r = r_M - \frac{(r_i - r_M)(-\sigma_M \pm \sigma)}{\sigma_i + \sigma_M}.$$

Example 11.8 Suppose the expected rate of return for investment i is $r_i = 0.07$ with standard deviation $\sigma_i = 0.25$ and the expected rate of return for the market is $r_M = 0.05$ with standard deviation $\sigma_M = 0.20$. Suppose the rates of return have a correlation of $\rho_{i,M} = 0.57$. If an investor is willing to hold a portfolio with a standard deviation in its rate of return at $\sigma = 0.22$ then the maximum rate of return can be found via the method outlined above. According to Eq. (11.31), the maximum rate of return on the portfolio is 0.0650248. This occurs when $x = 0.751242$ or when slightly more than 75% of investment capital is placed in investment i. The curve depicting r and x as parametric functions of σ is shown in Fig. 11.8.

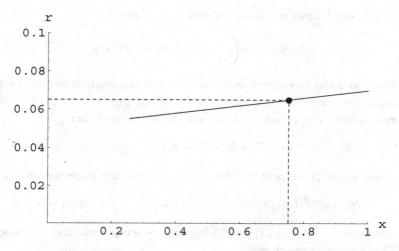

Fig. 11.8 This curve shows the relationship between the rate of return r, and the proportion of capital invested x.

11.7 Mean-Variance Analysis

This chapter concludes with a discussion of the stochastic process approach to determining the expected rate of return of a portfolio. If necessary, the reader may wish to review Chapter 5 before proceeding. In this section the portfolio will consist of positions in a security and a European call option on the security. The reader should not be confused by the portfolio. The investor will purchase a units of the security and b units of the call option, where either a or b could be positive or negative. For example, the investor might take a short position in the security and hedge this position with a long position in the call option. Likewise, a short position in the call option can be offset with a long position in the security.

We will assume the present value of the security is $S(0)$, the value of the option is C, the strike price is K, and the expiry date is T. The risk-free interest rate is r. Suppose the value of the security obeys a Brownian-type motion, stochastic process with drift parameter μ and volatility σ. In other words,

$$dS(t) = \mu S(t)\, dt + \sigma S(t)\, dW(t).$$

The reader will recall that Itô's Lemma (Lemma 5.4) implies the logarithm of the security follows a Wiener process of the form:

$$d(\ln S(t)) = \left(\mu - \frac{1}{2}\sigma^2\right) dt + \sigma \, dW(t).$$

At time $t = 0$ the investor devotes resources in the amount of $aS(0) + bC$ to create the portfolio. At the expiry date T the positions in the portfolio are zeroed out. The present value of any gain or loss is then

$$R = e^{-rT} \left(aS(T) + b(S(T) - K)^+\right) - (aS(0) + bC).$$

Using the linearity property of the expected value, the expected return is

$$\mathrm{E}\,[R] = e^{-rT} \left(a\mathrm{E}\,[S(T)] + b\mathrm{E}\,\left[(S(T) - K)^+\right]\right) - aS(0) - bC.$$

According to Eqs. (5.24) and (5.25) the expected value and variance of $\ln S(T)/S(0)$ are respectively

$$\mathrm{E}\,[\ln(S(T)/S(0))] = \left(\mu - \frac{1}{2}\sigma^2\right) T \qquad (11.32)$$

and

$$\mathrm{Var}\,(\ln(S(T)/S(0))) = \sigma^2 T. \qquad (11.33)$$

Now from the expressions in Eqs. (11.32) and (11.33) and by making use of Theorems 3.1 and 3.4 we have

$$\mathrm{E}\,[\ln S(T)] = \left(\mu - \frac{1}{2}\sigma^2\right) T + \ln S(0)$$
$$\mathrm{Var}\,(\ln S(T)) = \mathrm{Var}\,(\ln S(T)) = \sigma^2 T.$$

Thus by Lemma 3.1, $\mathrm{E}\,[S(T)] = S(0)e^{\mu T}$. Provided that $r < \mu$ the expression

$$e^{-rT}\mathrm{E}\,[S(T)] = S(0)e^{(-r+\mu)T} > S(0).$$

In order to determine a closed form expression for $\mathrm{E}\,[(S(T) - K)^+]$ we will make use of Corollary 3.2. $S(T)$ is a lognormal random variable with parameters $\ln S(0) + (\mu - \sigma^2/2)T$ (the drift parameter) and $\sigma\sqrt{T}$ (the volatility parameter). If we let

$$w = \frac{\ln(S(0)/K) + (\mu - \sigma^2/2)T}{\sigma\sqrt{T}}$$

then according to Corollary 3.2,

$$\mathrm{E}\left[(S(T) - K)^+\right] = S(0)e^{\mu T}\phi\left(w + \sigma\sqrt{T}\right) - K\phi(w).$$

Thus the expected value of the rate of return is

$$r = \mathrm{E}\left[R\right] \tag{11.34}$$
$$= aS(0)\left(e^{(\mu-r)T} - 1\right) + be^{-rT}\left(S(0)e^{\mu T}\phi\left(w + \sigma\sqrt{T}\right) - K\phi(w) - C\right).$$

The level sets of the expected return are lines in the ab-plane.

The variance in the rate of return can be calculated as

$$\begin{aligned}
\mathrm{Var}\,(R) &= \mathrm{Var}\left(e^{-rT}\left(aS(T) + b(S(T) - K)^+\right)\right) \\
&= e^{-2rT}\mathrm{Var}\left(aS(T) + b(S(T) - K)^+\right) \\
&= e^{-2rT}\left[a^2\mathrm{Var}\,(S(T)) + b^2\mathrm{Var}\left((S(T) - K)^+\right)\right. \\
&\quad\left. + 2ab\mathrm{Cov}\left(S(T), (S(T) - K)^+\right)\right]. \tag{11.35}
\end{aligned}$$

Referring to the second part of Lemma 3.1 we have

$$\mathrm{Var}\,(S(T)) = S^2(0)e^{2\mu T}\left(e^{\sigma^2 T} - 1\right).$$

According to Corollary 3.4,

$$\begin{aligned}
\mathrm{Var}\left((S(T) - K)^+\right) &= S^2(0)e^{(2\mu+\sigma^2)T}\phi(w + 2\sigma\sqrt{T}) \\
&\quad - 2KS(0)e^{\mu T}\phi(w + \sigma\sqrt{T}) + K^2\phi(w) \\
&\quad - \left(S(0)e^{\mu T}\phi(w + \sigma\sqrt{T}) - K\phi(w)\right)^2.
\end{aligned}$$

Lemma 11.2 implies that

$$\begin{aligned}
\mathrm{Cov}&\left(S(T), (S(T) - K)^+\right) \\
&= KS(0)e^{\mu T}\left(\phi(w) - \phi(w + \sigma\sqrt{T})\right) \\
&\quad + S^2(0)e^{2\mu T}\left(e^{\sigma^2 T}\phi(w + 2\sigma\sqrt{T}) - \phi(w + \sigma\sqrt{T})\right).
\end{aligned}$$

Putting these last three results together yields, at long last, the complicated, but complete, expression for Var (R).

$$\begin{aligned}
\text{Var}\,(R) = e^{-2rT}\Big[&a^2 S^2(0)e^{2\mu T}\left(e^{\sigma^2 T}-1\right) \\
&+ b^2\left(S^2(0)e^{(2\mu+\sigma^2)T}\phi(w+2\sigma\sqrt{T}) - 2KS(0)e^{\mu T}\phi(w+\sigma\sqrt{T})\right. \\
&+ K^2\phi(w) - \left.\left(S(0)e^{\mu T}\phi(w+\sigma\sqrt{T}) - K\phi(w)\right)^2\right) \\
&+ 2ab\left(S^2(0)e^{2\mu T}\left[e^{\sigma^2 T}\phi(w+2\sigma\sqrt{T}) - \phi(w+\sigma\sqrt{T})\right]\right. \\
&+ \left.\left. KS(0)e^{\mu T}\left[\phi(w) - \phi(w+\sigma\sqrt{T})\right]\right)\right]
\end{aligned} \tag{11.36}$$

The level sets of Var (R) are parabolas in the ab-plane.

Example 11.9 Suppose a share of a security has a current price of \$100 while the drift parameter and volatility of the price of the security are respectively 0.08 and 0.21 per annum. The risk free interest rate is 3.5% annually. The value of a European call option for a strike price of \$102 with an exercise time of 12 months on the security is \$9.075. An investor can use the preceding analysis to determine that the expected rate of return on a portfolio consisting of a position x in the security and a position y in the call option is given by Eq. (11.34) as

$$E\,[R] = 4.60279x + 3.12945y.$$

The line of zero return is given by the equation $y = -1.4708x$. Thus at a point in the second or fourth quadrant above this line the expected return will be positive. We look for a point in the second or fourth quadrant since typically x and y must be of opposite sign due to the fact that the investor will take opposite positions in the security and the option. A density plot in Fig. 11.9 shows the return in various subsets of the plane. If the investor decides that a return of $E\,[R] = 2$ is desirable they will then select x and y so that the variance in the return is minimized over all portfolios for which $E\,[R] = 2$. This is yet another constrained optimization exercise. According to Eq. (11.36) the variance in the return is

$$\text{Var}\,(R) = 493.329x^2 + 681.025xy + 271.757y^2$$

Rather than use the method of Lagrange multipliers to find the minimum of Var (R) we will take the more elementary approach of solving the equation $2 = 4.60279x + 3.12945y$ for y and substituting into the expression for

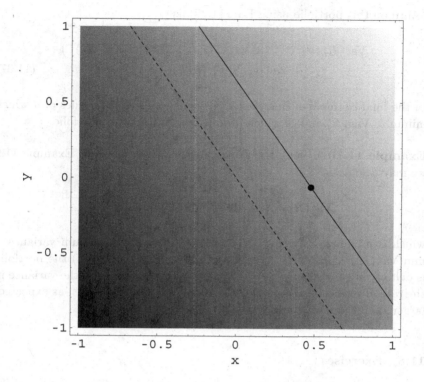

Fig. 11.9   The expected return of the portfolio is positive above the dashed diagonal line shown in the density plot. The solid line illustrates the locus of points at which the expected return equals 2.

variance. In this case we get the quadratic expression

$$\text{Var}\,(R) = 110.995 - 75.6514x + 79.5567x^2$$

which is minimized when $x = 0.475456$. Under these conditions $y = -0.0602088$ and the minimum variance is $\text{Var}\,(R) = 93.0107$.

It is instructive to understand that constructing a portfolio of minimum variance is different than constructing a hedged portfolio. Recall that a portfolio is delta neutral if for every option in the portfolio (either a long or short position) Δ units of the underlying security are held (in the opposite position). Considering a portfolio of the form $\mathcal{P} = C - aS$ (with $a > 0$ for the sake of simplicity) then we may rewrite the variance in the rate of return on this portfolio using Eq. (11.35) as

$$\text{Var}\,(R) = e^{-2rT}\left[a^2\text{Var}\,(S(T)) + \text{Var}\,\left((S(T) - K)^+\right)\right.$$
$$\left. - 2a\text{Cov}\,\left(S(T), (S(T) - K)^+\right)\right]. \tag{11.37}$$

In the final example of this chapter we will compare the values of a which minimize $\text{Var}\,(R)$ and Δ which produce a delta neutral portfolio.

Example 11.10   Using the same values as those given in Example 11.9 we may calculate

$$\text{Var}\,(R) = 493.329a^2 - 681.025a + 271.757$$

which is minimized when $a = 0.690234$. In fact the minimum variance is $\min(\text{Var}\,(R)) = 36.7241$. According to Eqs. (8.36) and (9.8) the portfolio is delta neutral when $a = \Delta = 0.570391$. For this value of a the variance in the rate of return on the portfolio is $\text{Var}\,(R) = 43.8095$ and is, as expected, larger than the minimum variance.

11.8   Exercises

(1) Using the definition of covariance and variance prove the first two statements of Theorem 11.1.
(2) Show that for any two random variables X and Y

$$\text{Var}\,(X + Y) = \text{Var}\,(X) + \text{Var}\,(Y) + 2\text{Cov}\,(X, Y).$$

If X and Y are independent random variables, show this formula reduces to the result mentioned in Theorem 2.7.
(3) Fill in the remaining details of the proof of Lemma 11.1 for the cases in which $\text{E}\,[X^2] = 0$, $\text{E}\,[X^2] = \infty$, $\text{E}\,[Y^2] = 0$, or $\text{E}\,[Y^2] = \infty$
(4) The data shown in the table below was originally published in the *New York Times*, Section 3, pg. 1, 05/31/1998. It lists the name of a corporate CEO, the CEO's corporation, the CEO's golf handicap, and a rating of the stock of the CEO's corporation. Determine the covariance and correlation between the CEO's golf handicap, and the corporation's stock rating.

Name	Corp.	Handicap	Rating
Terrence Murray	Fleet Financial	10.1	67
William T. Esrey	Sprint	10.1	66
Hugh L. McColl Jr.	Nationsbank	11	64
James E. Cayne	Bear Stearns	12.6	64
John R. Stafford	Amer. Home Prod.	10.9	58
John B. McCoy	Banc One	7.6	58
Frank C. Herringer	Transamerica	10.6	55
Ralph S. Larsen	Johnson&Johnson	16.1	54
Paul Hazen	Wells Fargo	10.9	54
Lawrence A. Bossidy	Allied Signal	12.6	51
Charles R. Shoemate	Bestfoods	17.6	49
James E. Perrella	Ingersoll-Rand	12.8	49
William P. Stiritz	Ralston Purina	13	48

(5) The table below lists the names, heights, and weights of a sample of *Playboy* centerfolds [Dean *et al.* (2001)]. This data will be used to illustrate the concept of covariance. Let X represent height and Y represent weight for each woman.

Name	Height (in.)	Weight (lb.)
Allias, Henriette	68.5	125
Anderson, Pamela	67	105
Broady, Eloise	68	125
Butler, Cher	67	123
Clark, Julie	65	110
Sloan, Tiffany	66	120
Stewart, Liz	67	116
Taylor, Tiffany	67	115
Witter, Cherie	69	117
York, Brittany	66	120

Determine the covariance and correlation between the heights and weights listed.

(6) Consider the match paired data in the table below.

X	0	1	1	1	1	1	1	1	1	1
Y	y	1	2	3	4	5	6	7	8	9

Show that while the functional relationship between X and Y is not linear, as $y \to -\infty$, the correlation between X and Y approaches $9/10$.

(7) Suppose that a company sells a European call option on a security. The standard deviation on the price of the call option is $\sigma_O = 0.045$. The company hedges its position by buying a different security for which the standard deviation in value is $\sigma_D = 0.037$. The correlation between the values of the two instruments is $\rho(O, D) = 0.86$. What is the optimal ratio of the number of shares of security purchased to the number of options sold so that the variance in the value of the portfolio is minimized?

(8) Show using the concepts of the stochastic calculus that the variance in the change in value of the portfolio $\mathcal{P} = C - nS$, where C is a European call option on the underlying security S, is minimized when $n = \Delta$, the Delta of the call option.

(9) Which of the following utility functions are concave on their domain?

    (a) $u(x) = \ln x$

    (b) $u(x) = (\ln x)^2$

    (c) $u(x) = \tan^{-1} x$

(10) The mean of a discrete random variable as defined in Eq. (2.4) is sometimes called the **arithmetic mean**. The **harmonic mean** is defined as

$$\mathcal{H} = \frac{1}{\sum_X \frac{P(X)}{X}}. \tag{11.38}$$

The **geometric mean** is defined as

$$\mathcal{G} = \prod_X X^{P(X)}. \tag{11.39}$$

Use Jensen's inequality (11.15) to show that $\mathcal{H} \leq \mathcal{G} \leq \mathrm{E}[X]$ and that equality holds only when $X_1 = X_2 = \cdots = X_n$. You may assume that all the random variables are positive.

(11) Verify inequality (11.16) for $f(x) = \tanh x$ and $\phi(t) = t$.

(12) Find the certainty equivalent for the choice between (a) flipping a fair coin and either winning \$10 or losing \$2, or (b) receiving an amount C with certainty. Assume that your utility function is $f(x) = x - x^2/50$.

(13) Suppose you must choose between receiving an amount C with certainty or playing a game in which a fair die is rolled. If a prime number results you win \$15, but if a non-prime number results you lose that amount of money. Assume that your utility function is $f(x) = x - x^2/2$.

(14) Suppose a risk-averse investor with utility function $u(x) = \ln x$ will invest a proportion α of their total capital x in an investment which will

pay them either $2\alpha x$ with probability p or nothing with probability $1-p$. The amount of capital not invested will earn interest for one time period at the simple rate $r = 11\%$. What proportion of their capital should the investor allocate if they wish to achieve the maximum expected utility?

(15) Consider the function $u(x) = 1 - e^{-bx}$ where $b > 0$.

(a) Show that $u(x)$ is concave.
(b) Show that $0 \le x_1 < x_2$ implies $u(x_1) < u(x_2)$.

(16) Suppose an investor whose utility function is $u(x) = 1 - e^{-x/100}$ has a total of \$1000 to invest in two securities. The expected rate of return for the first investment is $r_1 = 0.08$ with a standard deviation in the rate of return of $\sigma_1 = 0.03$. The expected rate of return for the second investment is $r_2 = 0.13$ with a standard deviation in the rate of return of $\sigma_2 = 0.09$. The correlation in the rates of return is $\rho = -0.26$. What is the optimal amount of money to place into each investment?

(17) Suppose that an investor will split a unit of wealth between the following securities possessing variances in their rates of return as listed in the table below. Assuming that the rates of return are uncorrelated, determine the proportion of the portfolio which will be allocated to each security so that the variance in the returned wealth is minimized.

Security	σ^2
A	0.24
B	0.41
C	0.27
D	0.16
E	0.33

(18) For any real scalar c show that for $r(\mathbf{w})$ and $\sigma^2(\mathbf{w})$ as defined in Eqs. (11.19) and (11.20) the following results hold.

$$r(c\mathbf{w}) = cr(\mathbf{w})$$
$$\sigma^2(c\mathbf{w}) = c^2\sigma^2(\mathbf{w})$$

(19) Prove Lemma 11.3.

(20) In this exercise we will extend the work that was done in exercise 17 by determining the optimal portfolio which minimizes the variance in the return under the condition that the portfolio is financed with money borrowed at the interest rate of 11%.

Security	r	σ^2
A	0.13	0.24
B	0.12	0.41
C	0.15	0.27
D	0.11	0.16
E	0.17	0.33

(21) Verify that the solutions to the set of Eqs. (11.26) and (11.28) are given by the values of λ and x in (11.29) and (11.30).

(22) Suppose the risk-free interest rate is 4.75% per year, the expected rate of return on the stock market is 7.65% per year, and the standard deviation in the return on the market is 22% per year. If the covariance in the expected returns on a particular stock and the market is 15%, determine the expected rate of return on the stock.

(23) Suppose a share of a security has a current price of $83 while the drift parameter and volatility of the price of the security are respectively 0.13 and 0.25 per annum. The risk free interest rate is 5.5% annually. Find the value of 3-month European call option on the security with a strike price of $86. Use the mean-variance analysis of Sec. 11.7 to determine the positions of the optimal portfolio with an expected return of $10.

Chapter 12

American Options

For the most part the preceding chapters have been concerned with European-style options, their properties, and their uses in hedging. For the reader who has come this far, the final chapter will explore the pricing and properties of American-style options. Recall that American-style options may be exercised at any time up to expiry. In this chapter we will see there are important differences between the European- and American-style options as well as nearly paradoxical similarities. We will develop a means of approximating the values of American puts and calls using the binomial approach outlined in Section 8.5. Before tackling the pricing procedure for American-style options, students should understand the binomial model presented in the earlier section.

12.1 Parity and American Options

In this chapter we will use the subscripts "a" and "e" to distinguish between American and European options respectively. Unless otherwise stated the underlying security is assumed to pay no dividends. Earlier an arbitrage argument was used to show that American and European options on the same underlying security with the same strike price and expiry time obey the following inequalities (see Section 7.1 and exercise 1 in that same chapter).

$$C_e \leq C_a \quad \text{and} \quad P_e \leq P_a$$

Intuitively these inequalities hold because the American options give the holder all the rights of the European options with the addition of the possibility of early exercise. There are market scenarios under which an American option may be exercised to yield a profit while the equivalent European option would expire out of the money. See Figure 12.1. Consider an Amer-

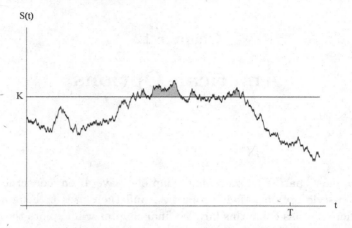

Fig. 12.1   The shaded region indicates the intervals before expiry in which the price of the underlying security $S(t)$ exceeds the strike price K. An American call option could be exercised profitably in these intervals, but a European call option could not, and would ultimately expire unused.

ican call versus a European call on the same underlying security with the same strike price K and expiry time T. If the random walk of the underlying security S is such that at a time $0 \le t < T$ it is the case that $S(t) - K - C_a > 0$, the American option could be exercised at time t and generate a positive profit. Since $t < T$ the European option could not be exercised. The European option would only generate a positive profit if $S(T) - K - C_e > 0$ and there is no guarantee that will occur. The European put and call prices also obey the Put-Call Parity Formula expressed in Eq. (7.1). However, the American options do not obey a Put-Call Parity Formula. In this section we will develop some bounds on the prices of the American options and explore a surprising relationship between C_a and C_e.

Assume that the American options are created at time $t = 0$ when the price of the underlying security is S. The risk-free interest rate is r compounded continuously. If the strike price is K then the values of the American put and call options with identical strike price and expiry $T > 0$ obey the following inequality.

$$C_a + K \ge S + P_a \tag{12.1}$$

To prove this, assume to the contrary that $C_a + K < S + P_a$. In this case an investor could short the security, sell the put, and buy the call.

This produces a cash flow of $S + P_a - C_a$. If this amount is positive it can be invested at the risk-free rate, otherwise it is borrowed at the risk-free rate. If the holder of the American put chooses to exercise the put at time $0 \leq t \leq T$, the investor can exercise the call option and purchase the security for K. At that time the investor's balance is

$$(S + P_a - C_a)e^{rt} - K > Ke^{rt} - K \geq 0.$$

If the American put expires out of the money, the investor will close their short position in the security at time T by exercising the call option. In this case

$$(S + P_a - C_a)e^{rT} - K > Ke^{rT} - K > 0.$$

Thus the investor receives a non-negative profit in either case, violating the principle of no arbitrage. Consequently the inequality in (12.1) holds.

Now it is also the case that

$$S + P_a \geq C_a + Ke^{-rT}. \tag{12.2}$$

A proof by contradiction and a no arbitrage argument will establish this inequality as well. Suppose $S + P_a < C_a + Ke^{-rT}$. An investor could sell an American call and buy the security and the American put. This generates a cash flow of $C_a - S - P_a$ at $t = 0$. If necessary the investor will borrow funds at the risk-free rate r compounded continuously. If the holder of the call decides to exercise it at any time $0 \leq t \leq T$, the investor may sell the security for the strike price K. Thus at time t the investor's asset balance is

$$(C_a - S - P_a)e^{rt} + K = (C_a + Ke^{-rt} - S - P_a)e^{rt}$$
$$\geq (C_a + Ke^{-rT} - S - P_a)e^{rt}$$

since $r > 0$. By assumption $S + P_a < C_a + Ke^{-rT}$, so the last expression above is positive. The investor has earned a risk-less positive profit, but this contradicts the no arbitrage assumption. Therefore the inequality in (12.2) is true.

By rearranging terms in (12.1) and (12.2) and combining the two inequalities we have proved the following theorem.

Theorem 12.1 *If the risk-free interest rate is r compounded continuously, if C_a and P_a are the values of American call and put options respectively both with strike price K and expiry T, and if the value of the* *underlying security is S, then*

$$S - K \leq C_a - P_a \leq S - Ke^{-rT}. \tag{12.3}$$

Perhaps one of the most unexpected results governing the value of American-style options is the equality holding between C_e and C_a. We have seen that $C_a \geq C_e$ for American and European call options with the same strike price, expiry time, and underlying security. We have also seen a simulation of an example for which the American call is in the money for a period before expiry and could be exercised to generate a positive profit while the European option would be out of the money at expiry (Figure 12.1). Since the American call, through possible early exercise, may generate a greater profit than its European counterpart, it is surprising that $C_a = C_e$. However, the usual type of no-arbitrage proof will establish the equality. Suppose that $C_a > C_e$. Since the American call is (supposedly) worth more, an investor could sell the American call and buy a European call with the same strike price K, expiry date T, and underlying security. The net cash flow $C_a - C_e > 0$ would be invested at the risk-free rate r. Assuming the interest is compounded continuously, at any time $0 \leq t \leq T$, the amount due is $(C_a - C_e)e^{rt}$. If the holder of the American call chooses to exercise the option at some time $t \leq T$, the investor may sell short a share of the security for amount K and add the proceeds to the amount invested at the risk-free rate. At time T the investor must close out the short position in the security and may use the European option to do so. Upon settlement of the short position the holdings of the investor are

$$(C_a - C_e)e^{rT} + K(e^{r(T-t)} - 1) > 0.$$

If the American option is not exercised, the investor has holdings of $(C_a - C_e)e^{rT} > 0$ at expiry. A rational investor will only exercise the European option if profit is increased, else the European option will be allowed to expired unused. In either case the investor earns a risk-free positive profit. This completes the proof of the next theorem.

Theorem 12.2 *If C_a and C_e are the values of American and European call options respectively on the same underlying security with identical strike prices and expiry times, then*

$$C_a = C_e. \tag{12.4}$$

Now that we have a means of determining the value of an American call on a non-dividend paying security we may use it to establish a bound on the value of an American call on the same underlying security with the same expiry and strike price.

Example 12.1 The price of a security is currently \$36, the risk-free interest rate is 5.5% compounded continuously, and the strike price of a six-month American call option worth \$2.03 is \$37. The range of no arbitrage values of a six-month American put on the same security with the same strike price can be found by making use of the inequality in (12.3).

$$36 - 37 \leq 2.03 - P_a \leq 36 - 37e^{-0.055(6/12)}$$
$$2.03 \leq P_a \leq 3.03$$

The American put may be worth more than the European put (all parameters of the option being equal) due to the possibility of early exercise. In the next section we will develop a binomial model approximation to the value of an American put.

12.2 American Puts Valued by a Binomial Model

As earlier in Section 8.5 we will assume for the purposes of pricing an American put on a non-dividend paying security that the

- strike price of the American put is K,
- expiry date of the American put is $T > 0$,
- price of the security at time t with $0 \leq t \leq T$ is $S(t)$,
- continuously compounded risk-free interest rate is r, and
- price of the security follows a geometric Brownian motion with variance σ^2.

If the interval $[0, T]$ is divided into n subintervals of length $\Delta t = T/n$ where $n \in \mathbb{N}$, we will assume that at time $k\Delta t$ with $k = 0, 1, \ldots, n - 1$ the security price is $S(k\Delta t)$. At time $(k + 1)\Delta t$ the price from the previous time step may have increased by a factor $u > 1$ or decreased by a factor $0 < d < 1$. The probability of an increase is p and naturally that is a decrease is $1 - p$. The discrete dynamics of the movement of the price of the security should approximate geometric Brownian motion for which the volatility in the security price is σ^2. Following the same line of reasoning as in Section 8.5 we obtain the following relationships.

$$u = \frac{1}{2}\left(e^{-r\Delta t} + e^{(r+\sigma^2)\Delta t} + \sqrt{\left(e^{-r\Delta t} + e^{(r+\sigma^2)\Delta t}\right)^2 - 4}\right) \quad (12.5)$$

$$d = \frac{1}{2}\left(e^{-r\Delta t} + e^{(r+\sigma^2)\Delta t} - \sqrt{\left(e^{-r\Delta t} + e^{(r+\sigma^2)\Delta t}\right)^2 - 4}\right) \quad (12.6)$$

$$p = \frac{e^{r\Delta t} - d}{u - d} \quad (12.7)$$

These values of the parameters of the discrete model insure that the discrete model is arbitrage-free and that the discrete version of the security has the same expectation and variance as the continuous version.

An American put is always worth at least as much as the payoff generated by immediate exercise. Thus $P_a(t) \geq (K - S(t))^+$ for any $0 \leq t \leq T$. The quantity $(K - S(t))^+$ is called the **intrinsic value** of the American put. Since we will make frequent reference to the intrinsic value of the option, we will define the function $Q(t) = (K - S(t))^+$ and use it where the intrinsic value is needed. The reader should bear in mind that the intrinsic value represents the payoff of an in the money put option. To develop a pricing formula for the American put, consider the coarsest possible time discretization of the interval $[0, T]$. For the moment we will assume the American put can only be exercised at $t = 0$ or at $t = T$. The value of the underlying security would evolve along one of the two branches shown in Fig. 12.2. The corresponding intrinsic values of the American put are shown in Fig. 12.3. At expiry the value of an American put is its intrinsic value. Thus we always have the relationship $P_a(T) = Q(T)$, assuming $t = T$ is the expiration date of the option. In this single step model, the American put will be worth the greater of its intrinsic value at $t = 0$ or the present value of the expected value of the American put at $t = T$. This can be expressed as

$$P_a(0) = \max\left\{Q(0), e^{-rT}\left[p(K - uS(0))^+ + (1 - p)(K - dS(0))^+\right]\right\}$$
$$= \max\left\{Q(0), e^{-rT}E\left[(K - S(T))^+\right]\right\}.$$

Consequently if the intrinsic value of the put at time $t = 0$ is greater than the present value of the expected value of the option at time $t = T$, the put could be exercised early. This approach to pricing the American put can be generalized to a multi-step evolution of the security price.

Suppose the interval $[0, T]$ has been partitioned into n subintervals of length $\Delta t = T/n$. The partition consists of the points $t_0 < t_1 < \cdots < t_n$

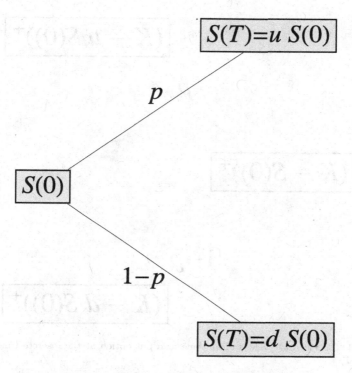

Fig. 12.2   A single-step discrete approximation to the evolution of the value of a security.

where $t_i = i\Delta t$. For convenience we will assign $Q(\infty) = 0$, indicating that an expired option has zero value. If $i = n$ then

$$P_a(n\Delta t) = P_a(T) = Q(T), \tag{12.8}$$

i.e. the value of the option is the intrinsic value. If $i = n-1$ then the value of the put is defined to be the greater of the intrinsic value at $t = (n-1)\Delta t$ and the present value at $t = (n-1)\Delta t$ of the expected value of the put at $t = n\Delta t$. In other words

$$P_a((n-1)\Delta t) = \max\{Q((n-1)\Delta t), e^{-r\Delta t}\mathrm{E}\,[Q(n\Delta t)]\}$$
$$= \max_{j\in\{n-1,n,\infty\}}\{e^{-r(j-n+1)\Delta t}Q(j\Delta t)\}.$$

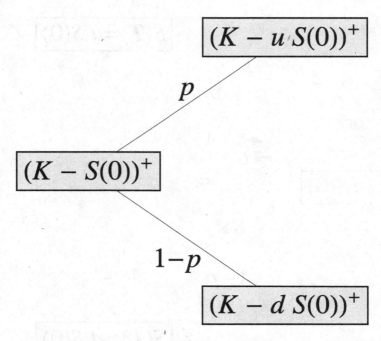

Fig. 12.3   The intrinsic values of an American put option at the discrete times $t = 0$ and $t = T$.

This relationship may be generalized to define the value of the American put at any time value in the partition of $[0, T]$. For $0 \le i < n$ we define

$$P_a(i\Delta t) = \max_{j \in \{i, i+1, \ldots, n, \infty\}} \{e^{-r(j-i)\Delta t} \mathrm{E}\left[Q(j\Delta t)\right]\}. \qquad (12.9)$$

Therefore, prior to expiry, the value of the American put is the greater of its intrinsic value and the discounted value of its expected value at the next time step. Equations (12.8) and (12.9) constitute a recursive algorithm for pricing an American put option. A binomial lattice of values of the underlying security determines the intrinsic values of the American put. At expiry the value of the put is given by Eq. (12.8). At every time step prior to the expiration of the option, the value of the American put is determined by Eq. (12.9).

Before exploring some of the properties of the binomial lattice formula for the price of an American put, we will work through the details of a two-step binomial model.

Example 12.2 Suppose the current price of a security is $32, the risk-free interest rate is 10% compounded continuously, and the volatility of Brownian motion for the security is 20%. The price of a two-month American put with a strike price of $34 on the security can be found as outlined below. The length of a time step will be $\Delta t = 1/12$. The parameters u and d given in Eqs. (12.5) and (12.6) and governing the proportional increase and decrease in the price of the security are

$$u \approx 1.0603 \quad \text{and} \quad d \approx 0.9431.$$

The probability of an increase in the price of the security occurring between time steps is given by the formula in Eq. (12.7).

$$p \approx 0.5567$$

The time until expiry will be divided into two single-month time steps and a binomial lattice of security prices will be created. The binomial lattice of security values is shown in Fig. 12.4. Now we may create a corresponding binomial lattice of intrinsic values for the American put. Each node in this lattice is merely the positive part of the payoff generating by exercise of the put. These intrinsic values are shown in Fig. 12.5.

Now we can begin to recursively calculate the values of the American put at the nodes of the binomial lattice. At the leaf (right-most) nodes, the price of the option is its intrinsic value according to Eq. (12.8). Thus a partial lattice of put values showing only those at expiry is shown in Fig. 12.6.

At $t = 1/12$, for the upper of the two nodes in the binomial lattice at this time step, the value of the American put according to Eq. (12.9) is

$$P_a(1/12) = \max\{(34 - 33.9316)^+, e^{-0.10/12}(p(0) + (1-p)2)\}$$
$$= \max\{0.0683805, 0.879248\}$$
$$= 0.879248.$$

In a similar fashion we may calculate the value of the American put at the lower of the two nodes in the lattice at $t = 1/2$. The partial binomial lattice of values of the American put for $t = 1/12$ and $t = 2/12$ is shown in Fig. 12.7.

Finally the value of the American put at $t = 0$ is

$$P_a(0) = \max\{(34 - 32)^+, e^{-0.10/12}[p(0.879248) + (1-p)(3.82166)]\}$$
$$= 2.16551.$$

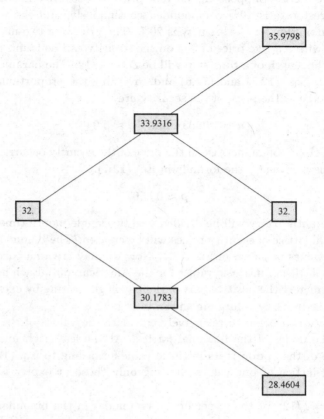

Fig. 12.4   A two-step discrete approximation to the evolution of the value of a security.

The binomial lattice of values of the American put is shown in Fig. 12.8.

The next section contains some results describing the properties of the American put option. One simple property is evident from the formulas for the put detailed in Eqs. (12.8) and (12.9), namely the value of the American put is at least as great as the intrinsic value of the option.

$$P_a(t) \geq Q(t)$$

A binomial lattice providing a side-by-side comparison of the value of the American put and the intrinsic value of the option for the previous example

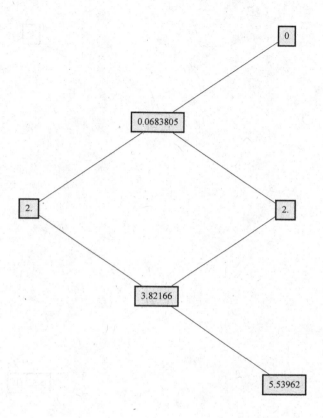

Fig. 12.5 The intrinsic values of an American put with a strike price of $34 on the security shown in Fig. 12.4.

is shown in Fig. 12.9.

12.3 Properties of the Binomial Pricing Formula

The pricing algorithm for an American-style put option given in Eqs. (12.8) and (12.9) is simple to implement in a spreadsheet or computer program. Its simplicity may hide some of the mathematical properties of the value of the option. In this section, brief descriptions and justifications of some

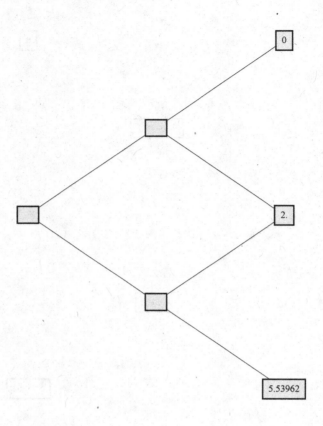

Fig. 12.6 The values of an American put with a strike price of $34 at expiry on the security shown in Fig. 12.4.

of the properties of the put option will be presented. These results will be helpful in deciding the optimal time to exercise an American put option.

Lemma 12.1 *Suppose the value of the security underlying an American put option follows a path through the n-step binomial lattice such that for some $i \in \{0, 1, \ldots, n-1\}$ we have $(K - dS(t_i))^+ = 0$, then $Q(t_i) = 0$.*

Proof. Since $d < 1 < u$ we have $dS(t_i) < S(t_i) < uS(t_i)$ which implies

$$(K - uS(t_i)) < (K - S(t_i)) < (K - dS(t_i)),$$

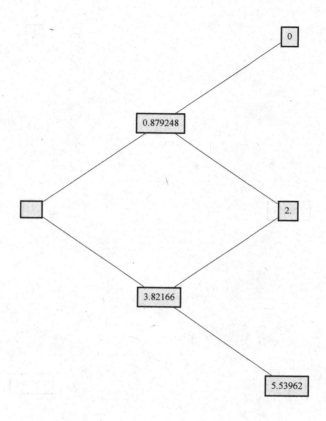

Fig. 12.7   More values of an American put with a strike price of $34 at expiry on the security shown in Fig. 12.4.

from which it follows that

$$(K - uS(t_i))^+ \leq (K - S(t_i))^+ \leq (K - dS(t_i))^+.$$

We have assumed $(K - dS(t_i))^+ = 0$, which implies $(K - S(t_i))^+ = Q(t_i) = 0$. □

This lemma can be thought of as stating that if the intrinsic value of an American put at the next downward movement of the price of the underlying security would be zero, then the current intrinsic value of the option is zero as well. Likewise the intrinsic value of the American put

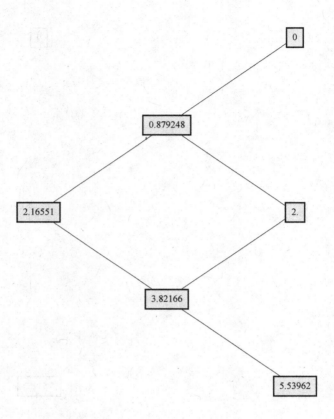

Fig. 12.8   The values of the American-style puts corresponding to the values of the securities in Fig. 12.4.

at higher values of the security price would be zero. The following lemma reveals the effect this has on the value of the value of the option.

Lemma 12.2   *Suppose the value of the security underlying an American put option follows a path through the n-step binomial lattice such that for some $i \in \{0, 1, \ldots, n-1\}$ we have $(K - dS(t_i))^+ = 0$, then $P_a(t_i) = e^{-r\Delta t}P_a(t_{i+1})$.*

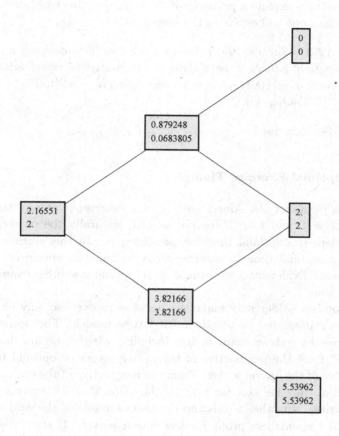

Fig. 12.9 The top number in each box represents the price of the American put option for the previous example while the bottom number is the intrinsic value of the option.

Proof. By Lemma 12.1 $Q(t_i) = 0$ and by Eq. (12.9)

$$P_a(t_i) = \max_{j \in \{i, i+1, \ldots, n, \infty\}} \{e^{-r(j-i)\Delta t} \mathrm{E}[Q(t_j)]$$

$$= \max_{j \in \{i+1, i+2, \ldots, n, \infty\}} \{e^{-r(j-i)\Delta t} \mathrm{E}[Q(t_j)]$$

$$= e^{-r\Delta t} \max_{j \in \{i+1, i+2, \ldots, n, \infty\}} \{e^{-r(j-[i+1])\Delta t} \mathrm{E}[Q(t_j)]$$

$$= e^{-r\Delta t} P_a(t_{i+1}).$$

$\square$

Now we may explore a property of the intrinsic value which holds when the American put will expire in the money.

Lemma 12.3 *Suppose the value of the security underlying an American put option follows a path through the n-step binomial lattice such that for some $i \in \{0, 1, \ldots, n-1\}$ we have $(K - uS(t_i))^+ > 0$, then $Q(t_i) > e^{-r\Delta t}\mathrm{E}\left[Q(t_{i+1})\right]$.*

Proof.   See exercise 5. $\square$

12.4   Optimal Exercise Time

Since the owner of the American put may exercise it at any time step between $t = 0$ and $t = T$ (or not at all), naturally the owner will be interested in the optimal time for exercising it. In this section we will determine optimal time to exercise an American put option on a stock that pays no dividends. The optimal exercise time is another example of a stopping time.

The option holder may make the decision to exercise only on the information represented in the binomial lattice model. The optimal time for exercise depends on many factors including whether we are thinking of "optimal" from the perspective of the option owner or optimal from the perspective of the option writer. From the perspective of the option holder, the optimal time for exercise may be the worst time for exercise for the option writer. In other words the optimal exercise for the option owner may yield the smallest profit for the option writer. If the option seller has determined the price of the put correctly, then even in the worst case scenario, there should be no arbitrage opportunity for the option buyer. For the option seller the optimal outcome may be for the option to expire unused. In Chapter 11 the concept of the utility function was introduced. The holder of an American option may act so as to maximize the expected utility of the option, which could lead to early exercise of both puts and calls.

To keep things simple we will consider for the moment only the value of the American option and not the investor's utility function. As long as the value of the American put exceeds the intrinsic value, the owner should not exercise the option. Mathematically this is stated as the inequality

$$P_a(t) > Q(t) \quad \Longrightarrow \quad \text{option owner should not exercise.}$$

To act otherwise is for the option holder to accept a smaller payoff than would be generated if the option were sold to another party. Care must be taken in the determination of the optimal time to exercise the put, since we have defined $P_a(T) = Q(T)$, *i.e.* the value of the put is the intrinsic value at expiry (see Eq. (12.8)). If the option has not been used prior to expiry, it would only be used at expiry if $P_a(T) > 0$. Ideally the owner would like to exercise the option at a time for which the payoff is as large as possible. In Fig. 12.9 this may appear to be at the second time step since a payoff of approximately 5.53962 appears there. However, this particular payoff can only be achieved if the security follows a path of two consecutive downward movements in price. This behavior is not certain to occur.

The American put should be exercised the first time the value of the option is equal to the intrinsic value, provided that value is positive. This allows for the case of an option expiring unused (in which case its value would always have been higher than the intrinsic value prior to expiry and its value at expiry was zero). With these conditions in mind we will define τ^* to be the **optimal stopping time** having value

$$\tau^* = \begin{cases} \infty & \text{if } P_a(t) > Q(t) \text{ for } t \in [0,T) \text{ and } P_a(T) = 0, \\ T & \text{if } P_a(t) > Q(t) \text{ for } t \in [0,T) \text{ and } P_a(T) > 0, \\ \min_{t \in [0,T)} \{t \mid P_a(t) = Q(t)\} & \text{otherwise.} \end{cases}$$

In the discrete setting it is convenient to define $i^* \in \{0, 1, \ldots, n, \infty\}$ as

$$i^* = \begin{cases} \infty & \text{if } P_a(i\Delta t) > Q(i\Delta t) \text{ for } i \in \{0, 1, \ldots, n-1\} \text{ and } P_a(n\Delta t) = 0, \\ n & \text{if } P_a(i\Delta t) > Q(i\Delta t) \text{ for } i \in \{0, 1, \ldots, n-1\} \text{ and } P_a(n\Delta t) > 0, \\ \min_{i \in \{0,1,\ldots,n-1\}} \{i \mid P_a(i\Delta t) = Q(i\Delta t)\} & \text{otherwise.} \end{cases}$$

$$(12.10)$$

Then $\tau^* = i^* \Delta t$.

We will give an informal proof that τ^* is the optimal stopping time below. The reader interested in a more complete proof should consult [Shreve (2004)].

Theorem 12.3 *Let i^* be defined as in Eq. (12.10) then $\tau^* = i^* \Delta t$ is the optimal exercise time for the owner of the American option and furthermore*

$$P_a(0) = e^{-r\tau^*} \mathrm{E}\left[P_a(\tau^*)\right].$$

Proof. For notational convenience we will define $\hat{P}_a(t)$ as

$$\hat{P}_a(t) = P_a(\min\{t, \tau^*\}). \tag{12.11}$$

Suppose we are at the ith time step of the binomial lattice with $i \in \{0, 1, \ldots, n-1\}$ and $i^* > i$. According to the definition of i^* given in Eq. (12.10), the value of the put exceeds its intrinsic value, or symbolically $P_a(i\Delta t) > Q(i\Delta t)$. According to Eq. (12.9)

$$
\begin{aligned}
\hat{P}_a(i\Delta t) &= P_a(i\Delta t) \quad \text{(since } i^* > i\text{)} \\
&= \max_{j \in \{i, i+1, \ldots, n, \infty\}} \{e^{-r(j-i)\Delta t} \mathrm{E}\left[Q(j\Delta t)\right]\} \\
&= e^{-r\Delta t} \max_{j \in \{i+1, i+2, \ldots, n, \infty\}} \{e^{-r(j-[i+1])\Delta t} \mathrm{E}\left[Q(j\Delta t)\right]\} \\
&\quad \text{(since } P_a(i\Delta t) > Q(i\Delta t)\text{)} \\
&= e^{-r\Delta t} P_a((i+1)\Delta t) \\
&= e^{-r\Delta t} \mathrm{E}\left[\hat{P}_a((i+1)\Delta t)\right].
\end{aligned}
\tag{12.12}
$$

On the other hand if $i^* \leq i$ then the put has been exercised and its value is set.

$$
\begin{aligned}
\hat{P}_a(i\Delta t) &= P_a(i^*\Delta t) \quad \text{(since } i^* \leq i\text{)} \\
&= p P_a(i^*\Delta t) + (1-p) P_a(i^*\Delta t) \\
&= p \hat{P}_a((i+1)\Delta t) + (1-p) \hat{P}_a((i+1)\Delta t) \\
&= \mathrm{E}\left[\hat{P}_a((i+1)\Delta t)\right].
\end{aligned}
\tag{12.13}
$$

The American put must be exercised on or before the nth time step or expire unused. Thus if we set $i = 0$, then if $i^* = 0$ we have

$$
P_a(0) = Q(0) = \mathrm{E}\left[Q(0)\right] = e^0 \mathrm{E}\left[Q(0)\right] = e^0 \mathrm{E}\left[P_a(0)\right],
$$

which satisfies Eq. (12.11). Note that $\mathrm{E}\left[Q(0)\right] = Q(0)$ since at $t = 0$ the option's payoff is known with certainty. If $i^* > 0$ then by Eq. (12.12)

$$
P_a(0) = \hat{P}_a(0) = e^{-r\Delta t} \mathrm{E}\left[\hat{P}_a(\Delta t)\right].
$$

Now if $i^* = 1$ then $\hat{P}_a(\Delta t) = P_a(\Delta t)$ and

$$
P_a(0) = e^{-r\Delta t} \mathrm{E}\left[P_a(\Delta t)\right],
$$

which once again satisfies Eq. (12.11). The other possibility is that $i^* > 1$. Another application of Eq. (12.12) gives us

$$
P_a(0) = e^{-r\Delta t} \mathrm{E}\left[e^{-r\Delta t} \mathrm{E}\left[\hat{P}_a(2\Delta t)\right]\right] = e^{-r(2\Delta t)} \mathrm{E}\left[\hat{P}_a(2\Delta t)\right].
$$

We are assuming that the risk-free interest rate is constant which allows it to be brought out of the expected value.

To finish the proof we must consider two cases: (1) i^* is finite, and (2) $i^* = \infty$. For the case that i^* is finite we may apply the reasoning above for $i = 0, 1, \ldots, i^* - 1$ and we obtain

$$P_a(0) = e^{-ri^*\Delta t} \mathrm{E}\left[P_a(i^*\Delta t)\right].$$

Now for $i = i^*, i^* + 1, \ldots, n$ we have from Eq. (12.13) $\hat{P}_a(i\Delta t) = \mathrm{E}\left[\hat{P}_a((i+1)\Delta t)\right]$. Thus

$$\mathrm{E}\left[\hat{P}_a(i\Delta t)\right] = \hat{P}_a(i^*\Delta t) = P_a(i^*\Delta t)$$

for all $i = i^*, i^* + 1, \ldots, n$. Finally if $i^* = \infty$ we know the payoff of the American put is zero and Eq. (12.11) is satisfied. $\qquad\square$

Knowledgeable readers may recognize the concept of the **martingale** used in the proof of Theorem 12.3. This chapter concludes with an example illustrating the result of Theorem 12.3.

Example 12.3 Suppose the current price of a non-dividend paying security is \$36, the volatility in the security price is 20% annually, and the risk-free interest rate is 6% annually. If an American put option with maturity occurring in 4 months is written its price may be approximated using the binomial model with $\Delta = 1/12$ (or one month). The binomial lattice is shown in Fig. 12.10. The upper number in each box is the value of the put at that node in the lattice. The lower number is the corresponding intrinsic value (payoff from immediate exercise). The first time these values are equal is at $t^* = 1/12$. Thus we may verify that

$$
\begin{aligned}
P_a(0) &= e^{-rt^*} \mathrm{E}\left[P_a(t^*)\right] \\
&= e^{-0.06/12}\left[(0.528557)(2.33928) + (1 - 0.528557)(6.03319)\right] \\
&= 4.06039.
\end{aligned}
$$

12.5 Exercises

(1) The price is a security is currently \$50, the risk-free interest rate is 6% compounded continuously, and the strike price of a three-month American put option worth \$9.75 is \$51. Find the range of no arbitrage

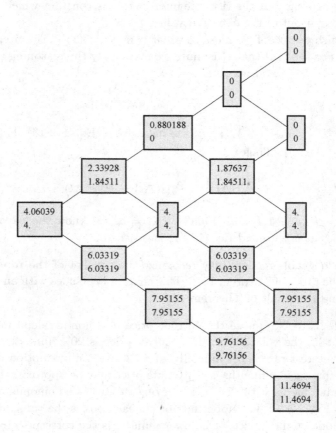

Fig. 12.10   The top number in each box represents the price of the American put option while the bottom number is the intrinsic value of the option.

values of a three-month American call on the same security with the same strike price.

(2) The price is a security is currently $93, the risk-free interest rate is 5.74% compounded continuously, and the strike price of a two-month American call option worth $11.77 is $90. Find the range of no arbitrage values of a two-month American put on the same security with the same strike price.

(3) The price of a security is $56, the risk-free interest rate is 12% compounded continuously, and the volatility of Brownian motion for the

security is 25%. Find the price of a two-month American put on the security with a strike price of $58.

(4) Verify that, in the absence of arbitrage, the expression for p in Eq. (12.7) lies in the interval $[0, 1]$ and thus is a valid value for a probability.

(5) Complete the proof of Lemma 12.3.

(6) Verify the conclusion of Theorem 12.3 for an American put option maturing in 3 months on a stock whose current price is $80. The strike price of the option in $78, the risk-free interest rate is 5% per annum, and the volatility of the stock price is 25%.

Appendix A

Sample Stock Market Data

In this appendix are end-of-day closing stock market prices for Sony Corporation stock. The data were collected from the website www.siliconinvestor.com and cover the one year period of August 13, 2001 until August 12, 2002.

Date	Close	Date	Close
Aug-12-2002	42.86	Jul-15-2002	50.3
Aug-9-2002	44.44	Jul-12-2002	50.9
Aug-8-2002	44.0	Jul-11-2002	51.2
Aug-7-2002	43.85	Jul-10-2002	50.47
Aug-6-2002	42.5	Jul-9-2002	52.75
Aug-5-2002	42.01	Jul-8-2002	51.98
Aug-2-2002	42.71	Jul-5-2002	53.17
Aug-1-2002	44.55	Jul-3-2002	51.75
Jul-31-2002	45.33	Jul-2-2002	50.0
Jul-30-2002	46.65	Jul-1-2002	51.55
Jul-29-2002	46.14	Jun-28-2002	53.1
Jul-26-2002	44.7	Jun-27-2002	50.3
Jul-25-2002	45.49	Jun-26-2002	48.95
Jul-24-2002	47.1	Jun-25-2002	49.41
Jul-23-2002	45.21	Jun-24-2002	50.1
Jul-22-2002	44.97	Jun-21-2002	48.63
Jul-19-2002	46.3	Jun-20-2002	50.23
Jul-18-2002	48.95	Jun-19-2002	50.06
Jul-17-2002	47.82	Jun-18-2002	51.8
Jul-16-2002	49.59	Jun-17-2002	52.78

Date	Close	Date	Close
Jun-14-2002	52.45	Apr-26-2002	55.5
Jun-13-2002	53.27	Apr-25-2002	56.0
Jun-12-2002	54.31	Apr-24-2002	53.98
Jun-11-2002	54.01	Apr-23-2002	53.92
Jun-10-2002	54.36	Apr-22-2002	53.93
Jun-7-2002	55.5	Apr-19-2002	53.25
Jun-6-2002	55.5	Apr-18-2002	53.87
Jun-5-2002	56.7	Apr-17-2002	53.8
Jun-4-2002	55.91	Apr-16-2002	53.7
Jun-3-2002	56.3	Apr-15-2002	52.17
May-31-2002	58.11	Apr-12-2002	50.86
May-30-2002	58.69	Apr-11-2002	51.7
May-29-2002	57.8	Apr-10-2002	52.72
May-28-2002	58.23	Apr-9-2002	51.31
May-24-2002	59.4	Apr-8-2002	52.35
May-23-2002	59.7	Apr-5-2002	51.5
May-22-2002	59.56	Apr-4-2002	52.71
May-21-2002	58.05	Apr-3-2002	52.46
May-20-2002	57.65	Apr-2-2002	51.15
May-17-2002	58.81	Apr-1-2002	51.1
May-16-2002	56.48	Mar-28-2002	51.7
May-15-2002	56.15	Mar-27-2002	51.45
May-14-2002	55.6	Mar-26-2002	50.85
May-13-2002	55.0	Mar-25-2002	50.85
May-10-2002	54.75	Mar-22-2002	52.3
May-9-2002	54.6	Mar-21-2002	53.16
May-8-2002	55.14	Mar-20-2002	52.01
May-7-2002	52.65	Mar-19-2002	55.21
May-6-2002	54.0	Mar-18-2002	53.66
May-3-2002	54.02	Mar-15-2002	55.25
May-2-2002	53.45	Mar-14-2002	55.09
May-1-2002	54.8	Mar-13-2002	52.4
Apr-30-2002	54.2	Mar-12-2002	53.9
Apr-29-2002	55.48	Mar-11-2002	56.2

Date	Close	Date	Close
Mar-8-2002	56.59	Jan-17-2002	45.51
Mar-7-2002	56.05	Jan-16-2002	44.5
Mar-6-2002	53.4	Jan-15-2002	45.91
Mar-5-2002	50.3	Jan-14-2002	47.02
Mar-4-2002	50.7	Jan-11-2002	47.55
Mar-1-2002	47.8	Jan-10-2002	48.92
Feb-28-2002	46.2	Jan-9-2002	48.0
Feb-27-2002	47.22	Jan-8-2002	47.82
Feb-26-2002	45.75	Jan-7-2002	49.31
Feb-25-2002	47.25	Jan-4-2002	49.57
Feb-22-2002	46.71	Jan-3-2002	47.25
Feb-21-2002	45.3	Jan-2-2002	45.95
Feb-20-2002	45.4	Dec-31-2001	45.1
Feb-19-2002	44.87	Dec-28-2001	46.25
Feb-15-2002	44.85	Dec-27-2001	44.0
Feb-14-2002	45.55	Dec-26-2001	44.11
Feb-13-2002	44.85	Dec-24-2001	44.36
Feb-12-2002	43.77	Dec-21-2001	44.36
Feb-11-2002	43.45	Dec-20-2001	45.5
Feb-8-2002	42.4	Dec-19-2001	46.75
Feb-7-2002	42.5	Dec-18-2001	47.0
Feb-6-2002	41.0	Dec-17-2001	45.55
Feb-5-2002	41.75	Dec-14-2001	46.2
Feb-4-2002	43.25	Dec-13-2001	45.28
Feb-1-2002	43.5	Dec-12-2001	47.3
Jan-31-2002	44.76	Dec-11-2001	46.95
Jan-30-2002	44.35	Dec-10-2001	46.61
Jan-29-2002	43.7	Dec-7-2001	48.5
Jan-28-2002	45.2	Dec-6-2001	49.86
Jan-25-2002	46.4	Dec-5-2001	49.7
Jan-24-2002	45.86	Dec-4-2001	46.8
Jan-23-2002	44.91	Dec-3-2001	45.99
Jan-22-2002	43.56	Nov-30-2001	47.7
Jan-18-2002	44.7	Nov-29-2001	47.0
Nov-28-2001	46.11	Oct-10-2001	37.0

Date	Close	Date	Close
Nov-27-2001	48.25	Oct-9-2001	34.31
Nov-26-2001	48.55	Oct-8-2001	35.7
Nov-23-2001	46.55	Oct-5-2001	35.25
Nov-21-2001	46.49	Oct-4-2001	35.3
Nov-20-2001	45.6	Oct-3-2001	34.4
Nov-19-2001	48.14	Oct-2-2001	33.72
Nov-16-2001	46.2	Oct-1-2001	33.74
Nov-15-2001	45.3	Sep-28-2001	33.2
Nov-14-2001	40.83	Sep-27-2001	37.19
Nov-13-2001	40.5	Sep-26-2001	36.25
Nov-12-2001	39.95	Sep-25-2001	37.05
Nov-9-2001	39.8	Sep-24-2001	38.7
Nov-8-2001	39.79	Sep-21-2001	36.75
Nov-7-2001	39.3	Sep-20-2001	37.56
Nov-6-2001	41.8	Sep-19-2001	37.25
Nov-5-2001	41.2	Sep-18-2001	36.35
Nov-2-2001	39.72	Sep-17-2001	37.0
Nov-1-2001	39.14	Sep-10-2001	40.4
Oct-31-2001	38.2	Sep-7-2001	41.48
Oct-30-2001	38.12	Sep-6-2001	42.2
Oct-29-2001	38.6	Sep-5-2001	42.7
Oct-26-2001	40.33	Sep-4-2001	44.0
Oct-25-2001	40.43	Aug-31-2001	44.9
Oct-24-2001	40.84	Aug-30-2001	45.92
Oct-23-2001	41.7	Aug-29-2001	47.33
Oct-22-2001	41.74	Aug-28-2001	47.75
Oct-19-2001	41.0	Aug-27-2001	47.72
Oct-18-2001	40.5	Aug-24-2001	47.91
Oct-17-2001	40.1	Aug-23-2001	46.49
Oct-16-2001	40.75	Aug-22-2001	48.55
Oct-15-2001	39.58	Aug-21-2001	47.1
Oct-12-2001	40.01	Aug-20-2001	48.25
Oct-11-2001	39.96	Aug-17-2001	47.9
Aug-16-2001	50.48	Aug-14-2001	50.1
Aug-15-2001	51.65	Aug-13-2001	49.16

Solutions to Chapter Exercises

B.1 The Theory of Interest

(1) The principal amount is $P = \$3659$, the annual interest rate is $r = 0.065$, the period is $t = 5$ years. Thus the final compound balance is

$$P(1 + rt) = 3659(1 + (5)(0.065)) = \$4848.18,$$

where the final balance has been rounded to the nearest cent.

(2) The principal amount is $P = \$3993$, the annual interest rate is $r = 0.043$, the period is $t = 2$ years. If there are n compounding periods per year, the account balance is

$$A = P\left(1 + \frac{r}{n}\right)^{nt}.$$

(a) If the interest is compounded monthly $n = 12$ and

$$A = 3993(1 + \frac{0.043}{12})^{24} = \$4350.93$$

(b) If the interest is compounded weekly $n = 52$ and

$$A = 3993(1 + \frac{0.043}{52})^{104} = \$4351.44$$

(c) If the interest is compounded daily $n = 365$ and

$$A = 3993(1 + \frac{0.043}{365})^{730} = \$4351.57$$

(d) If the interest is compounded continuously

$$A = Pe^{rt} = 3993e^{(0.043)(2)} = \$4351.60$$

(3) The period is $t = 1$, the interest rate is $r = 0.08$ compounded $n = 4$ times per period. The effective simple annual interest rate R is

$$R = -1 + (1 + \frac{0.08}{4})^4 \approx 0.08243.$$

(4) Since the competitor pays interest at rate $r = 0.0525$ compounded daily we will compute his effective simple annual interest rate and then determine the monthly compounded rate which matches it. The effective annual interest rate r_e is

$$r_e = -1 + (1 + \frac{0.0525}{365})^{365} \approx 0.05390.$$

Thus we wish to solve the following equation for i.

$$0.05390 = -1 + (1 + \frac{i}{12})^{12}$$
$$\sqrt[12]{1.05390} = 1 + \frac{i}{12}$$
$$0.05261 = i$$

Thus if you pay an interest rate compounded monthly of at least 5.261122% you will compete favorably.

(5) The continuously compounded interest-bearing account will increase in value more rapidly, thus we wish to solve the following equation.

$$1000e^{0.0475t} - 1000(1 + \frac{0.0475}{365})^{365t} = 1$$

This equation cannot be solved algebraically for t, so we use Newton's method to approximate the solution, $t \approx 42.66$ years.

(6) We focus on finding the limit $\lim_{h \to 0} (1 + h)^{1/h}$. If we let $y = (1 + h)^{1/h}$, then

$$\ln y = \ln \left((1 + h)^{1/h} \right)$$
$$= \frac{1}{h} \ln(1 + h)$$
$$e^{\ln y} = e^{\frac{1}{h} \ln(1+h)}$$
$$y = e^{\frac{\ln(1+h)}{h}}.$$

Therefore

$$\lim_{h \to 0} (1 + h)^{1/h} = \lim_{h \to 0} e^{\frac{\ln(1+h)}{h}}$$

$$= e^{\lim_{h \to 0} \frac{\ln(1+h)}{h}}$$

since the exponential function is continuous everywhere. The limit in the exponent is the definition of the derivative of $\ln x$ at $x = 1$. Thus the exponent limit is 1 and $\lim_{h \to 0} (1 + h)^{1/h} = e^1 = e$.

(7) To determine the preferable investment we must determine the present value of the payouts. Since the interest rate is $r = 0.0275$ compounded continuously we have

$$P_A = 200e^{-0.0275} + 211e^{-0.055} + 198e^{-0.0825} + 205e^{-0.11} = 760.25$$
$$P_B = 198e^{-0.0275} + 205e^{-0.055} + 211e^{-0.0825} + 200e^{-0.11} = 760.12$$

Thus investment A is preferable.

(8) If the price of the house is $200000 and your down payment is 20% then you will borrow $P = \$160000$. If the monthly payment on a $t = 30$ year, fixed rate mortgage should not exceed $x = \$1500$ then we can substitute these quantities into Eq. (1.9) and use Newton's Method to approximate r.

$$P = x \frac{n}{r} \left(1 - \left[1 + \frac{r}{n}\right]^{-nt}\right)$$
$$160000 = 1500 \frac{12}{r} \left(1 - \left[1 + \frac{r}{12}\right]^{-(12)(30)}\right)$$
$$r \approx 0.1080$$

Thus the interest rate must not exceed 10.80% annually.

(9) The effective annual real rate of interest is

$$r_i = \frac{0.0505 - 0.0202}{1 + 0.0202} = 0.0297.$$

The equivalent nominal annual rate compounded quarterly is then

$$\left(1 + \frac{r}{4}\right)^4 = 1 + 0.0297$$
$$r \approx 0.0294$$

(10) Differentiating the function $f(r)$ given in Eq. (1.11) yields

$$f'(r) = \sum_{i=1}^{n} A_i(-i)(1+r)^{-i-1} = -\sum_{i=1}^{n} A_i(i)(1+r)^{-(i+1)} < 0$$

for $-1 < r < \infty$. Thus if there exist two distinct rates of return r_1 and r_2 then by the Mean Value Theorem $f'(r) = 0$ for some r between r_1 and r_2, contradicting the inequality above.

(11) Using Eq. (1.11) we setup the equation

$$10000 = \frac{2000}{1+r} + \frac{3000}{(1+r)^2} + \frac{4000}{(1+r)^3} + \frac{3000}{(1+r)^4}$$

whose solution is approximated using Newton's Method. The rate of return $r \approx 0.0718$ or equivalently 7.18% per year.

(12) (a) The present value of initial investment and payout amounts for investment A is (assuming an annual interest rate of $r = 0.0433$), $-\$10.91$. The present value of the initial investment and payout amounts for investment B is $\$18.63$. Thus investment B is preferable if present value is the decision criterion.

  (b) The rate of return of investment A is $r_A \approx 0.0394$ while the rate of return of investment B is $r_B \approx 0.0499$. Thus investment B would be the preferable investment if rate of return were the deciding criterion.

B.2   Discrete Probability

(1) Assuming the regular tetrahedron is fair so that it equally likely to land on any of its four faces, the probability of it landing on 3 is $p = 1/4$.

(2) Let the ordered pair (m, n) represent the numeric outcomes of the rolling of the dice. The outcome of the first die is m and the outcome of the second die is n. If you are concerned about the problem of keeping track of which is the first die and which is the second, assume that the first die is painted green while the second is painted red. According to the Multiplication Rule the probability of the event (m, n) is

$$P(m, n) = P(\text{first die shows } m) P(\text{second die shows } n)$$

since the dice are independent of each other. Since each of the die has six equally likely simple outcomes, then $P(m, n) = 1/36$. The table of 36 outcomes is shown below.

$$
\begin{array}{cccccc}
(1,1) & (1,2) & (1,3) & (1,4) & (1,5) & (1,6) \\
(2,1) & (2,2) & (2,3) & (2,4) & (2,5) & (2,6) \\
(3,1) & (3,2) & (3,3) & (3,4) & (3,5) & (3,6) \\
(4,1) & (4,2) & (4,3) & (4,4) & (4,5) & (4,6) \\
(5,1) & (5,2) & (5,3) & (5,4) & (5,5) & (5,6) \\
(6,1) & (6,2) & (6,3) & (6,4) & (6,5) & (6,6)
\end{array}
$$

Since we are interested in the sums of the numbers shown on the upward faces of the dice we will tabulate them below.

$$
\begin{array}{cccccc}
2 & 3 & 4 & 5 & 6 & 7 \\
3 & 4 & 5 & 6 & 7 & 8 \\
4 & 5 & 6 & 7 & 8 & 9 \\
5 & 6 & 7 & 8 & 9 & 10 \\
6 & 7 & 8 & 9 & 10 & 11 \\
7 & 8 & 9 & 10 & 11 & 12
\end{array}
$$

Thus we can see the sample space of sums for the fair dice is the set $\{2, 3, 4, 5, 6, 7, 8, 9, 10, 11, 12\}$. Each of the entries in the previous table occurs with probability 1/36 and therefore the probabilities of the outcomes of rolling a pair of fair dice are

Outcome	Prob.	Outcome	Prob.
2	$\frac{1}{36}$	7	$\frac{1}{6}$
3	$\frac{1}{18}$	8	$\frac{5}{36}$
4	$\frac{1}{12}$	9	$\frac{1}{9}$
5	$\frac{1}{9}$	10	$\frac{1}{12}$
6	$\frac{5}{36}$	11	$\frac{1}{18}$
		12	$\frac{1}{36}$

(3) Let the probability that the batter strikes out in the first inning be denoted $P(1) = 1/3$. Let the probability that the batter strikes out in the fifth inning be denoted $P(5) = 1/4$. Let the probability that the batter strikes out in both innings be denoted $P(1 \wedge 5) = 1/10$. According to the Addition Rule the probability that the batter strikes out in either the first or the fifth inning is

$$
P(1 \vee 5) = P(1) + P(5) - P(1 \wedge 5) = \frac{1}{3} + \frac{1}{4} - \frac{1}{10} = \frac{29}{60}.
$$

(4) The eight possible outcomes of the experiment are listed in the table below.

Person	1	2	3
	Red	Red	Red
	Red	Red	Blue
	Red	Blue	Red
Hat	Blue	Red	Red
	Red	Blue	Blue
	Blue	Blue	Red
	Blue	Red	Blue
	Blue	Blue	Blue

In 6 of the outcomes two of the three people will see their companions wearing mis-matched hats and pass while the third will see their two companions wearing matching hats opposite in color to their own hat's color. Thus one of the three will guess the correct color in these six cases. In the remaining two cases all three hats are of the same color, so the strategy of guessing the opposite color would fail. Thus the probability of winning under the strategy is $6/8 = 3/4 = 0.75$.

(5) Since the cards are drawn without replacement the outcome of the second draw is dependent on the outcome of the first draw. Using the Multiplication Rule and conditional probability we have

$$P\,(2\ \text{aces}) = P\,(\text{2nd ace}|\text{1st ace})\,P\,(\text{1st ace})$$
$$= \left(\frac{3}{51}\right)\left(\frac{4}{52}\right)$$
$$= \frac{1}{221}.$$

(6) Since the cards are drawn without replacement the outcome of the second draw is dependent on the outcome of the first draw. Using the Multiplication Rule and conditional probability we have

$$\dot{P}\,(\text{2nd ace} \wedge \text{1st not ace}) = P\,(\text{2nd ace}|\text{1st not ace})\,P\,(\text{1st not ace})$$
$$= \left(\frac{4}{51}\right)\left(\frac{48}{52}\right)$$
$$= \frac{16}{221}.$$

(7) Since the cards are drawn without replacement the outcome of the fourth draw is dependent on the outcome of the third draw which

is dependent on the outcome of the second draw which is dependent on the outcome of the first draw. Using the Multiplication Rule and conditional probability we have

$$P\,(4\text{ aces}) = P\,(4\text{th ace}|3\text{rd ace}) \cdot P\,(3\text{rd ace}|2\text{nd ace}) \cdot$$
$$P\,(2\text{nd ace}|1\text{st ace}) \cdot P\,(1\text{st ace})$$
$$= \left(\frac{1}{49}\right)\left(\frac{2}{50}\right)\left(\frac{3}{51}\right)\left(\frac{4}{52}\right)$$
$$= \frac{1}{270725}.$$

(8) If $n \in \{2, \dots, 49\}$ is the position of the first ace drawn then

$$P\,(n) = \frac{48}{52} \cdot \frac{47}{51} \cdots \frac{50-n}{54-n} \cdot \frac{4}{53-n}.$$

When $n = 1$, $P\,(1) = 1/13$. By calculating $P\,(2), P\,(3), \dots, P\,(49)$ we obtain the following probabilities for the first ace appearing on the nth draw.

n	$P(n)$	n	$P(n)$
2	$\frac{16}{221}$	26	$\frac{8}{833}$
3	$\frac{376}{5525}$	27	$\frac{92}{10829}$
4	$\frac{17296}{270725}$	28	$\frac{2024}{270725}$
5	$\frac{3243}{54145}$	29	$\frac{253}{38675}$
6	$\frac{3036}{54145}$	30	$\frac{44}{7735}$
7	$\frac{2838}{54145}$	31	$\frac{38}{7735}$
8	$\frac{1892}{38675}$	32	$\frac{228}{54145}$
9	$\frac{1763}{38675}$	33	$\frac{57}{15925}$
10	$\frac{328}{7735}$	34	$\frac{48}{15925}$
11	$\frac{164}{4165}$	35	$\frac{8}{3185}$
12	$\frac{152}{4165}$	36	$\frac{16}{7735}$
13	$\frac{703}{20825}$	37	$\frac{1}{595}$
14	$\frac{8436}{270725}$	38	$\frac{4}{2975}$
15	$\frac{222}{7735}$	39	$\frac{22}{20825}$
16	$\frac{12}{455}$	40	$\frac{44}{54145}$
17	$\frac{11}{455}$	41	$\frac{33}{54145}$
18	$\frac{352}{15925}$	42	$\frac{24}{54145}$
19	$\frac{5456}{270725}$	43	$\frac{12}{38675}$
20	$\frac{992}{54145}$	44	$\frac{8}{38675}$
21	$\frac{899}{54145}$	45	$\frac{1}{7735}$
22	$\frac{116}{7735}$	46	$\frac{4}{54145}$
23	$\frac{522}{38675}$	47	$\frac{2}{54145}$
24	$\frac{36}{2975}$	48	$\frac{4}{270725}$
25	$\frac{9}{833}$	49	$\frac{1}{270725}$

The draw with the highest probability of being the first ace is the first draw.

(9) The outcomes of different spins of a roulette wheel are assumed to be independent, thus the probability that any spin of the wheel has an outcome of black is $P(\text{black}) = 9/19$.

(10) The outcomes of different spins of a roulette wheel are assumed to be independent, thus the probability that any spin of the wheel has an outcome of 00 is $P(00) = 1/38$.

(11) To be a probability distribution function

$$\sum_{x=1}^{10} f(x) = \sum_{x=1}^{10} \frac{c}{x} = c \sum_{x=1}^{10} \frac{1}{x} = \frac{7381c}{2520} = 1,$$

thus $c = 2520/7381$.

(12)

$$P\,(0\ \text{black}) = \frac{5}{20} \cdot \frac{4}{19} \cdot \frac{3}{18} = \frac{1}{114}$$

$$P\,(1\ \text{black}) = \frac{15}{20} \cdot \frac{5}{19} \cdot \frac{4}{18} + \frac{5}{20} \cdot \frac{15}{19} \cdot \frac{4}{18} + \frac{5}{20} \cdot \frac{4}{19} \cdot \frac{15}{18} = \frac{5}{38}$$

$$P\,(2\ \text{black}) = \frac{15}{20} \cdot \frac{14}{19} \cdot \frac{5}{18} + \frac{15}{20} \cdot \frac{5}{19} \cdot \frac{14}{18} + \frac{5}{20} \cdot \frac{15}{19} \cdot \frac{14}{18} = \frac{35}{76}$$

$$P\,(3\ \text{black}) = \frac{15}{20} \cdot \frac{14}{19} \cdot \frac{13}{18} = \frac{91}{228}$$

(13) This situation can be thought of as a binomial experiment with 25 trials and the probability of success on a single trial of 0.0016. Thus the probability that a manufacturing run will be rejected is

$$P\,(X \geq 2) = \sum_{n=2}^{25} P\,(X = n)$$

$$= \sum_{n=2}^{25} \binom{25}{n}(0.0016)^n(1 - 0.0016)^{25-n}$$

$$\approx 0.00075.$$

(14) The probability of a child being born female is $p = 100/205 = 20/41$. Thus the expected number of female children in a family of 6 total children is

$$\sum_{x=0}^{6}(x \cdot P\,(x)) = \sum_{x=0}^{6}\left(x \cdot \binom{6}{x}\left(\frac{20}{41}\right)^{x}\left(1 - \frac{20}{41}\right)^{6-x}\right)$$

$$= \frac{120}{41} \approx 2.93.$$

(15) Using the probabilities calculated in exercise 8 and the definition of expected value we have the expected position of the first ace,

$$\mu = \sum_{n=1}^{49}(n \cdot P\,(n)) = \frac{53}{5} = 10.6.$$

(16) By the definition of expected value

$$E\left[aX + b\right] = \sum_X \left((aX + b)P\left(X\right)\right)$$

$$= a\sum_X \left(XP\left(X\right)\right) + b\sum_X P\left(X\right)$$

$$= aE\left[X\right] + b$$

(17) Using the probabilities calculated in exercise 8 and the expected value calculated in exercise 15, the variance in the position of the first ace drawn is

$$\sum_{n=1}^{49}\left(n^2 \cdot P\left(n\right)\right) - \mu^2 = \frac{901}{5} - \frac{2809}{25} = \frac{1696}{25} = 67.84.$$

This implies the standard deviation in the appearance of the first ace is approximately 8.2365.

(18) By the definition of variance

$$\begin{aligned}
\text{Var}\left(aX + b\right) &= E\left[(aX + b)^2\right] - \left(E\left[aX + b\right]\right)^2 \\
&= E\left[a^2 X^2 + 2abX + b^2\right] - \left(aE\left[X\right] + b\right)^2 \\
&= a^2 E\left[X^2\right] + 2abE\left[X\right] + b^2 - a^2\left(E\left[X\right]\right)^2 - 2abE\left[X\right] - b^2 \\
&= a^2 E\left[X^2\right] - a^2\left(E\left[X\right]\right)^2 \\
&= a^2\text{Var}\left(X\right)
\end{aligned}$$

B.3 Normal Random Variables and Probability

(1) The probability distribution function for X is given by the piecewise defined function

$$f(x) = \begin{cases} \frac{1}{5} & \text{if } -4 \le x \le 1 \\ 0 & \text{otherwise.} \end{cases}$$

Therefore

$$P\left(X \ge 0\right) = \int_0^\infty f(x)\,dx = \int_0^1 \frac{1}{5}\,dx = \frac{1}{5}.$$

(2) We know that for any random variable X with probability distribution

function $f(x)$ we must have $\int_{-\infty}^{\infty} f(x)\, dx = 1$, thus

$$\int_{-\infty}^{\infty} f(x)\, dx = \int_{1}^{\infty} \frac{c}{x^3}\, dx$$

$$= c \lim_{M \to \infty} \int_{1}^{M} \frac{1}{x^3}\, dx$$

$$= c \lim_{M \to \infty} \left(\frac{-1}{2x^2} \Big|_{1}^{M} \right)$$

$$= c \lim_{M \to \infty} \left(\frac{-1}{2M^2} + \frac{1}{2} \right)$$

$$1 = \frac{c}{2}$$

$$c = 2.$$

(3)

$$1 = \int_{-\infty}^{\infty} f(x)\, dx$$

$$= \int_{-\infty}^{a} f(x)\, dx + \int_{a}^{\infty} f(x)\, dx$$

$$= P(X < a) + P(X \geq a)$$

$$1 - P(X < a) = P(X \geq a)$$

(4) By definition, the mean of a continuously distributed random variable is

$$\mu = \int_{-\infty}^{\infty} x f(x)\, dx.$$

Thus for the given probability distribution function

$$\mu = \int_{-\infty}^{\infty} \frac{x}{\pi(1+x^2)}\, dx$$

$$= \int_{-\infty}^{0} \frac{x}{\pi(1+x^2)}\, dx + \int_{0}^{\infty} \frac{x}{\pi(1+x^2)}\, dx$$

$$= \lim_{M\to-\infty} \int_{M}^{0} \frac{x}{\pi(1+x^2)}\, dx + \lim_{N\to\infty} \int_{0}^{N} \frac{x}{\pi(1+x^2)}\, dx$$

$$= \lim_{M\to-\infty} \left(\frac{1}{2\pi}\ln(1+x^2)\Big|_{M}^{0} \right) + \lim_{N\to\infty} \left(\frac{1}{2\pi}\ln(1+x^2)\Big|_{0}^{N} \right)$$

$$= \lim_{M\to-\infty} \left(-\frac{1}{2\pi}\ln(1+M^2) \right) + \lim_{N\to\infty} \left(\frac{1}{2\pi}\ln(1+N^2) \right).$$

The improper integral above does not converge and hence the mean of the random variable does not exist.

(5)

$$E\left[aX + b\right] = \int_{-\infty}^{\infty} (ax + b)f(x)\, dx$$

$$= a \int_{-\infty}^{\infty} xf(x)\, dx + b \int_{-\infty}^{\infty} f(x)\, dx$$

$$= aE\left[X\right] + b$$

(6) ***Proof.*** Let X_1, X_2, ..., X_k be continuous random variables with probability distribution function $f(x_1, x_2, \ldots, x_k)$. By the definition of expected value

$$E\left[X_1 + X_2 + \cdots X_k\right]$$

$$= \int_{-\infty}^{\infty} \int_{-\infty}^{\infty} \cdots \int_{-\infty}^{\infty} (x_1 + x_2 + \cdots + x_k)f(x_1, x_2, \ldots, x_k)\, dx_1\, dx_2 \ldots dx_k$$

$$= \int_{-\infty}^{\infty} \int_{-\infty}^{\infty} \cdots \int_{-\infty}^{\infty} x_1 f(x_1, x_2, \ldots, x_k)\, dx_1\, dx_2 \ldots dx_k$$

$$+ \int_{-\infty}^{\infty} \int_{-\infty}^{\infty} \cdots \int_{-\infty}^{\infty} x_2 f(x_1, x_2, \ldots, x_k)\, dx_1\, dx_2 \ldots dx_k$$

$$+ \cdots + \int_{-\infty}^{\infty} \int_{-\infty}^{\infty} \cdots \int_{-\infty}^{\infty} x_k f(x_1, x_2, \ldots, x_k)\, dx_1\, dx_2 \ldots dx_k$$

$$= \int_{-\infty}^{\infty} x_1 f_1(x_1)\, dx_1 + \int_{-\infty}^{\infty} x_2 f_2(x_2)\, dx_2 + \cdots + \int_{-\infty}^{\infty} x_k f_k(x_k)\, dx_k$$

$$= E\left[X_1\right] + E\left[X_2\right] + \cdots + E\left[X_k\right].$$

To keep the notation compact we have written the marginal probability density of X_i as $f_i(x_i)$ for $i = 1, 2, \ldots, k$, where

$$f_i(x_i) = \int_{-\infty}^{\infty} \int_{-\infty}^{\infty} \cdots \int_{-\infty}^{\infty} f(x_1, x_2, \ldots, x_k)\, dx_1 \ldots dx_{i-1} dx_{i+1} \ldots dx_k.$$

$\square$

(7) **Proof.** Let X_1, X_2, $\ldots$, X_k be pairwise independent, continuous random variables with probability distribution function $f(x_1, x_2, \ldots, x_k)$. Since the component random variables are assumed to be pairwise independent the joint probability distribution can be rewritten as

$$f(x_1, x_2, \ldots, x_k) = f_1(x_1) f_2(x_2) \cdots f_k(x_k).$$

Thus by the definition of expected value

$$\mathrm{E}\left[X_1 X_2 \cdots X_k\right]$$
$$= \int_{-\infty}^{\infty} \int_{-\infty}^{\infty} \cdots \int_{-\infty}^{\infty} x_1 x_2 \cdots x_k f(x_1, x_2, \ldots, x_k)\, dx_1\, dx_2 \ldots dx_k$$
$$= \int_{-\infty}^{\infty} \int_{-\infty}^{\infty} \cdots \int_{-\infty}^{\infty} x_1 x_2 \cdots x_k f_1(x_1) f_2(x_2) \cdots f_k(x_k)\, dx_1\, dx_2 \ldots dx_k$$
$$= \left(\int_{-\infty}^{\infty} x_1 f_1(x_1)\, dx_1\right) \left(\int_{-\infty}^{\infty} x_2 f_2(x_2)\, dx_2\right) \cdots \left(\int_{-\infty}^{\infty} x_k f_k(x_k)\, dx_k\right)$$
$$= \mathrm{E}\left[X_1\right] \mathrm{E}\left[X_2\right] \cdots \mathrm{E}\left[X_k\right].$$

$\square$

(8) **Proof.** Let X be a continuous random variable with probability distribution function $f(x)$. Suppose that $\mathrm{E}\left[X\right] = \mu$. By the definition of variance

$$\mathrm{Var}\left(X\right) = \int_{-\infty}^{\infty} (x - \mu)^2 f(x)\, dx$$
$$= \int_{-\infty}^{\infty} x^2 f(x)\, dx - 2\mu \int_{-\infty}^{\infty} x f(x)\, dx + \mu^2 \int_{-\infty}^{\infty} f(x)\, dx$$
$$= \mathrm{E}\left[X^2\right] - 2\mu^2 + \mu^2$$
$$= \mathrm{E}\left[X^2\right] - \mu^2$$

$\square$

(9) **Proof.** Let X be a continuous random variable with probability distribution function $f(x)$. Suppose $a, b \in \mathbb{R}$.

$$\text{Var}\,(aX + b) = \text{E}\left[(aX + b)^2\right] - (\text{E}\,[aX + b])^2$$

$$= \int_{-\infty}^{\infty} (ax + b)^2 f(x)\, dx - (a\text{E}\,[X] + b)^2$$

$$= a^2 \int_{-\infty}^{\infty} x^2 f(x)\, dx + 2ab \int_{-\infty}^{\infty} x f(x)\, dx + b^2 \int_{-\infty}^{\infty} f(x)\, dx$$

$$\quad - a^2\,(\text{E}\,[X])^2 - 2ab\text{E}\,[X] - b^2$$

$$= a^2\text{E}\left[X^2\right] + 2ab\text{E}\,[X] + b^2 - a^2\,(\text{E}\,[X])^2 - 2ab\text{E}\,[X] - b^2$$

$$= a^2 \left(\text{E}\left[X^2\right] - (\text{E}\,[X])^2\right)$$

$$= a^2 \text{Var}\,(X)$$

$$\square$$

(10) **Proof.** Let X_1, X_2, $\ldots$, X_k be pairwise independent, continuous random variables with probability distribution function $f(x_1, x_2, \ldots, x_k)$. Let the mean of X_i be μ_i for $i = 1, 2, \ldots, k$. By the definition of variance

$$\text{Var}\,(X_1 + X_2 + \cdots + X_k)$$

$$= \text{E}\left[(X_1 + X_2 + \cdots + X_k)^2\right] - (\text{E}\,[X_1 + X_2 + \cdots + X_k])^2$$

$$= \int_{-\infty}^{\infty} \cdots \int_{-\infty}^{\infty} (x_1 + \cdots + x_k)^2 f(x_1, \ldots, x_k)\, dx_1 \ldots dx_k$$

$$\quad - (\mu_1 + \cdots + \mu_k)^2$$

$$= \int_{-\infty}^{\infty} \cdots \int_{-\infty}^{\infty} (x_1 + \cdots + x_k)^2 f_1(x_1) \cdots f_k(x_k)\, dx_1 \ldots dx_k$$

$$\quad - (\mu_1 + \cdots + \mu_k)^2$$

$$= \int_{-\infty}^{\infty} \int_{-\infty}^{\infty} \cdots \int_{-\infty}^{\infty} (x_1^2 + x_2^2 + \cdots + x_k^2 + 2x_1 x_2$$

$$+ \cdots + 2x_{k-1} x_k) f_1(x_1) f_2(x_2) \cdots f_k(x_k) \, dx_1 \, dx_2 \ldots dx_k$$

$$- (\mu_1^2 + \mu_2^2 + \cdots + \mu_k^2 + 2\mu_1 \mu_2 + \cdots + 2\mu_{k-1} \mu_k)$$

$$= \mathrm{E}\left[X_1^2\right] + \mathrm{E}\left[X_2^2\right] + \cdots + \mathrm{E}\left[X_k^2\right]$$

$$+ 2\mathrm{E}\left[X_1 X_2\right] + \cdots + 2\mathrm{E}\left[X_{k-1} X_k\right]$$

$$- (\mu_1^2 + \mu_2^2 + \cdots + \mu_k^2 + 2\mu_1 \mu_2 + \cdots + 2\mu_{k-1} \mu_k)$$

$$= \mathrm{E}\left[X_1^2\right] + \mathrm{E}\left[X_2^2\right] + \cdots + \mathrm{E}\left[X_k^2\right]$$

$$+ 2\mathrm{E}\left[X_1\right]\mathrm{E}\left[X_2\right] + \cdots + 2\mathrm{E}\left[X_{k-1}\right]\mathrm{E}\left[X_k\right]$$

$$- (\mu_1^2 + \mu_2^2 + \cdots + \mu_k^2 + 2\mu_1 \mu_2 + \cdots + 2\mu_{k-1} \mu_k)$$

$$= \mathrm{E}\left[X_1^2\right] + \mathrm{E}\left[X_2^2\right] + \cdots + \mathrm{E}\left[X_k^2\right] + 2\mu_1 \mu_2 + \cdots + 2\mu_{k-1} \mu_k$$

$$- (\mu_1^2 + \mu_2^2 + \cdots + \mu_k^2 + 2\mu_1 \mu_2 + \cdots + 2\mu_{k-1} \mu_k)$$

$$= \mathrm{E}\left[X_1^2\right] - \mu_1^2 + \mathrm{E}\left[X_2^2\right] - \mu_2^2 + \cdots + \mathrm{E}\left[X_k^2\right] - \mu_k^2$$

$$= \mathrm{Var}\left(X_1\right) + \mathrm{Var}\left(X_2\right) + \cdots + \mathrm{Var}\left(X_k\right)$$

$\square$

(11) The expected value of X is calculated as

$$\mathrm{E}\left[X\right] = \int_{-\infty}^{\infty} x f(x) \, dx$$

$$= \int_{-1}^{2} \frac{2}{5} x |x| \, dx$$

$$= \frac{2}{5} \left(\int_{-1}^{0} (-x^2) \, dx + \int_{0}^{2} x^2 \, dx \right)$$

$$\mu = \frac{14}{15}.$$

The variance is calculated as

$$\text{Var}(X) = \int_{-\infty}^{\infty} x^2 f(x)\, dx - \mu^2$$

$$= \int_{-1}^{2} \frac{2}{5} x^2 |x|\, dx - \frac{196}{225}$$

$$= \frac{2}{5} \left(\int_{-1}^{0} (-x^3)\, dx + \int_{0}^{2} x^3\, dx \right) - \frac{196}{225}$$

$$= \frac{373}{450}.$$

(12) **Proof.**   Suppose $m, n \in \mathbb{Z}$, then

$$n - m \quad \text{is even} \Longleftrightarrow n - m + 2m = n + m \quad \text{is even.} \qquad \square$$

(13) If $f(x)$ is three times continuously differentiable at $x = x_0$ then we can write $f(x)$ as

$$f(x) = f(x_0) + f'(x_0)(x - x_0) + f''(x_0)\frac{(x - x_0)^2}{2} + f'''(z)\frac{(x - x_0)^3}{6}$$

where z lies between x and x_0. Hence

$$f(x_0 + h) = f(x_0) + f'(x_0)h + f''(x_0)\frac{h^2}{2} + f'''(z_1)\frac{h^3}{6}$$

$$f(x_0 - h) = f(x_0) - f'(x_0)h + f''(x_0)\frac{h^2}{2} - f'''(z_2)\frac{h^3}{6}$$

Subtracting the second equation from the first and solving for $f'(x_0)$ yields

$$f'(x_0) = \frac{f(x_0 + h) - f(x_0 - h)}{2h} - (f'''(z_1) + f'''(z_2))\frac{h^2}{12}.$$

Thus when h is small,

$$f'(x_0) \approx \frac{f(x_0 + h) - f(x_0 - h)}{2h}.$$

(14) The limit as stated is indeterminate of the form $0 \cdot \infty$, thus with some algebra and l'Hôpital's Rule we obtain

$$
\lim_{M \to \infty} (M - K) \int_M^\infty f(t)\, dt = \lim_{M \to \infty} \frac{\int_M^\infty f(t)\, dt}{(M - K)^{-1}}
$$

$$
= \lim_{M \to \infty} \frac{-f(M)}{-(M - K)^{-2}}
$$

$$
= \lim_{M \to \infty} (M - K)^2 f(M)
$$

$$
= 0.
$$

The last equality is true since f is a probability density function for a random variable whose variance is finite.

(15) The improper integral can be evaluated as follows.

$$
\frac{\sigma}{\sqrt{2\pi}} \int_{(K-\mu)/\sigma}^\infty t e^{-t^2/2}\, dt = \lim_{M \to \infty} \frac{\sigma}{\sqrt{2\pi}} \int_{(K-\mu)/\sigma}^M t e^{-t^2/2}\, dt
$$

$$
= \lim_{M \to \infty} \left(\frac{\sigma}{\sqrt{2\pi}} - e^{-t^2/2} \Big|_{(K-\mu)/\sigma}^M \right)
$$

$$
= \frac{\sigma}{\sqrt{2\pi}} \lim_{M \to \infty} \left(-e^{-M^2/2} + e^{-(K-\mu)^2/2\sigma^2} \right)
$$

$$
= \frac{\sigma}{\sqrt{2\pi}} e^{-(K-\mu)^2/2\sigma^2}
$$

(16) Using the method of integration by substitution, where $u = x^2/(4kt)$ we obtain

$$
\int \frac{x}{2\sqrt{k\pi t}} e^{-\frac{x^2}{4kt}}\, dx = \sqrt{\frac{kt}{\pi}} \int e^{-u}\, du
$$

$$
= -\sqrt{\frac{kt}{\pi}} e^{-u} + C
$$

$$
= -\sqrt{\frac{kt}{\pi}} e^{-\frac{x^2}{4kt}} + C
$$

where C is a constant of integration.

(17) Using the technique of integration by parts with

$$
u = \frac{x}{\sqrt{k\pi t}} \qquad\qquad v = -kt e^{-\frac{x^2}{4kt}}
$$

$$
du = \frac{1}{\sqrt{k\pi t}}\, dx \qquad dv = \frac{x}{2} e^{-\frac{x^2}{4kt}}\, dx
$$

we obtain

$$\int \frac{x^2}{2\sqrt{k\pi t}} e^{-\frac{x^2}{4kt}}\, dx = -\frac{ktx}{\sqrt{k\pi t}} e^{-\frac{x^2}{4kt}} + \int \frac{kt}{\sqrt{k\pi t}} e^{-\frac{x^2}{4kt}}\, dx$$

$$= \sqrt{\frac{kt}{\pi}}\left(-xe^{-\frac{x^2}{4kt}} + \int e^{-\frac{x^2}{4kt}}\, dx \right)$$

(18) Assuming that $k > 0$ and $t > 0$ we have

$$\lim_{M\to\infty} 2ktMe^{-M^2/4kt} = \lim_{M\to\infty} \frac{2ktM}{e^{M^2/4kt}}.$$

The limit on the right-hand side of the equation is indeterminate of the form ∞/∞ and thus we apply l'Hôpital's Rule.

$$\lim_{M\to\infty} \frac{2ktM}{e^{M^2/4kt}} = \lim_{M\to\infty} \frac{2kt}{\frac{M}{2kt}e^{M^2/4kt}}$$

$$= \lim_{M\to\infty} \frac{(2kt)^2}{Me^{M^2/4kt}}$$

$$= 0.$$

(19) The probability distribution function for the standard normal random variable is

$$f(x) = \frac{1}{\sqrt{2\pi}} e^{-x^2/2}$$

and thus

(a) $P\left(-1 < X < 1\right) = \dfrac{1}{\sqrt{2\pi}} \displaystyle\int_{-1}^{1} e^{-x^2/2}\, dx \approx 0.682689,$

(b) $P\left(-2 < X < 2\right) = \dfrac{1}{\sqrt{2\pi}} \displaystyle\int_{-2}^{2} e^{-x^2/2}\, dx \approx 0.9545,$

(c) $P\left(-3 < X < 3\right) = \dfrac{1}{\sqrt{2\pi}} \displaystyle\int_{-3}^{3} e^{-x^2/2}\, dx \approx 0.9973,$

(d) $P\left(1 < X < 3\right) = \dfrac{1}{\sqrt{2\pi}} \displaystyle\int_{1}^{3} e^{-x^2/2}\, dx \approx 0.157305.$

(20) **Proof.** Suppose X is a normally distributed random variable with mean μ and variance σ^2. We will let $z = g(x) = (x - \mu)/\sigma$.

$$E[Z] = E[g(X)]$$

$$= \frac{1}{\sigma\sqrt{2\pi}} \int_{-\infty}^{\infty} \frac{x - \mu}{\sigma} e^{-\frac{(x-\mu)^2}{2\sigma^2}}\, dx$$

Making the substitution $u = (x - \mu)/\sigma$ yields

$$E[Z] = \frac{1}{\sqrt{2\pi}} \int_{-\infty}^{\infty} u e^{-\frac{u^2}{2}} \, du$$
$$= 0.$$

Therefore

$$\text{Var}(Z) = E[Z^2] - (E[Z])^2$$
$$= E\left[(g(X))^2\right]$$
$$= \frac{1}{\sigma\sqrt{2\pi}} \int_{-\infty}^{\infty} \left(\frac{x - \mu}{\sigma}\right)^2 e^{-\frac{(x-\mu)^2}{2\sigma^2}} \, dx$$
$$= \frac{1}{\sqrt{2\pi}} \int_{-\infty}^{\infty} u^2 e^{-\frac{u^2}{2}} \, du$$
$$= 1,$$

where we made the substitution $u = (x - \mu)/\sigma$. $\qquad\square$

(21) Think of the consecutive years as year 1 and 2. The rainfalls are normally distributed random variables, call them X_1 and X_2 in each year with means $\mu_1 = \mu_2 = 14$ and standard deviations $\sigma_1 = \sigma_2 = 3.2$. Let the random variable X be the sum of the rainfalls in two consecutive years, then $\mu = E[X] = E[X_1] + E[X_2] = \mu_1 + \mu_2 = 28$. Likewise

$$\sigma^2 = \text{Var}(X) = \text{Var}(X_1) + \text{Var}(X_2) = \sigma_1^2 + \sigma_2^2 = (3.2)^2 + (3.2)^2 = 20.48,$$

assuming that rainfalls in different years are independent. Now let $Z = (X - \mu)/\sigma$ and then

$$P(X > 30) = P\left(Z > \frac{30 - 28}{\sqrt{20.48}}\right)$$
$$= P(Z > 0.44192)$$
$$= 1 - P(Z \le 0.44192)$$
$$= 1 - \phi(0.44192)$$
$$\approx 0.329266$$

(22) Let the random variable X represent the ratio of consecutive days

selling prices of the security.

$$P(X > 1) = P(\ln X > 0)$$
$$= P\left(Z > \frac{0 - \mu}{\sigma}\right)$$
$$= P\left(Z > \frac{0 - 0.01}{0.05}\right)$$
$$= P(Z > -0.20)$$
$$= \phi(0.20)$$
$$\approx 0.57926$$

Thus the probability of a one-day increase in the price of the security is approximately 0.57926. Consequently the probability of a one-day decrease in the price of the security is approximately $1 - 0.57926 = 0.42074$. The probability of a four-day decrease in the selling price of the security is

$$P(X^4 < 1) = P(4 \ln X < 0)$$
$$= P\left(Z < \frac{0 - 4\mu}{2\sigma}\right)$$
$$= P\left(Z < \frac{0 - 0.04}{0.10}\right)$$
$$= P(Z < -0.40)$$
$$= \phi(-0.40)$$
$$\approx 0.344578.$$

(23) Since X is uniformly distributed on $[a, b]$ its probability distribution function is

$$f(x) = \begin{cases} \frac{1}{b-a} & \text{if } a \le x \le b, \\ 0 & \text{otherwise.} \end{cases}$$

We will make use of Theorem 3.9.

$$\int_x^\infty f(t)\, dt = \begin{cases} 1 & \text{if } x < a \\ \frac{b-x}{b-a} & \text{if } a \le x \le b \\ 0 & \text{if } x > b \end{cases}$$

Thus

$$E\left[(X-K)^+\right] = \int_K^\infty \left(\int_x^\infty f(t)\,dt\right)dx$$

$$= \begin{cases} \int_K^a 1\,dx + \int_a^b \frac{b-x}{b-a}\,dx & \text{if } K < a \\ \int_K^b \frac{b-x}{b-a}\,dx & \text{if } a \le K \le b \\ 0 & \text{if } K > b \end{cases}$$

$$= \begin{cases} \frac{a+b}{2} - K & \text{if } K < a \\ \frac{(b-K)^2}{2(b-a)} & \text{if } a \le K \le b \\ 0 & \text{if } K > b \end{cases}$$

(24) The result follows graphically from the symmetry of the standard normal distribution about the y-axis. See Fig. 3.2. However, if we wish to establish this result using the definition of the standard normal distribution, we have

$$\phi(-x) = \frac{1}{\sqrt{2\pi}} \int_{-\infty}^{-x} e^{-t^2/2}\,dt.$$

Now we make the substitution $u = -t$.

$$\frac{1}{\sqrt{2\pi}} \int_{-\infty}^{-x} e^{-t^2/2}\,dt = -\frac{1}{\sqrt{2\pi}} \int_\infty^x e^{-u^2/2}\,du$$

$$= \frac{1}{\sqrt{2\pi}} \int_x^\infty e^{-u^2/2}\,du$$

$$= 1 - \frac{1}{\sqrt{2\pi}} \int_{-\infty}^x e^{-u^2/2}\,du$$

$$\phi(-x) = 1 - \phi(x)$$

(25) **Proof.** Using the definition of variance and Eqs. (3.15) and (3.17) we have

$$\text{Var}\left((X-K)^+\right)$$

$$= E\left[\left((X-K)^+\right)^2\right] - \left(E\left[(X-K)^+\right]\right)^2$$

$$= \left((\mu-2K)^2 + \sigma^2\right)\phi\left(\frac{\mu-2K}{\sigma}\right) + \frac{(\mu-2K)\sigma}{\sqrt{2\pi}}e^{-(\mu-2K)^2/2\sigma^2}$$

$$- \left(\frac{\sigma}{\sqrt{2\pi}}e^{-(\mu-K)^2/2\sigma^2} + (\mu-K)\phi\left(\frac{\mu-K}{\sigma}\right)\right)^2.$$

$\square$

(26) **Proof.**   Using the definition of variance and Eqs. (3.16) and (3.19) we have

$$\mathrm{Var}\left((X-K)^+\right)$$
$$= \mathrm{E}\left[\left((X-K)^+\right)^2\right] - \left(\mathrm{E}\left[(X-K)^+\right]\right)^2$$
$$= e^{2(\mu+\sigma^2)}\phi(w+2\sigma) - 2Ke^{\mu+\sigma^2/2}\phi(w+\sigma) + K^2\phi(w)$$
$$- \left(e^{\mu+\sigma^2/2}\phi(w+\sigma) - K\phi(w)\right)^2$$

where $w = (\mu - \ln K)/\sigma$.   □

B.4   The Arbitrage Theorem

(1) Converting from the odds against to probabilities that the outcomes may occur produces a table of probabilities like the following:

Outcome	Probability
A	$\frac{1}{3}$
B	$\frac{1}{4}$
C	$\frac{1}{2}$

Since the probabilities sum to $\frac{13}{12} > 1$ the Arbitrage Theorem guarantees the existence of a betting strategy which yields a positive payoff regardless of the outcome. Suppose we wager x, y, and z on outcomes A, B, and C respectively. The payoffs under different winning scenarios are listed in the next table.

Winning Outcome	Payoff
A	$2x - y - z$
B	$-x + 3y - z$
C	$-x - y + z$

We need only find values for x, y, and z which make the three expressions in the column on the right positive. The graphic below shows the "cone" where all three payoffs are positive. Notice that each amount wagered must be negative (this just means we take the wagers from others). By inspection we can see that $x = -3$, $y = -2.5$, and $z = -5$ is one solution to this set of inequalities.

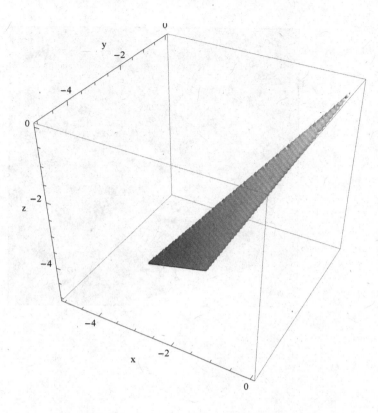

(2) The region in convex and unbounded. There are three corners to the region located at $(0, 6)$, $(3/2, 3/2)$, and $(6, 0)$. The region is on the next page.

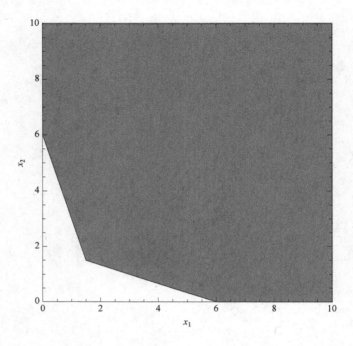

(3) The cost function will be minimized at the smallest value of k for which the graph of $x_1 + x_2 = k$ intersects the feasible region sketched in exercise 2. This occurs when $(x_1, x_2) = (3/2, 3/2)$. Thus the minimum cost is $k = 3$.

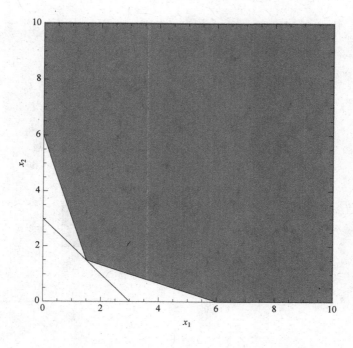

(4) $0 \leq x_1 \leq 2$ and $0 \leq x_2 \leq 3$

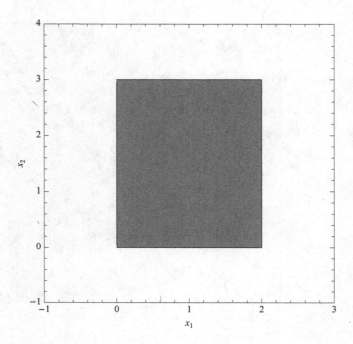

(5) We can think of the solution as a system of inequalities:

$$x_1 \geq 0$$
$$x_2 \geq 0$$
$$x_3 \geq 0$$
$$x_4 \geq 0$$
$$x_1 + x_3 = 2$$
$$x_2 + x_4 = 3$$

where x_3 and x_4 are the newly introduced slack variables. The last two equations can be re-written in matrix/vector form as

$$A\mathbf{x} = \begin{bmatrix} 1 & 0 & 1 & 0 \\ 0 & 1 & 0 & 1 \end{bmatrix} \begin{bmatrix} x_1 \\ x_2 \\ x_3 \\ x_4 \end{bmatrix} = \begin{bmatrix} 2 \\ 3 \end{bmatrix} = \mathbf{b}$$

(6) Let vector $\mathbf{b} = \langle 1, 1, 2 \rangle$ and $\mathbf{y} = \langle y_1, y_2, y_3 \rangle$. The problem as stated

can be thought of as a dual problem:

Minimize: $\mathbf{b}^T\mathbf{y}$ subject to $A^T\mathbf{y} = \begin{bmatrix} 1 & 2 & 3 \end{bmatrix}\mathbf{y} \geq [15] = \mathbf{c}$.

This is equivalent to the primal problem:

Maximize: $\mathbf{c}^T\mathbf{x} = 15\mathbf{x}$ subject to $A\mathbf{x} \leq \mathbf{b}$,

where $\mathbf{x} = \langle x_1 \rangle$. From the inequality constraint in the primal problem we know that $x_1 \leq 1/2$. Thus the maximum value of $15x_1$ is $15/2$. Thus the minimum value of $\mathbf{b}^T\mathbf{y} = 15/2$. According to Theorem 4.3 this minimum occurs when $\mathbf{y} = \langle 0, \frac{15}{2}, 0 \rangle$.

(7) Let vector $\mathbf{b} = \langle 0, 7, 9, 0 \rangle$ and $\mathbf{y} = \langle y_1, y_2, y_3, y_4 \rangle$. The problem as stated can be thought of as the dual problem:

Minimize: $\mathbf{b}^T\mathbf{y}$ subject to $A^T\mathbf{y} = \begin{bmatrix} 1 & 1 & 1 & 0 \\ 0 & 1 & 1 & 2 \end{bmatrix}\mathbf{y} \geq \begin{bmatrix} 5 \\ 1 \end{bmatrix} = \mathbf{c}$.

This is equivalent to the primal problem:

Maximize: $\mathbf{c}^T\mathbf{x} = 5x_1 + x_2$ subject to $A\mathbf{x} \leq \mathbf{b}$.

The constraint inequality for the primal problem is equivalent to the following system of linear inequalities.

$$x_1 \leq 0$$
$$x_1 + x_2 \leq 7$$
$$x_1 + x_2 \leq 9$$
$$2x_2 \leq 0$$

The first and last of these inequalities imply strict inequality in the second and third. Thus by Theorem 4.3 $y_2 = y_3 = 0$. Therefore the dual problem can be restated as the following:

Minimize: $\langle 0, 0 \rangle \cdot \langle y_1, y_4 \rangle$ subject to $\begin{bmatrix} 1 & 0 \\ 0 & 2 \end{bmatrix}\begin{bmatrix} y_1 \\ y_4 \end{bmatrix} \geq \begin{bmatrix} 5 \\ 1 \end{bmatrix}$.

Thus the minimum value of the cost function is 0 and occurs where $\mathbf{y} = \langle 5, 0, 0, 1/2 \rangle$.

(8) In this case we can introduce a slack variable x_4 to convert the inequality constraint in the problem to an equality constraint. Thus we want to think of the linear problem as:

Minimize $x_1 + x_2 + x_3$ subject to $2x_1 + x_2 = 4$ and $x_3 + x_4 = 6$ with $x_i \geq 0$ for $i = 1, 2, 3, 4$.

If we define the following vectors and matrix

$$\mathbf{c} = \langle 1, 1, 1, 0 \rangle, \ \mathbf{x} = \langle x_1, x_2, x_3, x_4 \rangle, \ A = \begin{bmatrix} 2 & 1 & 0 & 0 \\ 0 & 0 & 1 & 1 \end{bmatrix}, \quad \text{and } \mathbf{b} = \langle 4, 6 \rangle,$$

then the linear problem is the one of minimizing $\mathbf{c}^T \mathbf{x}$ subject to $A\mathbf{x} = \mathbf{b}$ and $\mathbf{x} \geq \mathbf{0}$. The dual to this problem is the one of maximizing $\mathbf{b}^T \mathbf{y}$ subject to $A^T \mathbf{y} \leq \mathbf{c}$ where $\mathbf{y} = \langle y_1, y_2 \rangle$. This is equivalent to maximizing $4y_1 + 6y_2$ subject to the system of inequalities

$$2y_1 \leq 1$$
$$y_1 \leq 1$$
$$y_2 \leq 1$$
$$y_2 \leq 0.$$

(9) Paying attention to the inequality constraints of the linear problem indicates that $y_1 \leq 1/2$ and $y_2 \leq 0$. Thus by Theorem 4.3, $x_2 = x_3 = 0$. We can also see that the maximum of $\mathbf{b}^T \mathbf{y}$ will be 2. This is also the minimum of the dual problem. However, we can also solve the dual problem directly after re-writing it as

Minimize: $\langle 1, 0 \rangle \cdot \langle x_1, x_4 \rangle$ subject to $\begin{bmatrix} 2 & 0 \\ 0 & 1 \end{bmatrix} \begin{bmatrix} x_1 \\ x_4 \end{bmatrix} = \begin{bmatrix} 4 \\ 6 \end{bmatrix}$.

From the constraint equation we see that now $\mathbf{x} = \langle 2, 0, 0, 6 \rangle$ and the minimum value of the cost function is 2.

(10) Written in matrix/vector form this linear problem becomes:

Maximize $\mathbf{b}^T \mathbf{y} = \langle 2, 0, 4 \rangle \cdot \langle y_1, y_2, y_3 \rangle$ subject to

$$A^T \mathbf{y} = \begin{bmatrix} 1 & 1 & 0 \\ 0 & 1 & 2 \end{bmatrix} \begin{bmatrix} y_1 \\ y_2 \\ y_3 \end{bmatrix} \leq \begin{bmatrix} 1 \\ 1 \end{bmatrix} = \mathbf{c}.$$

Thus the dual problem is

Minimize $\mathbf{c}^T \mathbf{x} = x_1 + x_2$ subject to

$$A\mathbf{x} = \begin{bmatrix} 1 & 0 \\ 1 & 1 \\ 0 & 2 \end{bmatrix} \begin{bmatrix} x_1 \\ x_2 \end{bmatrix} = \begin{bmatrix} 2 \\ 0 \\ 4 \end{bmatrix} = \mathbf{b}.$$

(11) We can see from the constraint equation for the dual problem that $\mathbf{x} = \langle x_1, x_2 \rangle = \langle 2, 2 \rangle$. Thus the minimum value of the cost function is

$\mathbf{c}^T\mathbf{x} = \langle 1,1 \rangle \cdot \langle 2,2 \rangle = 4$. Therefore the maximum of the dual problem is likewise 4. Since $x_1 + x_2 = 4 > 0$ then $y_2 = 0$ by Theorem 4.3 and we may re-write the dual problem as "maximize $2y_1 + 4y_3 = 4$ subject to $y_1 \leq 1$ and $2y_3 \leq 1$." This maximum occurs at $\mathbf{y} = \langle 1,0,1/2 \rangle$.

(12) Assume that $a < b$ and suppose $a \geq 0$ and $b \leq 0$. Then $b \leq 0 \leq a$ which implies $b \leq a$ a contradiction.

B.5   Random Walks and Brownian Motion

(1) If f is a continuous, twice differentiable function whose second derivative is 0 on the interval $[0, A]$, then f must be a linear function of the form $f(x) = ax + b$ where a and b are constants. Since $f(0) = 0$ then $b = 0$. Since $f(A) = 1$ then $a = 1/A$. Thus we have $f(x) = x/A$. This agrees with our earlier derivation of exit probabilities for the discrete, symmetric random walk.

(2) If g is a continuous, twice differentiable function whose second derivative is -2 on the interval $[0, A]$, then g must be a quadratic function of the form $g(x) = ax^2 + bx + c$ where a, b, and c are constants. Since $g(0) = 0$ then $c = 0$. Since $g''(x) = -2$ then $a = -1$ which implies $g(x) = x(b - x)$. Since $g(A) = 0$ then $b = A$. Thus we have $g(x) = x(A - x)$. This agrees with our earlier derivation of the conditional stopping times for the discrete, symmetric random walk.

(3) If $f(x)$ is four times continuously differentiable at $x = x_0$ then we can write $f(x)$ as

$$f(x) = f(x_0) + f'(x_0)(x - x_0) + f''(x_0)\frac{(x - x_0)^2}{2} + f'''(x_0)\frac{(x - x_0)^3}{6}$$
$$+ f^{(4)}(z)\frac{(x - x_0)^4}{24}$$

where z lies between x and x_0. Hence

$$f(x_0 + h) = f(x_0) + f'(x_0)h + f''(x_0)\frac{h^2}{2} + f'''(x_0)\frac{h^3}{6} + f^{(4)}(z_1)\frac{h^4}{24}$$
$$f(x_0 - h) = f(x_0) - f'(x_0)h + f''(x_0)\frac{h^2}{2} - f'''(x_0)\frac{h^3}{6} + f^{(4)}(z_2)\frac{h^4}{24}$$

Adding the two equations and solving for $f''(x_0)$ yields

$$f''(x_0) = \frac{f(x_0 + h) - 2f(x_0) + f(x_0 - h)}{h^2} - \left(f^{(4)}(z_1) + f^{(4)}(z_2) \right)\frac{h^2}{24}.$$

Thus when h is small,

$$f''(x_0) \approx \frac{f(x_0 + h) - 2f(x_0) + f(x_0 - h)}{h^2}.$$

(4) This is an example of a symmetric random walk. Because of the spatial homogeneity property of the random walk, the desired probability is equivalent to determining the probability that a symmetric random walk initially in state $S(0) = 50$ will attain a value of 75 while avoiding the absorbing boundary at 0. Using Theorem 5.3, the probability is $\mathcal{P}_{75}(50) = 50/75 = 2/3$.

(5) As in exercise 4 the exit time is equivalent to the exit time of a symmetric random walk on the discrete interval $[0, 75]$ which begins in the state $S(0) = 50$. Using Theorem 5.4, the probability is $\Omega_{0,75}(50) = 50(75 - 50) = 1250$.

(6) As in exercise 4 the conditional exit time is equivalent to the conditional exit time of a symmetric random walk on the discrete interval $[0, 75]$ which begins in the state $S(0) = 50$. Using Eq. 5.10 with $A = 75$ and $i = 50$ we have

$$\Omega_{75}(50) = \frac{1}{3}(75^2 - 50^2) = 3125/3 \approx 1041.67.$$

(7) Let $P = P(0)e^{\mu t}$, then

$$\begin{aligned}
\frac{dP}{dt} &= \frac{d}{dt}\left(P(0)e^{\mu t}\right) \\
&= P(0)\frac{d}{dt}\left(e^{\mu t}\right) \\
&= P(0)e^{\mu t}\frac{d}{dt}\left(\mu t\right) \\
&= P(0)e^{\mu t}\mu \\
&= \mu P.
\end{aligned}$$

(8) Various results are possible due to the fact that we are simulating a stochastic process. Using the data collected in Appendix A, the closing stock price on the last day for which data was collected was \$42.86. From the data the estimated drift parameter and volatility are $\mu = -0.000555$ day^{-1} and $\sigma = 0.028139$ day^{-1} respectively. If we substitute these values in Eq. (5.36) and set $P(t_0) = 42.86$ with $\Delta t = 1$ we can iteratively generate 252 (or any other amount for that matter) new days' closing prices. The random walk generated is pictured below.

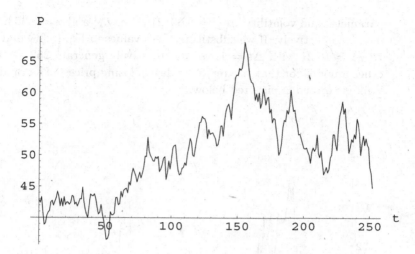

(9) From the website `finance.yahoo.com` the closing prices of stock for Continental Airlines, Inc. (symbol CAL) were captured for the time period of 06/15/2004 until 06/15/2005. A histogram of $\Delta X = \ln P(t_{i+1}) - \ln P(t_i)$ is shown below. The mean and standard deviation of ΔX are $\mu_{\Delta X} = 0.00121091$ day^{-1} and $\sigma_{\Delta X} = 0.0332183$ day^{-1} respectively.

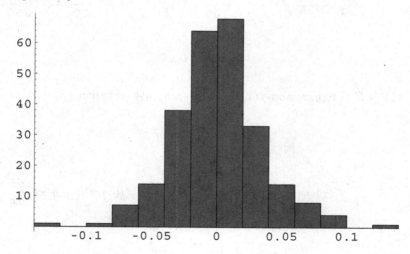

(10) Various results are possible due to the fact that we are simulating a stochastic process. The closing stock price on the last day for which data was collected was \$10.21. From the data the estimated drift

parameter and volatility are $\mu = 0.00121091$ day^{-1} and $\sigma = 0.0332183$ day^{-1} respectively. If we substitute these values in Eq. (5.36) and set $P(t_0) = 10.21$ with $\Delta t = 1$ we can iteratively generate 252 (or any other amount for that matter) new days' closing prices. The random walk generated is pictured below.

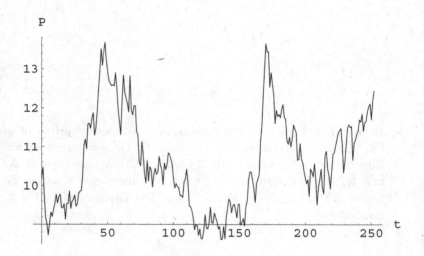

(11) If $F \equiv F(y, z)$ and $f(x) = F(y_0 + xh, z_0 + xk)$ then

$$f'(x) = F_y(y_0 + xh, z_0 + xk)\frac{dy}{dx} + F_z(y_0 + xh, z_0 + xk)\frac{dz}{dx}$$
$$= F_y(y_0 + xh, z_0 + xk)\frac{d}{dx}[y_0 + xh]$$
$$+ F_z(y_0 + xh, z_0 + xk)\frac{d}{dx}[z_0 + xk]$$
$$= F_y(y_0 + xh, z_0 + xk)h + F_z(y_0 + xh, z_0 + xk)k$$
$$f'(0) = F_y(y_0, z_0)h + F_z(y_0, z_0)k.$$

Differentiating once again produces

$$f''(x) = F_{yy}(y_0 + xh, z_0 + xk)h\frac{dy}{dx} + F_{yz}(y_0 + xh, z_0 + xk)h\frac{dz}{dx}$$
$$+ F_{zy}(y_0 + xh, z_0 + xk)k\frac{dy}{dx} + F_{zz}(y_0 + xh, z_0 + xk)k\frac{dz}{dx}$$
$$= hF_{yy}(y_0 + xh, z_0 + xk)\frac{d}{dx}[y_0 + xh]$$
$$+ hF_{yz}(y_0 + xh, z_0 + xk)\frac{d}{dx}[z_0 + xk]$$
$$+ kF_{zy}(y_0 + xh, z_0 + xk)\frac{d}{dx}[y_0 + xh]$$
$$+ kF_{zz}(y_0 + xh, z_0 + xk)\frac{d}{dx}[z_0 + xk]$$
$$= h^2 F_{yy}(y_0 + xh, z_0 + xk) + 2hk F_{yz}(y_0 + xh, z_0 + xk)$$
$$+ k^2 F_{zz}(y_0 + xh, z_0 + xk)$$
$$f''(0) = h^2 F_{yy}(y_0, z_0) + 2hk F_{yz}(y_0, z_0) + k^2 F_{zz}(y_0, z_0).$$

By definition the Taylor remainder of order 3 for $f(x)$ is $R_3(x) = \frac{f'''(\alpha)}{3!}x^3$ for some α between 0 and x. If we notice the binomial coefficient pattern among the partial derivatives in the first and second derivatives of $f(x)$ then we can readily see that

$$f'''(x) = h^3 F_{yyy}(y_0 + xh, z_0 + xk) + 3h^2 k F_{yyz}(y_0 + xh, z_0 + xk)$$
$$+ 3hk^2 F_{yzz}(y_0 + xh, z_0 + xk) + k^3 F_{zzz}(y_0 + xh, z_0 + xk).$$

From this expression we may directly determine the Taylor remainder of order 3.

(12) If we apply Itô's Lemma with $a(P, t) = \mu P$ and $b(P, t) = \sigma P$ and $Y = P^n$, then

$$dY = \left(\mu P n P^{n-1} + \frac{1}{2}(\sigma P)^2 n(n-1)P^{n-2}\right) dt + \sigma P n P^{n-1} \, dW(t)$$
$$= \left(n\mu P^n + \frac{n(n-1)\sigma^2}{2}P^n\right) dt + n\sigma P^n \, dW(t)$$

(13) If we apply Itô's Lemma with $a(P, t) = \mu P$ and $b(P, t) = \sigma P$ and

$Y = \ln P$, then

$$dY = \left(\mu P \left[\frac{1}{P} \right] + \frac{1}{2}(\sigma P)^2 \left[\frac{-1}{P^2} \right] \right) dt + \sigma P \left[\frac{1}{P} \right] dW(t)$$

$$= \left(\mu - \frac{\sigma^2}{2} \right) dt + \sigma \, dW(t)$$

(14)

$$E\left[X^4 \right] = \frac{1}{\sigma \sqrt{2\pi}} \int_{-\infty}^{\infty} x^4 e^{-x^2/2\sigma^2} \, dx$$

$$= \frac{1}{\sigma \sqrt{2\pi}} \int_{-\infty}^{\infty} x^3 \left(x e^{-x^2/2\sigma^2} \right) \, dx$$

If we use integration by parts with

$$u = x^3 \qquad\qquad v = -\sigma^2 e^{-x^2/2\sigma^2}$$
$$du = 3x^2 \, dx \qquad\qquad dv = x e^{-x^2/2\sigma^2} \, dx$$

then the integral for the expected value above becomes

$$E\left[X^4 \right] = \frac{1}{\sigma \sqrt{2\pi}} \left(-\sigma^2 x^3 e^{-x^2/2\sigma^2} \right) \Big|_{-\infty}^{\infty} + \frac{3\sigma^2}{\sigma \sqrt{2\pi}} \int_{-\infty}^{\infty} x^2 e^{-x^2/2\sigma^2} \, dx$$

$$= \frac{3\sigma^2}{\sigma \sqrt{2\pi}} \int_{-\infty}^{\infty} x^2 e^{-x^2/2\sigma^2} \, dx$$

$$= \frac{3\sigma^2}{\sigma \sqrt{2\pi}} \int_{-\infty}^{\infty} x \left(x e^{-x^2/2\sigma^2} \right) \, dx.$$

Again using integration by parts with

$$u = x \qquad\qquad v = -\sigma^2 e^{-x^2/2\sigma^2}$$
$$du = dx \qquad\qquad dv = x e^{-x^2/2\sigma^2} \, dx$$

then the integral for the expected value above becomes

$$E\left[X^4 \right] = \frac{3\sigma^2}{\sigma \sqrt{2\pi}} \left[\left(-\sigma^2 e^{-x^2/2\sigma^2} \right) \Big|_{-\infty}^{\infty} + \int_{-\infty}^{\infty} \sigma^2 e^{-x^2/2\sigma^2} \, dx \right]$$

$$= 3\sigma^4 \left(\frac{1}{\sigma \sqrt{2\pi}} \int_{-\infty}^{\infty} e^{-x^2/2\sigma^2} \, dx \right)$$

$$= 3\sigma^4.$$

B.6 Forwards and Futures

(1) Let the bid price be B, then the ask price is $A = B + 0.25$. The net cost to the investor is

$$1000B - 1000A = -250.$$

The round trip cost is $250.

(2) Suppose the buyer pays the seller $S < S(0)$, then the new owner may then instantly sell the stock for $S(0)$ guaranteeing a profit of $S(0) - S > 0$.

On the other hand the seller initially possesses a portfolio worth $S(0)$. If the buyer is willing to pay the seller $S > S(0)$, then the seller may borrow the stock from another investor and sell it to the buyer for S. Then the buyer immediately re-purchases the stock for $S(0)$ and returns it to the other investor. The seller is left with a profit of $S - S(0) > 0$.

(3) The future value of $17 one month hence at continuously compounded interest rate 5.05% is

$$F = 17e^{0.0505/12} \approx 17.0717.$$

(4) The forward contract is worth

$$F = 23e^{0.0475(3/12)} \approx 23.2748.$$

(5) The value of the prepaid forward is

$$F = 97 - 2.50e^{-0.0365(6/12)} - 2.75e^{-0.0365(12/12)} \approx 91.8938.$$

The value of a forward contract on the dividend paying stock is

$$F = 91.8938e^{0.0365(12/12)} \approx 95.3099.$$

(6) The value of three-month prepaid forward on the investment is

$$F = 195e^{-0.0195(3/12)} \approx 194.052.$$

The value of a three-month forward contract on the investment is

$$F = 195e^{(0.0455-0.0195)(3/12)} \approx 196.272.$$

(7) The equation

$$990 = 1000e^{(0.0505-r)(1/2)}$$

implies $r \approx 0.0706$. Thus the continuous dividend rate is 7.06%.

(8) Using inequality (6.1)

$$(74 - 2(2))e^{0.03(6/12)} \le F \le (75 + 2(2))e^{0.04(6/12)}$$
$$71.0579 \le F \le 80.5959.$$

(9) The initial margin is in the amount of

$$(100)(950)(0.125) = 11875,$$

which after one month of interest compounded continuously at 6% per annum will be 11934.52. A margin call will be issued only if the price of the security at the one-month mark satisfies

$$11934.52 + (S(1) - 950)(100) \le S(1)(100)(0.125)$$
$$S(1) \le 949.32.$$

(10) If the current price of the stock is $S(0)$ then the value of the forward contract is $S(0)e^{rT}$. However, the price of a forward contract is not paid until it matures. Hence the present value of the forward contract is $S(0)e^{rT}e^{-rT} = S(0)$. If this amount is lent at interest rate r continuously compounded for time T, then the payoff of entering into the long forward contract and lending the present value of the contract is

$$\underbrace{S(T) - S(0)e^{rT}}_{\text{forward}} + \underbrace{S(0)e^{rt} - S(0)}_{\text{loan}} = S(T) - S(0).$$

If the stock is purchased at $t = 0$ and sold at time $t = T$ the payoff is

$$S(T) - S(0).$$

Hence the two investments have the same payoff.

(11)

Day	No. of Contracts	Futures Price	Price Change	Margin Balance	Margin Call
0	1500	850.00	—	191,250.00	—
1	1500	774.67.	−75.33	78,302.40	95,997.60
2	1500	779.39	4.73	181,438.67	0.00
3	1500	778.42	−0.97	180,027.86	0.00
4	1500	749.56	−28.86	136,791.47	31,860.23
5	1500	742.87	−6.69	158,659.45	8,486.48
6	1500	735.64	−7.23	156,345.58	9,173.43
7	1500	741.59	5.95	174,489.36	0.00
8	1500	759.88	18.29	201,965.75	0.00
9	1500	766.25	6.38	211,583.59	0.00
10	1500	805.36	39.11	270,308.23	0.00

The profit to the holder of the long position in the futures contract is the difference between the final margin balance and the future value of the initial margin.

$$270,308.23 - 191,250e^{10(0.10)/365} = 78,533.54.$$

B.7 Options

(1) Suppose the value of an American put option is P_a and that $P_a < P_e$, the value of a European put option on the same underlying security with identical strike price and exercise time. An investor could sell the European put option and buy the American put option and invest the net positive cash flow of $P_e - P_a$ in a risk-free bond at interest rate r. At expiry, in either case whether the owner of the European put exercises the option or not, and the investor does likewise, the investor has a portfolio with value $(P_e - P_a)e^{rT} > 0$.

(2) The net payoff will be the quantity $(S(T) - K)^+ - C$ where $K = 60$ and $C = 10$. The payoff is plotted below.

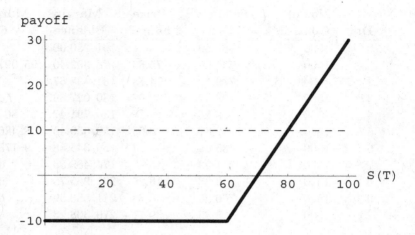

(3) We have already established that $C_e \leq C_a$ for European and American style options. Thus we need only show that $C_a(t) \leq S(t)$ for $0 \leq t \leq T$, where the expiry time is T. Let the strike price be K. If we suppose there is a time t^* at which $C_a(t^*) > S(t^*)$ then an investor could create a portfolio consisting of a short call and a long position in one share of the security. This portfolio generates a positive cash flow to the investor of $C_a(t^*) - S(t^*)$ which will be placed in a risk-free savings account earning interest at rate r compounded continuously. For any time $t^* \leq t \leq T$ the investor can close out their long position in the security and make a profit of

$$(C_a(t^*) - S(t^*))\, e^{r(t-t^*)} + \min\{K, S(t)\} > 0.$$

(4) Since $C_e \geq S - Ke^{-rT}$ then when $S = 29$, $K = 26$, $r = 0.06$, and $T = 1/4$,

$$C_e \geq 29 - 26e^{-(0.06)(0.25)}$$
$$\geq 3.39$$

(5) Here $S = 31$, $T = 1/4$, $C_e = 3$, $K = 31$, and $r = 0.10$ then according to the Put-Call Parity Formula in Eq. (7.1):

$$P_e + S = C_e + Ke^{-rT}$$
$$P_e + 31 = 3 + 31e^{-(0.10)(0.25)}$$
$$P_e = 2.23$$

(6) Here $S = 31$, $T = 1/4$, $C_e = 3$, $P_e = 2.25$, and $r = 0.10$ then according to the Put-Call Parity Formula in Eq. (7.1):

$$P_e + S = C_e + Ke^{-rT}$$
$$2.25 + 31 = 3 + Ke^{-(0.10)(0.25)}$$
$$K = 31.02$$

(7) Here $T = 1/6$, $K = 14$, $S = 11$, and $r = 0.07$ then according to the Put-Call Parity Formula in Eq. (7.1):

$$P_e = C_e - S + Ke^{-rT}$$
$$\geq -S + Ke^{-rT}$$
$$= -11 + 14e^{-(0.07)/6}$$
$$= 2.84$$

(8) An investor could take a long position in the put option and the security and short position in the call option while borrowing, at the risk-free rate, the strike price of the options. This generates an initial cash flow of

$$K + C_e - P_e - S = 30 + 3 - 1 - 31 = \$1.$$

This amount could be invested at the risk-free rate until the strike time arrives. At $t = 3$ months if $S(3) > 30 = K$, then the call option will be exercised and the investor cancels all positions with a portfolio worth

$$1e^{(0.10)(0.25)} + 30 - 30e^{(0.10)(0.25)} = \$0.27 > 0.$$

If the call option finishes out of the money, the investor exercises the put option and finishes with a portfolio worth again

$$1e^{(0.10)(0.25)} + 30 - 30e^{(0.10)(0.25)} = \$0.27 > 0.$$

(9) If $f_1(S,t)$ and $f_2(S,t)$ both satisfy Eq. (8.5) and if c_1 and c_2 are real

numbers then

$$(c_1 f_1 + c_2 f_2)_t + rS(c_1 f_1 + c_2 f_2)_S$$
$$+ \frac{1}{2}\sigma^2 S^2 (c_1 f_1 + c_2 f_2)_{SS} - r(c_1 f_1 + c_2 f_2)$$
$$= c_1 \left(f_{1,t} + rS f_{1,S} + \frac{1}{2}\sigma^2 S^2 f_{1,SS} - r f_1 \right)$$
$$+ c_2 \left(f_{2,t} + rS f_{2,S} + \frac{1}{2}\sigma^2 S^2 f_{2,SS} - r f_2 \right)$$
$$= c_1(0) + c_2(0)$$
$$= 0 \,.$$

Thus $f = c_1 f_1 + c_2 f_2$ is a solution to Eq. (8.5) as well.

(10) At time $t = T$ the payoff of the European put option will be

$$\max(0, K - S(T)) = (K - S(T))^+,$$

thus $F(S, T) = (K - S(T))^+$. If the stock is worthless (*i.e.*, along the boundary where $S = 0$) then the put option will be exercised and $F(0, t) = Ke^{-r(T-t)}$. As the price of the stock increases toward infinity the put option will not be exercised and thus $\lim_{S \to \infty} F(S, t) = 0$.

B.8 Solution of the Black-Scholes Equation

(1)

$$\hat{f}(w) = \int_{-\infty}^{\infty} f(x) e^{-iwx}\, dx$$
$$= \int_0^{\infty} e^{-ax} e^{-iwx}\, dx$$
$$= \lim_{M \to \infty} \int_0^M e^{-(a+iw)x}\, dx$$
$$= \frac{-1}{a + iw} \lim_{M \to \infty} e^{-(a+iw)x} \Big|_0^M$$
$$= \frac{-1}{a + iw} \lim_{M \to \infty} \left(e^{-(a+iw)M} - 1 \right)$$

Since $0 \leq |e^{-iwM}|e^{-aM} \leq e^{-aM}$ then by the Squeeze Theorem (Theorem 2.7 of [Smith and Minton (2002)]),

$$\lim_{M \to \infty} \left(e^{-(a+iw)M} - 1 \right) = -1 \quad \Longrightarrow \quad \hat{f}(w) = \frac{1}{a+iw}.$$

(2)

$$\begin{aligned}
f(x) &= \frac{1}{2\pi} \int_{-\infty}^{\infty} \hat{f}(w) e^{iwx} \, dw \\
&= \frac{1}{2\pi} \int_{0}^{\infty} e^{-aw} e^{iwx} \, dw \\
&= \frac{1}{2\pi} \lim_{M \to \infty} \int_{0}^{M} e^{-(a-ix)w} \, dw \\
&= \frac{1}{2\pi} \lim_{M \to \infty} \frac{-1}{a-ix} e^{-(a-ix)w} \Big|_{0}^{M} \\
&= \frac{-1}{2\pi(a-ix)} \lim_{M \to \infty} \left(e^{-(a-ix)M} - 1 \right)
\end{aligned}$$

Since $0 \leq |e^{-ixM}|e^{-aM} \leq e^{-aM}$ then by the Squeeze Theorem (Theorem 2.7 of [Smith and Minton (2002)]),

$$\lim_{M \to \infty} \left(e^{-(a-ix)M} - 1 \right) = -1 \quad \Longrightarrow \quad f(x) = \frac{1}{2\pi(a-ix)}.$$

(3)

$$\begin{aligned}
f(x) &= \frac{1}{2\pi} \int_{-\infty}^{\infty} \hat{f}(w) e^{iwx} \, dw \\
&= \frac{1}{2\pi} \int_{-\infty}^{\infty} e^{-aw^2} e^{iwx} \, dw \\
&= \frac{1}{2\pi} e^{-x^2/4a} \int_{-\infty}^{\infty} e^{-a\left(w - \frac{ix}{2a}\right)^2} \, dw \\
&= \frac{1}{2\pi} e^{-x^2/4a} \int_{-\infty}^{\infty} e^{-az^2} \, dz \\
&= \frac{1}{2\sqrt{\pi a}} e^{-x^2/4a}
\end{aligned}$$

The last step follows since

$$\int_{-\infty}^{\infty} e^{-az^2} \, dz = \sqrt{\frac{\pi}{a}}.$$

(4)

$$\frac{\partial}{\partial t}\left(F(S,t)\right) = K\frac{\partial}{\partial t}\left(v(x,\tau)\right)$$

$$= K\left(\frac{\partial v}{\partial x}\frac{dx}{dt} + \frac{\partial v}{\partial \tau}\frac{d\tau}{dt}\right)$$

$$= K\left(\frac{\partial v}{\partial x}\cdot 0 + \frac{\partial v}{\partial \tau}\cdot\left[\frac{-\sigma^2}{2}\right]\right)$$

$$= -\frac{K\sigma^2}{2}\frac{\partial v}{\partial \tau}$$

(5)

$$\frac{\partial^2 F}{\partial S^2} = \frac{\partial}{\partial S}\left(\frac{\partial F}{\partial S}\right)$$

$$= \frac{\partial}{\partial S}\left(e^{-x}v_x\right) \quad \text{(by Eq. (8.12))}$$

$$= -e^{-x}v_x\frac{dx}{dS} + e^{-x}\left(v_{xx}\frac{dx}{dS} + v_{x\tau}\frac{d\tau}{dS}\right)$$

$$= -e^{-x}v_x\cdot\frac{1}{S} + e^{-x}\left(v_{xx}\cdot\frac{1}{S} + v_{x\tau}\cdot 0\right)$$

$$= -\frac{e^{-x}}{S}(v_x - v_{xx})$$

$$= \frac{e^{-2x}}{K}(v_{xx} - v_x)$$

(6)

$$rF = F_t + \frac{1}{2}\sigma^2 S^2 F_{SS} + rSF_S$$

$$rKv = -\frac{K\sigma^2}{2}v_\tau + \frac{1}{2}\sigma^2 S^2\frac{e^{-2x}}{K}(v_{xx} - v_x) + rSe^{-x}v_x$$

$$rv = -\frac{\sigma^2}{2}v_\tau + \frac{1}{2}\sigma^2(v_{xx} - v_x) + rv_x$$

$$\frac{\sigma^2}{2}v_\tau = -rv + \frac{1}{2}\sigma^2(v_{xx} - v_x) + rv_x$$

$$v_\tau = v_{xx} + (k-1)v_x - kv$$

(7)

$$\lim_{x \to \infty} u(x, \tau) = \left[\lim_{x \to \infty} e^{(k-1)x/2+(k+1)^2\tau/4} \right] \left[\lim_{x \to \infty} v(x, \tau) \right]$$
$$= \lim_{x \to \infty} e^{(k-1)x/2+(k+1)^2\tau/4} \left(e^x - e^{-k\tau} \right)$$

Dropping the limit we see that

$$e^{(k-1)x/2+(k+1)^2\tau/4} \left(e^x - e^{-k\tau} \right)$$
$$= e^{(k+1)x/2+(k+1)^2\tau/4} - e^{(k-1)x/2+(k+1)^2\tau/4-k\tau}$$
$$= e^{\frac{(k+1)}{2}[x+(k+1)\tau/2]} - e^{(k-1)x/2+[(k+1)^2-4k]\tau/4}$$
$$= e^{\frac{(k+1)}{2}[x+(k+1)\tau/2]} - e^{(k-1)x/2+(k-1)^2\tau/4}$$
$$= e^{\frac{(k+1)}{2}[x+(k+1)\tau/2]} - e^{\frac{(k-1)}{2}[x+(k-1)\tau/2]}.$$

(8)

$$\left(e^{(k+1)(x+\sqrt{2\tau}y)/2} - e^{(k-1)(x+\sqrt{2\tau}y)/2} \right)^+ > 0$$

if and only if

$$e^{(k+1)(x+\sqrt{2\tau}y)/2} > e^{(k-1)(x+\sqrt{2\tau}y)/2}$$
$$e^{x+\sqrt{2\tau}y} > 1$$
$$x + \sqrt{2\tau}y > 0$$
$$y > -\frac{x}{\sqrt{2\tau}}$$

(9) By the result of exercise 8 we know that

$$\frac{1}{\sqrt{2\pi}} \int_{-\infty}^{\infty} \left(e^{(k+1)(x+\sqrt{2\tau}y)/2} - e^{(k-1)(x+\sqrt{2\tau}y)/2} \right)^+ e^{-y^2/2} \, dy$$
$$= \frac{1}{\sqrt{2\pi}} \int_{-x/\sqrt{2\tau}}^{\infty} \left(e^{(k+1)(x+\sqrt{2\tau}y)/2} - e^{(k-1)(x+\sqrt{2\tau}y)/2} \right) e^{-y^2/2} \, dy$$
$$= \frac{1}{\sqrt{2\pi}} \int_{-x/\sqrt{2\tau}}^{\infty} e^{[(k+1)(x+\sqrt{2\tau}y)-y^2]/2} - e^{[(k-1)(x+\sqrt{2\tau}y)-y^2]/2} \, dy$$

Completing the square in the exponents present in the integrand yields

$$y^2 - \sqrt{2\tau}(k \pm 1)y - (k \pm 1)x$$
$$= \left(y^2 - \sqrt{2\tau}(k \pm 1)y + \frac{(k \pm 1)^2\tau}{2}\right) - \frac{(k \pm 1)^2\tau}{2} - (k \pm 1)x$$
$$= \left(y - \frac{(k \pm 1)\sqrt{2\tau}}{2}\right)^2 - \frac{(k \pm 1)^2\tau}{2} - (k \pm 1)x$$

Thus the integral in Eq. (8.33) becomes

$$\frac{1}{\sqrt{2\pi}} \left[\int_{-x/\sqrt{2\tau}}^{\infty} e^{-(y-\frac{(k+1)\sqrt{2\tau}}{2})^2/2} e^{\frac{(k+1)^2\tau}{4}+(k+1)x/2}\, dy \right.$$
$$\left. - \int_{-x/\sqrt{2\tau}}^{\infty} e^{-(y-\frac{(k-1)\sqrt{2\tau}}{2})^2/2} e^{\frac{(k-1)^2\tau}{4}+(k-1)x/2}\, dy \right]$$
$$= \frac{e^{(k+1)^2\tau/4+(k+1)x/2}}{\sqrt{2\pi}} \int_{-x/\sqrt{2\tau}}^{\infty} e^{-(y-\frac{(k+1)\sqrt{2\tau}}{2})^2/2}\, dy$$
$$- \frac{e^{(k-1)^2\tau/4+(k-1)x/2}}{\sqrt{2\pi}} \int_{-x/\sqrt{2\tau}}^{\infty} e^{-(y-\frac{(k-1)\sqrt{2\tau}}{2})^2/2}\, dy$$

(10) If we let $w = -y + \frac{1}{2}(k+1)\sqrt{2\tau}$, then $dw = -dy$. When $y = -x/\sqrt{2\tau}$ then $w = x/\sqrt{2\tau} + \frac{1}{2}(k+1)\sqrt{2\tau}$ and

$$\frac{1}{\sqrt{2\pi}} \int_{-x/\sqrt{2\tau}}^{\infty} e^{-(y-\frac{1}{2}(k+1)\sqrt{2\tau})^2/2}\, dy$$
$$= \frac{1}{\sqrt{2\pi}} \int_{-\infty}^{x/\sqrt{2\tau}+\frac{1}{2}(k+1)\sqrt{2\tau}} e^{-w^2/2}\, dw$$
$$= \phi\left(\frac{x}{\sqrt{2\tau}} + \frac{1}{2}(k+1)\sqrt{2\tau}\right).$$

(11) If we let $w' = -y + \frac{1}{2}(k-1)\sqrt{2\tau}$, then $dw' = -dy$. When $y = -x/\sqrt{2\tau}$ then $w' = x/\sqrt{2\tau} + \frac{1}{2}(k-1)\sqrt{2\tau}$ and

$$\frac{1}{\sqrt{2\pi}} \int_{-x/\sqrt{2\tau}}^{\infty} e^{-(y-\frac{1}{2}(k-1)\sqrt{2\tau})^2/2}\, dy$$
$$= \frac{1}{\sqrt{2\pi}} \int_{-\infty}^{x/\sqrt{2\tau}+\frac{1}{2}(k-1)\sqrt{2\tau}} e^{-w'^2/2}\, dw'$$
$$= \phi\left(\frac{x}{\sqrt{2\tau}} + \frac{1}{2}(k-1)\sqrt{2\tau}\right).$$

(12) We verify the initial condition by evaluating a limit.

$$\lim_{\tau \to 0^+} u(x, \tau) = \lim_{\tau \to 0^+} \left(e^{\frac{(k+1)}{2}[x+(k+1)\tau/2]} \phi \left(\frac{x}{\sqrt{2\tau}} + \frac{1}{2}(k+1)\sqrt{2\tau} \right) \right.$$

$$\left. - e^{\frac{(k-1)}{2}[x+(k-1)\tau/2]} \phi \left(\frac{x}{\sqrt{2\tau}} + \frac{1}{2}(k-1)\sqrt{2\tau} \right) \right)$$

$$= \left(e^{(k+1)x/2} - e^{(k-1)x/2} \right) \lim_{\tau \to 0^+} \phi \left(\frac{x}{\sqrt{2\tau}} \right)$$

If $x < 0$ then $\lim_{\tau \to 0^+} \phi(x/\sqrt{2\tau}) = 0$. If $x = 0$ then $e^{(k+1)x/2} - e^{(k-1)x/2} = 0$. If $x > 0$ then $\lim_{\tau \to 0^+} \phi(x/\sqrt{2\tau}) = 1$. Hence we see that for any x,

$$\lim_{\tau \to 0^+} u(x, \tau) = \left(e^{(k+1)x/2} - e^{(k-1)x/2} \right)^+.$$

(13)

$$\frac{x}{\sqrt{2\tau}} + \frac{1}{2}(k+1)\sqrt{2\tau} = \frac{\ln(S/K)}{\sqrt{\sigma^2(T-t)}} + \frac{1}{2}\left(\frac{2r}{\sigma^2} + 1 \right) \sqrt{\sigma^2(T-t)}$$

$$= \frac{\ln(S/K) + (r + \sigma^2/2)(T-t)}{\sigma\sqrt{T-t}}$$

$$\frac{x}{\sqrt{2\tau}} + \frac{1}{2}(k-1)\sqrt{2\tau} = \frac{x}{\sqrt{2\tau}} + \frac{1}{2}(k+1)\sqrt{2\tau} - \sqrt{2\tau}$$

$$= \frac{\ln(S/K) + (r + \sigma^2/2)(T-t)}{\sigma\sqrt{T-t}} - \sigma\sqrt{T-t}$$

(14) Using the parameter values specified $w \approx 0.536471$ and thus $C_e \approx$ \$5.05739.

(15) Using the parameter values specified $w \approx 0.16662$ and thus $P_e \approx$ \$6.40141.

(16) One way to find the volatility is to approximate it using Newton's Method (Sec. 3.2 of [Smith and Minton (2002)]). We wish to find a root of the equation $C_e(\sigma) = 2.50$. The graph below shows that the solution is near 0.40 which we will use as the initial approximation to the solution for Newton's Method.

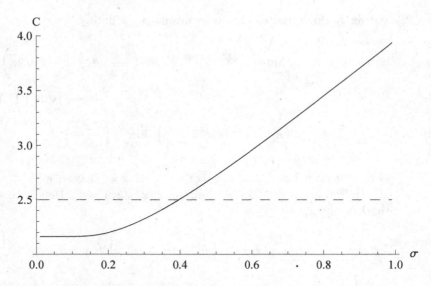

Newton's Method approximates the solution as $\sigma \approx 0.396436$.

(17) If $f(S, t) = S$, then $f_S = 1$, $f_{SS} = 0$, and $f_t = 0$. Thus

$$rf = f_t + \frac{1}{2}\sigma^2 S^2 f_{SS} + rS f_S$$

$$rS = 0 + \frac{1}{2}\sigma^2 S^2 \cdot 0 + rS$$

$$rS = rS.$$

Therefore the price of the security itself solves the Black-Scholes partial differential equation Eq. (8.5).

(18) If $f(S, t) = e^{rt}$, then $f_S = 0$, $f_{SS} = 0$, and $f_t = re^{rt}$. Thus

$$rf = f_t + \frac{1}{2}\sigma^2 S^2 f_{SS} + rS f_S$$

$$re^{rt} = re^{rt} + \frac{1}{2}\sigma^2 S^2 \cdot 0 + rS \cdot 0$$

$$re^{rt} = re^{rt}.$$

Therefore a unit of currency earning the risk-free interest rate r compounded continuously solves the Black-Scholes partial differential Eq. (8.5).

(19) If $Y = \ln S$ then by Itô's Lemma 5.4 Y obeys the stochastic process

$$dY = \left(\mu - \frac{\sigma^2}{2}\right) dt + \sigma \, dW(t).$$

Integrating both sides with respect to t produces

$$Y(t) = Y(0) + \left(\mu - \frac{\sigma^2}{2}\right)t + \sigma W(t)$$

where $Y(0) = \ln S(0)$. Then exponentiating both sides yields

$$S(t) = S(0)e^{(\mu - \sigma^2/2)t + \sigma W(t)}.$$

(20) Constructing a binomial lattice of stock values using $n = 4$ and

$$u = 1.07484, \quad d = 0.930374, \quad \text{and} \quad p = 0.545466,$$

we can summarize the values of the stock, the payoffs of the option, and binomial proababilities of achieving those payoffs.

S	$(S-K)^+$	P
32.218	0	0.0426839
37.2206	0	0.204893
43	1	0.368824
49.6768	7.6768	0.295073
57.3903	15.3903	0.0885262

Thus the value of the call option is approximated as

$$C \approx \frac{(1)(0.368824) + (7.6768)(0.295073) + (15.3903)(0.088562)}{e^{(0.11)(4/12)}} = 3.8526.$$

(21) Constructing a binomial lattice of stock values using $n = 3$ and

$$u = 1.09995, \quad d = 0.909134, \quad \text{and} \quad p = 0.502427,$$

we can summarize the values of the stock, the payoffs of the option, and binomial proababilities of achieving those payoffs.

S	$(K-S)^+$	P
72.1365	27.8635	0.123188
87.2769	12.7231	0.373171
105.595	0	0.376812
127.758	0	0.126829

Notice that the middle column is the payoff for a European put option. Thus the value of the put option is approximated as

$$P \approx \frac{(27.8635)(0.123188) + (12.7231)(0.373171)}{e^{(0.06)(3/12)}} = 8.05857.$$

B.9 Derivatives of Black-Scholes Option Prices

(1) Using Eq. (9.3) with $S = 300$, $K = 310$, $T = 1/4$, $t = 0$, $r = 0.03$, and $\sigma = 0.25$ we have

$$w = -0.139819$$
$$\Theta = -33.281$$

(2) To find Θ for a European put option we will make use of the Put/Call parity formula (Eq. (7.1)) and Eq. (9.3). According to the Put/Call parity formula,

$$\frac{\partial P}{\partial t} = \frac{\partial C}{\partial t} + Kre^{-r(T-t)}.$$

Substituting the expression already found for $\frac{\partial C}{\partial t}$ in the right-hand side of this equation yields

$$\frac{\partial P}{\partial t} = \frac{Se^{-w^2/2} - Ke^{-r(T-t)-(w-\sigma\sqrt{T-t})^2/2}}{2\sigma\sqrt{2\pi(T-t)}}\left(\frac{\ln(S/K)}{T-t} - r - \sigma^2/2\right)$$

$$- Ke^{-r(T-t)}\left(r\phi(w-\sigma\sqrt{T-t}) + \frac{\sigma e^{-(w-\sigma\sqrt{T-t})^2/2}}{2\sqrt{2\pi(T-t)}}\right)$$

$$+ Kre^{-r(T-t)}$$

$$= \frac{Se^{-w^2/2} - Ke^{-r(T-t)-(w-\sigma\sqrt{T-t})^2/2}}{2\sigma\sqrt{2\pi(T-t)}}\left(\frac{\ln(S/K)}{T-t} - r - \sigma^2/2\right)$$

$$+ Kre^{-r(T-t)}\left(1 - \phi(w-\sigma\sqrt{T-t})\right)$$

$$- \frac{\sigma Ke^{-r(T-t)-(w-\sigma\sqrt{T-t})^2/2}}{2\sqrt{2\pi(T-t)}}$$

$$= \frac{Se^{-w^2/2}}{2\sigma\sqrt{2\pi(T-t)}}\left(\frac{\ln(S/K)}{T-t} - r - \sigma^2/2\right)$$

$$+ Kre^{-r(T-t)}\phi(\sigma\sqrt{T-t} - w)$$

$$- \frac{Ke^{-r(T-t)-(w-\sigma\sqrt{T-t})^2/2}}{2\sqrt{2\pi(T-t)}}\left(\frac{\ln(S/K)}{\sigma(T-t)} - r/\sigma - \sigma/2 + \sigma\right)$$

$$= \frac{Se^{-w^2/2}}{2\sigma\sqrt{2\pi(T-t)}}\left(\frac{\ln(S/K)}{T-t} - r - \sigma^2/2\right)$$

$$+ Kre^{-r(T-t)}\phi(\sigma\sqrt{T-t} - w)$$

$$- \frac{Ke^{-r(T-t)-(w-\sigma\sqrt{T-t})^2/2}}{2\sigma\sqrt{2\pi(T-t)}}\left(\frac{\ln(S/K)}{T-t} - r + \sigma^2/2\right).$$

(3) Using Eq. (9.5) with $S = 275$, $K = 265$, $T = 1/3$, $t = 0$, $r = 0.02$, and $\sigma = 0.20$ we have

$$w = 0.436257$$
$$\Theta = -15.3073$$

(4) According to Eq. (8.36)

$$w = \frac{\ln(S/K) + (r + \sigma^2/2)(T-t)}{\sigma\sqrt{T-t}},$$

which implies that

$$\frac{\partial w}{\partial S} = \frac{\partial}{\partial S}\left[\frac{\ln(S/K)}{\sigma\sqrt{T-t}}\right]$$

$$= \frac{1}{\sigma\sqrt{T-t}}\cdot\frac{1}{S}$$

$$= \frac{1}{\sigma S\sqrt{T-t}}.$$

(5) Using Eq. (9.8) with $S = 150$, $K = 165$, $T = 5/12$, $t = 0$, $r = 0.025$, and $\sigma = 0.22$ we have

$$w = -0.526797$$
$$\Delta = 0.299167$$

(6) Using Eq. (9.9) with $S = 125$, $K = 140$, $T = 2/3$, $t = 0$, $r = 0.055$, and $\sigma = 0.15$ we have

$$w = -0.564706$$
$$\Delta = -0.713863$$

(7) Using Eq. (9.11) with $S = 180$, $K = 175$, $T = 1/3$, $t = 0$, $r = 0.0375$, and $\sigma = 0.30$ we have

$$w = 0.321416$$
$$\Gamma = 0.0121519$$

(8) According to Eq. (8.36)

$$w = \frac{\ln(S/K) + (r + \sigma^2/2)(T-t)}{\sigma\sqrt{T-t}},$$

which implies that

$$\frac{\partial w}{\partial \sigma} = \frac{\partial}{\partial \sigma} \left[\frac{\ln(S/K) + (r + \sigma^2/2)(T - t)}{\sigma\sqrt{T - t}} \right]$$

$$= \frac{\sigma(T - t)\sigma\sqrt{T - t} - (\ln(S/K) + (r + \sigma^2/2)(T - t))\sqrt{T - t}}{\sigma^2(T - t)}$$

$$= \frac{\sigma^2(T - t) - (\ln(S/K) + (r + \sigma^2/2)(T - t))}{\sigma^2\sqrt{T - t}}$$

$$= \sqrt{T - t} - \frac{1}{\sigma} \left(\frac{\ln(S/K) + (r + \sigma^2/2)(T - t)}{\sigma\sqrt{T - t}} \right)$$

$$= \sqrt{T - t} - \frac{w}{\sigma}$$

(9) Using Eq. (9.14) with $S = 300$, $K = 305$, $T = 1/2$, $t = 0$, $r = 0.0475$, and $\sigma = 0.25$ we have

$$w = 0.129235$$
$$V = 83.9247$$

(10) Using Eq. (9.19) with $S = 270$, $K = 272$, $T = 1/6$, $t = 0$, $r = 0.0375$, and $\sigma = 0.15$ we have

$$w = 0.012164$$
$$\rho = -23.4071$$

(11) Using Eq. (9.19) with $S = 305$, $K = 325$, $T = 1/3$, $t = 0$, $r = 0.0255$, and $\sigma = 0.35$ we have

$$w = -0.171209$$
$$\rho = 38.0758$$

B.10   Hedging

(1) If the call option is in the money then $S > K$. Start by considering the following limit.

$$\lim_{t \to T^-} w = \lim_{t \to T^-} \left(\frac{\ln(S/K) + (r + \sigma^2/2)(T - t)}{\sigma\sqrt{T - t}} \right)$$

$$= \lim_{t \to T^-} \left(\frac{\ln(S/K)}{\sigma\sqrt{T - t}} + \frac{(r + \sigma^2/2)}{\sigma}\sqrt{T - t} \right)$$

$$= +\infty$$

We know the limit is $+\infty$ since $\ln(S/K) > 0$ for an in the money call option.

Now since $\Delta_C = \phi(w)$ and $\phi(w)$ is a continuous function then

$$\lim_{t \to T^-} \Delta_C = \lim_{w \to \infty} \phi(w)$$
$$= 1.$$

(2) If the call option is out of the money then $S < K$. Start by considering the following limit.

$$\lim_{t \to T^-} w = \lim_{t \to T^-} \left(\frac{\ln(S/K) + (r + \sigma^2/2)(T - t)}{\sigma\sqrt{T - t}} \right)$$
$$= \lim_{t \to T^-} \left(\frac{\ln(S/K)}{\sigma\sqrt{T - t}} + \frac{(r + \sigma^2/2)}{\sigma}\sqrt{T - t} \right)$$
$$= -\infty$$

We know the limit is $-\infty$ since $\ln(S/K) < 0$ for an out of the money call option.

Now since $\Delta_C = \phi(w)$ and $\phi(w)$ is a continuous function then

$$\lim_{t \to T^-} \Delta = \lim_{w \to -\infty} \phi(w)$$
$$= 0.$$

(3) These calculations are similar to those used to create Table 10.1.

Week	S	Δ	Shares Held	Share Cost	Interest Cost	Cumulative Cost
0	45.00	0.408940	2044.70	92011.6	0.000000	92011.6
1	44.58	0.366642	1833.21	81724.6	79.6599	82663.0
2	46.55	0.526428	2632.14	122526	71.5663	119924
3	47.23	0.581970	2909.85	137432	103.826	133144
4	47.62	0.614837	3074.18	146393	115.271	141085
5	47.28	0.583262	2916.31	137883	122.146	133743
6	49.60	0.782988	3914.94	194181	115.790	183391
7	50.07	0.824866	4124.33	206505	158.773	194034
8	47.79	0.635479	3177.39	151848	167.987	148948
9	48.33	0.698867	3494.34	168881	128.953	164395
10	48.81	0.761174	3805.87	185765	142.326	179743
11	51.36	0.954430	4772.15	245098	155.614	229527
12	52.06	0.986327	4931.64	256741	198.715	238029
13	51.98	0.995748	4978.74	258795	206.075	240683
14	54.22	1.000000	5000.00	271100	208.374	242044
15	54.31	1.000000	5000.00	271550	209.552	242254

(4) These calculations are similar to those used to create Table 10.2.

Week	S	Δ	Shares Held	Share Cost	Interest Cost	Cumulative Cost
0	45.00	0.408940	2044.70	92011.6	0.00000	92011.6
1	44.58	0.366642	1833.21	81724.6	79.6599	82663.0
2	45.64	0.447836	2239.18	102196	71.5663	101263
3	44.90	0.374621	1873.11	84102.5	87.6693	84914.0
4	43.42	0.238290	1191.45	51732.7	73.5151	55389.9
5	42.23	0.140594	702.968	29686.3	47.9543	34809.3
6	41.18	0.073032	365.159	15037.3	30.1364	20928.5
7	41.52	0.073130	365.651	15181.8	18.1190	20967.0
8	41.94	0.075921	379.604	15920.6	18.1524	21570.4
9	42.72	0.097654	488.269	20858.9	18.6747	26231.2
10	44.83	0.254169	1270.84	56971.9	22.7099	61336.7
11	44.93	0.235189	1175.94	52835.1	53.1028	57125.9
12	44.19	0.114050	570.252	25199.4	49.4573	30409.9
13	41.77	0.001621	8.10264	338.447	26.3277	6955.25
14	39.56	0.000000	0.00000	0.00007	6.02157	6640.73
15	40.62	0.000000	0.00000	0.00000	5.74928	6645.48

(5) In this case the put option will not be exercised since the final price of the stock exceeds the strike price. In other words, the put option finishes out of the money. At the time the Put option is issued its price is $2.73256. Delta for a European Put option is always negative, thus the financial institution will take a short position in the stock in creating the hedge. The revenue generated by the sale of the Put option and the short position in the stock will be invested in a bond at the risk-free interest rate $r = 0.045$ yr^{-1}.

Week	S	Δ	Shares Shorted	Share Value	Interest Earned	Cumulative Earnings
0	45.00	−0.591060	2955.30	132988	0.00000	132988
1	44.58	−0.633358	3166.79	141175	115.136	142532
2	46.55	−0.473572	2367.86	110224	123.398	105465
3	47.23	−0.418030	2090.15	98717.8	3073	92440.2
4	47.62	−0.385163	1925.82	91707.4	80.0309	84694.6
5	47.28	−0.416738	2083.69	98516.8	73.3252	92232.2
6	49.60	−0.217012	1085.06	53819.1	79.8508	42780.1
7	50.07	−0.175134	875.670	43844.8	37.0372	32332.9
8	47.79	−0.364521	1822.61	87102.3	27.9925	77614.9
9	48.33	−0.301133	1505.66	72768.7	67.1958	62364.3
10	48.81	−0.238826	1194.13	58285.4	53.9924	47212.3
11	51.36	−0.045570	227.848	11702.3	40.8744	−2375.01
12	52.06	−0.013673	68.3635	3559.01	−2.05618	−10679.8
13	51.98	−0.004252	21.2594	1105.06	−9.24617	−13137.6
14	54.22	0.000000	0.00051	0.02738	−11.3740	−14301.6
15	54.31	0.000000	0.00000	0.00000	−12.3817	−14314.0

Since the buyer of the option pays for the option at the beginning of week 0, the net cost to the financial institution is

$$(5000)(2.73256)e^{0.045(15/52)} - 14314.0 = -472.69.$$

(6) In this exercise the Put option will be exercised since the final stock price is smaller than the strike price.

Week	S	Δ	Shares Shorted	Share Value	Interest Earned	Cumulative Earnings
0	45.00	−0.591060	2955.30	132988	0.00000	132988
1	44.58	−0.633358	3166.79	141175	115.136	142532
2	45.64	−0.552164	2760.82	126004	123.398	124127
3	44.90	−0.625379	3126.89	140397	107.464	140671
4	43.42	−0.761710	3808.55	165367	121.787	170390
5	42.23	−0.859406	4297.03	181464	147.517	191166
6	41.18	−0.926968	4634.84	190863	165.504	205253
7	41.52	−0.926870	4634.35	192418	177.691	205400
8	41.94	−0.924079	4620.40	193779	177.827	204993
9	42.72	−0.902346	4511.73	192741	177.473	200528
10	44.83	−0.745831	3729.16	167178	173.609	165619
11	44.93	−0.764811	3824.06	171815	143.386	170026
12	44.19	−0.885950	4429.75	195751	147.202	196939
13	41.77	−0.998379	4991.90	208512	170.502	220590
14	39.56	−1.000000	5000.00	197800	190.987	221102
15	40.62	−1.000000	5000.00	203100	191.421	221293

At the strike time the financial institution must honor the put option and cancel its short position in the stock. Thus the financial institution purchases 5000 shares of the stock at the strike price of $47 per share. This final transaction changes the holdings of the financial institution by

$$221293 - (5000)(47) + (5000)(2.73256)e^{0.045(15/52)} = \$134.31.$$

(7) Let $S = 85$, $K = 88$, $r = 0.055$, and $\sigma = 0.17$. We will let the number of four-month options be w_4 and the number of six-month options be w_6. If $\mathcal{P}$ represents the value of the portfolio consisting of a short position in the four-month options and a long position in the six-month options, then the value of the portfolio is $\mathcal{P} = w_4 C(S, \frac{1}{3}) - w_6 C(S, \frac{1}{2})$. Note that $C(85, \frac{1}{3}) = 1.2972$ and $C(85, \frac{1}{2}) = 2.59313$. Gamma for the portfolio is

$$\Gamma = \Gamma_4 w_4 - \Gamma_6 w_6$$
$$= 0.0474901 w_4 - 0.0390443 w_6.$$

Thus the portfolio is Gamma-neutral whenever $w_4 = 0.822155 w_6$.

(8) Using the results of exercise 7 and including a position of x shares of the security, the value of the portfolio is now $\mathcal{P} = Sx + w_4 C(S, \frac{1}{3}) -$

$w_6 C(S, \frac{1}{2})$. Delta for the portfolio is

$$\Delta = x + \Delta_4 w_4 - \Delta_6 w_6$$
$$= x + 0.45322 w_4 - 0.500131 w_6$$
$$= x + (0.45322)(0.822155) w_6 - 0.500131 w_6$$
$$= x - 0.127514 w_6.$$

Thus the portfolio is both Gamma- and Delta-neutral whenever $x = 0.127514 w_6$.

(9) Let $S = 95$, $r = 0.045$, and $\sigma = 0.23$. We will let the number of three-month options be w_3 and the number of five-month options be w_5. If $\mathcal{P}$ represents the value of the portfolio consisting of a short position in the three-month options and a long position in the five-month options, then the value of the portfolio is $\mathcal{P} = w_3 C(S, \frac{1}{4}) - w_5 C(S, \frac{5}{12})$. Note that $C(95, \frac{1}{4}) = 1.74863$ and $C(95, \frac{5}{12}) = 3.08418$. Gamma for the portfolio is

$$\Gamma = \Gamma_3 w_3 - \Gamma_5 w_5$$
$$= 0.0365043 w_3 - 0.0282844 w_5.$$

Thus the portfolio is Gamma-neutral whenever $w_3 = 0.774825 w_5$.

(10) Using the results of exercise 9 and including a position of x shares of the security, the value of the portfolio is now $\mathcal{P} = Sx + w_3 C(S, \frac{1}{4}) - w_5 C(S, \frac{5}{12})$. Delta for the portfolio is

$$\Delta = x + \Delta_3 w_3 - \Delta_5 w_5$$
$$= x + 0.489693 w_3 - 0.496454 w_5$$
$$= x + (0.489693)(0.774825) w_5 - 0.496454 w_5$$
$$= x - 0.117028 w_5.$$

Thus the portfolio is both Gamma- and Delta-neutral whenever $x = 0.117028 w_5$.

B.11 Optimizing Portfolios

(1) Proof of statement 1:

$$\mathrm{Cov}(X, X) = \mathrm{E}\left[X^2\right] - (\mathrm{E}[X])^2$$
$$= \mathrm{Var}(X) \quad \text{(by Theorem (2.6))}$$

Proof of statement 2:

$$\text{Cov}(X, Y) = \text{E}[XY] - \text{E}[X]\text{E}[Y] \quad \text{(by Eq. (11.2))}$$
$$= \text{E}[YX] - \text{E}[Y]\text{E}[X]$$
$$= \text{Cov}(Y, X) \quad \text{(by Eq. (11.2))}$$

(2) Let X and Y be random variables, then by Theorem 11.1

$$\text{Var}(X + Y) = \text{Cov}(X + Y, X + Y)$$
$$= \text{E}[X^2 + 2XY + Y^2] - (\text{E}[X] + \text{E}[Y])(\text{E}[X] + \text{E}[Y])$$
$$= \text{E}[X^2] + 2\text{E}[XY] + \text{E}[Y^2]$$
$$- (\text{E}[X])^2 - 2\text{E}[X]\text{E}[Y] - (\text{E}[Y])^2$$
$$= \text{E}[X^2] - (\text{E}[X])^2 + \text{E}[Y^2] - (\text{E}[Y])^2$$
$$+ 2(\text{E}[XY] - \text{E}[X]\text{E}[Y])$$
$$= \text{Var}(X) + \text{Var}(Y) + 2\text{Cov}(X, Y).$$

Recall that if X and Y are independent random variables $\text{E}[XY] = \text{E}[X]\text{E}[Y]$ and thus $\text{E}[XY] - \text{E}[X]\text{E}[Y] = 0$. Therefore, if X and Y are independent random variables

$$\text{Var}(X + Y) = \text{Var}(X) + \text{Var}(Y).$$

(3) Suppose that $\text{E}[X^2] = 0$, then by definition

$$\text{E}[X^2] = \sum_{i=1}^{n} x_i^2 \cdot \text{P}(X = x_i).$$

If $x_i \neq 0$ for some i, then $\text{P}(X = x_i) = 0$. Therefore we must have $\text{P}(X = 0) = 1$. In other words $X = 0$ which implies $XY = 0$ and thus

$$0 = (\text{E}[XY])^2 \leq \text{E}[X^2]\text{E}[Y^2] = 0.$$

A similar statement can be made when $\text{E}[Y^2] = 0$.
Suppose that $\text{E}[X^2] = \infty$, then

$$(\text{E}[XY])^2 < \infty = \text{E}[X^2]\text{E}[Y^2].$$

A similar statement can be made when $\text{E}[Y^2] = \infty$.

(4) Let X be the golf handicap of the CEO and let Y be the stock rating of the CEO's corporation. Then we may calculate the following

quantities:

$$E[X] = 11.9923$$
$$E[Y] = 56.6923$$
$$E[XY] = 671.069$$
$$\text{Var}(X) = 6.8741$$
$$\text{Var}(Y) = 45.5641$$

Thus we have

$$\text{Cov}(X,Y) = 671.069 - (11.9923)(56.6923) = -8.80207$$
$$\rho(X,Y) = \frac{-8.80207}{\sqrt{(6.8741)(45.5641)}} = -0.497354.$$

(5) Let X be the height of the model and let Y be the weight of the model. Then we may calculate the following quantities:

$$E[X] = 67.05$$
$$E[Y] = 117.6$$
$$E[XY] = 7887.85$$
$$\text{Var}(X) = 1.46944$$
$$\text{Var}(Y) = 41.8222$$

Thus we have

$$\text{Cov}(X,Y) = 7887.75 - (67.05)(117.6) = 2.77$$
$$\rho(X,Y) = \frac{2.77}{\sqrt{(1.46944)(41.8222)}} = 0.353346.$$

(6) From the data we calculate the following quantities:

$$E[X] = \frac{9}{10}$$
$$E[Y] = \frac{y+45}{10}$$
$$E[XY] = \frac{9}{2}$$
$$\text{Var}(X) = \frac{1}{10}$$
$$\text{Var}(Y) = \frac{y^2}{10} - y + \frac{55}{6}$$

Thus we have

$$\text{Cov}\,(X,Y) = \frac{9}{2} - \left(\frac{9}{10}\right)\left(\frac{y+45}{10}\right) = \frac{9}{100}(5-y)$$

$$\rho\,(X,Y) = \frac{\frac{9}{100}(5-y)}{\sqrt{\left(\frac{1}{10}\right)\left(\frac{y^2}{10} - y + \frac{55}{6}\right)}} = \frac{9}{10}\frac{5-y}{\sqrt{y^2 - 10y + \frac{275}{3}}}$$

As $y \to -\infty$, $\rho\,(X,Y) \to 9/10$. Not quite 1, but by increasing the sample size, the limit can be increased as well.

(7) If n is the ratio of the number of shares of security D purchased compared to the number of call options C sold, then the value of the portfolio is $\mathcal{P} = C - nD$ which implies the variance in the value of the portfolio is

$$\begin{aligned}\text{Var}\,(\mathcal{P}) &= n^2\text{Var}\,(D) - 2n\text{Cov}\,(C,D) + \text{Var}\,(C)\\ &= (0.037)^2 n^2 - 2n(0.86)(0.045)(0.037) + (0.045)^2\\ &= 0.001369 n^2 - 0.0028638 n + 0.002025.\end{aligned}$$

This quadratic function is minimized when $n = 1.04595$.

(8) From Eq. (11.8) we know that

$$n = \frac{\text{Cov}\,(\Delta C, \Delta S)}{\text{Var}\,(\Delta S)}.$$

Assuming that S is governed by an Itô process of the form

$$dS = \mu S\,dt + \sigma S\,dW$$

then over a short time interval of length ΔT

$$\begin{aligned}\text{Var}\,(\Delta S) &= \text{E}\left[(\Delta S)^2\right] - \text{E}[\Delta S]^2\\ &= \text{E}\left[(\mu S\,\Delta T + \sigma S\,\Delta W)^2\right] - \text{E}\left[(\mu S\,\Delta T + \sigma S\,\Delta W)\right]^2\\ &\approx \sigma^2 S^2 \Delta T.\end{aligned}$$

Note that we have retained expressions containing powers of ΔT that are 1 or less. If C is described by an Itô process of the form

$$dC = \left(\mu S\Delta + \frac{1}{2}\sigma^2 S^2\Gamma + \Theta\right)dt + (\sigma S\Delta)\,dW$$

then over a short time interval of length ΔT

$$\text{Cov}(\Delta C, \Delta S) = \text{E}\left[(\Delta C)(\Delta S)\right] - \text{E}\left[\Delta C\right]\text{E}\left[\Delta S\right]$$
$$= (\sigma^2 S^2 \Delta)\Delta T - \mu S\left(\mu S\Delta + \frac{1}{2}\sigma^2 S^2 \Gamma + \Theta\right)(\Delta T)^2$$
$$\approx (\sigma^2 S^2 \Delta)\Delta T.$$

Again, we have kept only the expressions containing the powers of ΔT which are 1 or less. Now substituting the expressions just found for the variance and covariance into the formula for n we obtain

$$n = \frac{(\sigma^2 S^2 \Delta)\Delta T}{\sigma^2 S^2 \Delta T} = \Delta.$$

(9) According to Theorem 11.4, a twice differentiable function $u(x)$, is concave when $u''(x) \leq 0$.

(a) $u(x) = \ln x$

$$u'(x) = \frac{1}{x}$$
$$u''(x) = -\frac{1}{x^2} < 0$$

for $0 < x < \infty$.

(b) $u(x) = (\ln x)^2$

$$u'(x) = \frac{2\ln x}{x}$$
$$u''(x) = \frac{2(1 - \ln x)}{x^2}$$

This function is not concave on its whole domain. It is concave only for x such that $\ln x \geq 1$, in other words only for $x \geq e$.

(c) $u(x) = \tan^{-1} x$

$$u'(x) = \frac{1}{1 + x^2}$$
$$u''(x) = -\frac{2x}{(1 + x^2)^2}$$

This function is not concave on its whole domain. It is concave only for $x \geq 0$.

(10) According to Jensen's inequality (11.15), since $f(x) = \ln x$ is a concave function then

$$\ln\left(\sum_X X\mathrm{P}(X)\right) \geq \sum_X (\ln X)\mathrm{P}(X)$$

$$= \sum_X \ln\left(X^{\mathrm{P}(X)}\right)$$

$$= \ln\left(\prod_X X^{\mathrm{P}(X)}\right).$$

Since the exponential function is monotone then

$$\mathrm{E}[X] = e^{\ln(\sum_X X\mathrm{P}(x))} \geq e^{\ln(\prod_X X^{\mathrm{P}(X)})} = \mathcal{G}.$$

Now if $X > 0$ then $T = \frac{1}{X} > 0$. By the previous result

$$\prod_X T^{\mathrm{P}(X)} \leq \mathrm{E}[T]$$

$$= \sum_X T\mathrm{P}(X)$$

$$\frac{1}{\prod_X X^{\mathrm{P}(X)}} \leq \sum_X \frac{\mathrm{P}(X)}{X}$$

$$\prod_X X^{\mathrm{P}(X)} \geq \frac{1}{\sum_X \frac{\mathrm{P}(X)}{X}}$$

$$\mathcal{G} \geq \mathcal{H}.$$

Thus we have shown that $\mathcal{H} \leq \mathcal{G} \leq \mathrm{E}[X]$.
Now suppose that all the random variable X takes on only one value (with probability 1). Then we have

$$X = \mathcal{H} = \mathcal{G} = \mathrm{E}[X].$$

Before finishing the exercise we must establish another inequality, an obvious one.

$$\mathrm{E}[X] = \sum_X X\mathrm{P}(X) \leq \max_X\{X\}$$

Equality holds if and only if X takes on only one value. From this inequality follows

$$\sum_X \frac{P(X)}{X} \le \max_X\{\frac{1}{X}\}$$

$$\frac{1}{\sum_X \frac{P(X)}{X}} \ge \min_X\{X\}.$$

Again equality holds if and only if X takes on only one value. Thus the following string of inequalities is true.

$$\min_X\{X\} \le \mathcal{H} \le \mathcal{G} \le E[X] \le \max_X\{X\}$$

Now if $\min_X\{X\} = \mathcal{H}$ then X takes on only one value and we have $\min_X\{X\} = \max_X\{X\}$ and thus

$$\min_X\{X\} = \mathcal{H} = \mathcal{G} = E[X] = \max_X\{X\}$$

(11) $\phi(t) = t$.

$$\int_0^1 f(\phi(t))\,dt = \int_0^1 \tanh t\,dt$$
$$= \ln(\cosh 1)$$
$$\approx 0.433781$$
$$< 0.462117$$
$$\approx \tanh\left(\frac{1}{2}\right)$$
$$= \tanh\left(\int_0^1 t\,dt\right)$$
$$= f\left(\int_0^1 \phi(t)\,dt\right)$$

(12) The certainty equivalent is the solution C of the equation

$$\frac{1}{2}(f(10) + f(-2)) = f(C)$$
$$\frac{74}{25} = C - \frac{C^2}{50}$$
$$C = 25 \pm 3\sqrt{53}$$

The certainty equivalent will be the smaller of the two roots of the quadratic equation, *i.e.*, $C = 25 - 3\sqrt{53} \approx 3.15967$.

(13) The certainty equivalent is the solution C of the equation

$$\frac{1}{2}f(15) + \frac{1}{2}f(-15) = f(C)$$

$$-\frac{225}{2} = C - \frac{C^2}{2}$$

$$C = 1 \pm \sqrt{226}$$

The certainty equivalent will be the smaller of the two roots of the quadratic equation, *i.e.*, $C = 1 - \sqrt{226} \approx -14.0333$.

(14) If we let X represent the random variable of the investor's return on the investment, then

$$X = \begin{cases} 2\alpha x + (1+r)(1-\alpha)x & \text{with probability } p, \\ (1+r)(1-\alpha)x & \text{with probability } 1-p. \end{cases}$$

If the investor's utility function is $u(x) = \ln x$, then the expected value of the utility function is

$$\text{E}\left[u(X)\right] = pu(2\alpha x + (1+r)(1-\alpha)x) + (1-p)u((1+r)(1-\alpha)x).$$

This function is maximized when

$$0 = \frac{d}{d\alpha}\text{E}\left[u(X)\right]$$

$$= \frac{p-1}{1-\alpha} + \frac{p(1-r)}{\alpha(1-r)+1+r}$$

$$\alpha = \frac{2p-r-1}{1-r}.$$

(15) Let $u(x) = 1 - e^{-bx}$ where $b > 0$, then $u'(x) = be^{-bx}$ and $u''(x) = -b^2e^{-bx}$. Thus $u''(x) \leq 0$ for all x. Hence $u(x)$ is concave.

The function $u(x)$ is monotone increasing since $u'(x) > 0$ for all x. Therefore if $0 \leq x_1 < x_2$ then $u(x_1) < u(x_2)$.

(16) Let W represent the wealth returned to the investor and let y represent the amount of money invested in the first investment $(0 \leq y \leq 1000)$.

$$\text{E}\left[W\right] = 1000 + 0.08y + 0.13(1000 - y)$$

$$= 1130 - 0.05y$$

$$\text{Var}\left(W\right) = (0.03)^2y^2 + (0.09)^2(1000 - y)^2$$

$$+ 2y(1000 - y)(0.03)(0.09)(-0.26)$$

$$= 0.010404y^2 - 17.604y + 8100$$

The investor's utility function, $u(x) = 1 - e^{-x/100}$, is maximized when $E[W] - \text{Var}(W)/200$ is maximized.

$$E[W] - \text{Var}(W)/200 = -0.00005202y^2 + 0.03802y + 1089.5$$

This quadratic expression is maximized when $y \approx 365.436$.

(17) According to Theorem 11.7

$$\alpha_A = \frac{\frac{1}{0.24}}{\frac{1}{0.24} + \frac{1}{0.41} + \frac{1}{0.27} + \frac{1}{0.16} + \frac{1}{0.33}} = 0.212697$$

$$\alpha_B = \frac{\frac{1}{0.41}}{\frac{1}{0.24} + \frac{1}{0.41} + \frac{1}{0.27} + \frac{1}{0.16} + \frac{1}{0.33}} = 0.124505$$

$$\alpha_C = \frac{\frac{1}{0.27}}{\frac{1}{0.24} + \frac{1}{0.41} + \frac{1}{0.27} + \frac{1}{0.16} + \frac{1}{0.33}} = 0.189064$$

$$\alpha_D = \frac{\frac{1}{0.16}}{\frac{1}{0.24} + \frac{1}{0.41} + \frac{1}{0.27} + \frac{1}{0.16} + \frac{1}{0.33}} = 0.319045$$

$$\alpha_E = \frac{\frac{1}{0.33}}{\frac{1}{0.24} + \frac{1}{0.41} + \frac{1}{0.27} + \frac{1}{0.16} + \frac{1}{0.33}} = 0.154689$$

(18) If c is a real number then

$$r(c\mathbf{w}) = E[R(c\mathbf{w})]$$
$$= E\left[\sum_{i=1}^{n} cw_i(R_i - r)\right] \quad \text{(by Eq. (11.18))}$$
$$= cE[R(\mathbf{w})] \quad \text{(by Theorem 2.3)}$$
$$= cr(\mathbf{w}).$$

Likewise

$$\sigma^2(c\mathbf{w}) = \text{Var}(R(c\mathbf{w}))$$
$$= \text{Var}\left(\sum_{i=1}^{n} cw_i(R_i - r)\right) \quad \text{(by Eq. (11.18))}$$
$$= c^2\text{Var}(R(\mathbf{w})) \quad \text{(by exercise 18 of Chap. 2)}$$
$$= c^2\sigma^2(\mathbf{w})$$

(19) In this case, the variance in the rate of return is to be minimized subject to the constraint that the expected value of the rate of return must be one. We will again use Lagrange multipliers to find the minimum

variance. We must solve the following system of two simultaneous equations.

$$\nabla \left(\sigma^2(\mathbf{w}) \right) = \lambda \nabla \left(r(\mathbf{w}) \right)$$
$$r(\mathbf{w}) = 1$$

The set of equations above are equivalent to the following set of $n+1$ equations.

$$2w_i \sigma_i^2 = \lambda(r_i - r) \quad \text{for } i = 1, 2, \ldots, n,$$
$$\sum_{i=1}^{n} w_i(r_i - r) = 1$$

If we solve the ith equation for w_i for $i = 1, 2, \ldots, n$ and substitute into the last equation we have

$$\lambda \sum_{i=1}^{n} \frac{(r_i - r)^2}{2\sigma_i^2} = 1.$$

This implies that

$$\lambda = \frac{1}{\sum_{i=1}^{n} \frac{(r_i - r)^2}{2\sigma_i^2}}.$$

Therefore

$$w_i = \frac{\frac{r_i - r}{\sigma_i^2}}{\sum_{i=1}^{n} \frac{(r_i - r)^2}{\sigma_i^2}}.$$

(20) According to Lemma 11.3

$$\alpha_A = 4.44549$$
$$\alpha_B = 1.30112$$
$$\alpha_C = 7.9031$$
$$\alpha_D = 0.0$$
$$\alpha_E = 9.69925$$

(21) We must show that the set of equations below is satisfied by the expressions in Eqs. (11.29) and (11.30). This can be accomplished by

direct substitution into the equations.

$$r_i - r_M = \lambda \left(2 \left[(\rho_{i,M}\sigma_i - \sigma_M)\sigma_M + x(\sigma_i^2 - 2\rho_{i,m}\sigma_i\sigma_M + \sigma_M^2) \right] \right)$$
$$\sigma^2 = x^2\sigma_i^2 + (1-x)^2\sigma_M^2 + 2x(1-x)\rho_{i,M}\sigma_i\sigma_M$$

(22) Substituting $r_f = 0.0475$, $r_M = 0.0765$, $\sigma_M = 0.22$, and $\text{Cov}\,(i, M) = 0.15$ into Eq. (11.27) we obtain

$$r_i - r_f = \frac{\text{Cov}\,(i, M)}{\sigma_M^2}(r_M - r_f)$$

$$r_i - 0.0475 = \frac{0.15}{(0.22)^2}(0.0765 - 0.0475)$$

$$r_i = 0.0475 + (3.09917)(0.029)$$

$$= 0.137376$$

(23) Using the values $S(0) = 83$, $\mu = 0.13$, $\sigma = 0.25$, $r = 0.055$, $T = 1/4$, $t = 0$, and $K = 86$ we can determine the value of the European call option as $C = 3.33212$.

Suppose a portfolio consists of a position x in the security and a position y in the option. According to Eq. (11.34) the expected return of the portfolio is a linear function of x and y given by

$$\text{E}\,[R] = 1.57093x + 0.808234y.$$

The variance in the expected rate of return is a quadratic function of x and y described in detail by Eq. (11.36). In this example the quadratic function takes the form

$$\text{Var}\,(R) = 112.631x^2 + 121.696xy + 43.0376y^2.$$

We would like to find the minimum variance portfolio having $\text{E}\,[R] = 10$. If we set the expected rate of return to $10 and solve for y we obtain

$$y = 12.3727 - 1.94366x.$$

Substituting this expression into the variance of R produces the quadratic function of x given below.

$$\text{Var}\,(R) = 38.6829x^2 - 564.248x + 6588.31$$

The minimum value occurs when $x = 7.29325$ and thus $y = -1.80294$.

B.12 American Options

(1) If $r = 0.06$, $S = 50$, $K = 51$, $T = 3/12$, and $P_a = 9.75$ then according to inequality (12.3)

$$50 - 51 \leq C_a - 9.75 \leq 50 - 51e^{-0.06(3/12)}$$
$$8.75 \leq C_a \leq 9.51$$

(2) If $r = 0.0574$, $S = 93$, $K = 90$, $T = 2/12$, and $C_a = 11.77$ then according to inequality (12.3)

$$93 - 90 \leq 11.77 - P_a \leq 93 - 90e^{-0.0574(2/12)}$$
$$0.56 \leq P_a \leq 8.77$$

(3) The binomial model parameters will be assigned as follows: $\Delta t = 1/12$. The proportional increase and decrease factors in the price of the security between time steps are respectively

$$u = e^{0.25\sqrt{1/12}} \approx 1.07484$$
$$d = e^{-0.25\sqrt{1/12}} \approx 0.930374.$$

The probability of an increase in the price of the security occurring between time steps is

$$p \approx 0.550973.$$

The binomial lattice of security prices resembles the following.

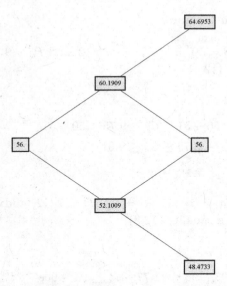

Using the algorithm described in Eq. (12.9), the lattice of put values is shown below.

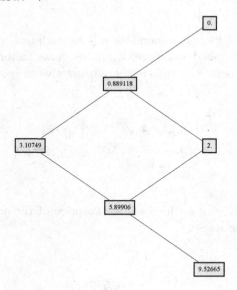

(4) It is sufficient to show that $u \geq e^{r\Delta t}$. To the contrary suppose that $u < e^{r\Delta t}$ then an investor could sell short the security at time $t = 0$ and invest the income in a risk-free bond at rate r compounded continuously. At $t = \Delta t$, the bond will be worth $S(0)e^{r\Delta t}$ while the maximum value of the security will be $uS(0) < e^{r\Delta}S(0)$. Thus the investor may close out the short position in the security and earn an amount

$$(e^{r\Delta t} - u)S(0) > 0$$

without risk. Consequently arbitrage is present.

(5) By assumption $(K - uS(t_i))^+ > 0$ so $(K - uS(t_i))^+ = K - uS(t_i)$ and $(K - dS(t_i))^+ = K - dS(t_i)$.

$$
\begin{aligned}
e^{-r\Delta t}\mathrm{E}\left[Q(t_{i+1})\right] &= e^{-r\Delta t}\left[p(K - uS(t_i)) + (1 - p)(K - dS(t_i))\right] \\
&= e^{-r\Delta t}\left[K - (pu + (1 - p)d)S(t_i)\right] \\
&= e^{-r\Delta t}\left[K - (p(u - d) + d)S(t_i)\right] \\
&= e^{-r\Delta t}\left[K - e^{r\Delta t}S(t_i)\right] \quad \text{(by Eq. (12.7))} \\
&= e^{-r\Delta t}K - S(t_i) \\
&< K - S(t_i) \\
&= (K - S(t_i))^+ \quad \text{(since } (K - uS(t_i))^+ > 0\text{)} \\
&= Q(t_i)
\end{aligned}
$$

(6) Assigning the values $S = 80$, $K = 78$, $r = 0.05$, $\Delta t = 1/12$, $n = 3$, and $\sigma = 0.25$ we have from Eqs. (12.5), (12.6), and (12.7)

$$u \approx 1.07521, \quad d \approx 0.93005, \quad \text{and} \quad p \approx 0.510642.$$

The value of the American put and the intrinsic value are listed in the table below.

i	$S(i\Delta t)$	$(K - S(i\Delta t))^+$	$P_a(i\Delta t)$
3	99.4424	0	0
	86.0159	0	0
	74.4040	3.59604	3.59604
	64.3588	13.6412	13.6412
2	92.4864	0	0
	80	0	1.75243
	69.1994	8.80063	8.80063
1	86.0169	0	0.854001
	74.4040	3.59604	5.1799
0	80	0	2.95856

The earliest time that the value of the option equals its intrinsic value is $i^* = 2$ corresponding to two months into the life of the option. We can verify that

$$2.95856 = e^{-0.05(2/12)} \left[p^2(0) + 2p(1 - p)(1.75243) + (1 - p)^2(8.80063) \right].$$

Bibliography

Bleecker, D. and Csordas, G. (1996). *Basic Partial Differential Equations*, International Press, Cambridge, Massachusetts, USA.

Boyce, W. E. and DiPrima, R. C. (2001). *Elementary Differential Equations and Boundary Value Problems*, 7th edition, John Wiley and Sons, Inc., New York, USA.

Broverman, S. A. (2004). *Mathematics of Investment and Credit*, 3rd edition, ACTEX Publications, Winsted, CT, USA.

Burden, R. L. and Faires, J. D. (2005). *Numerical Analysis*, 8th edition, Thomson Brooks/Cole, Belmont, CA, USA.

Chawla, M. M. (2006), Accurate computation of the Greeks, *International Journal of Applied Mathematics*, **7**, 4, pp. 379–388.

Chawla, M. M. and Evans, D. J. (2005), High-accuracy finite difference methods for the valuation of options, *International Journal of Computer Mathematics*, **82**, 9, pp. 1157–1165.

Churchill, R. V., Brown, J. W. and Verhey, R. F. (1976). *Complex Variables and Applications*, 3rd edition, McGraw-Hill Book Company, New York, USA.

Courant, R. and Robbins, H. (1969). *What is Mathematics?*, Oxford University Press, Oxford, UK.

Cox, J.C., Ross, S. and Rubinstein, M. (1979), Option pricing: a simplified approach, *Journal of Financial Economics*, **7**, pp. 229–264.

Dean, J., Corvin, J. and Ewell, D. (2001). List of *Playboy*'s playmates of the month with data sheet stats, Available at http://www3.sympatico.ca/jimdean/pmstats.txt.

DeGroot, M. H. (1975). *Probability and Statistics*, Behavioral Science: Quantitative Methods. Addison-Wesley Publishing Company, Reading, Massachusetts, USA.

Durrett, R. (1996). *Stochastic Calculus: A Practical Introduction*, CRC Press, Boca Raton, FL, USA.

Franklin, J. (1980). *Methods of Mathematical Economics*, Springer-Verlag, New York, NY, USA.

Gale, D., Kuhn, H. W. and Tucker, A. W. (1951). Linear programming and the theory of games, in T.C. Koopmans, editor, *Activity Analysis of Production and Allocation*, Wiley, New York, NY, USA, pp. 317–329.

Gardner, M. (October 1959). "Mathematical Games" column, *Scientific American*, pp. 180–182.

Goldstein, L. J., Lay, D. C. and Schneider, D. I. (1999). *Brief Calculus and Its Applications*, 8th edition, Prentice Hall, Upper Saddle River, NJ, USA.

Golub, G. H. and Van Loan, C. F. (1989). *Matrix Computations*, 2nd edition, Johns Hopkins University Press, Baltimore, Maryland, USA.

Greenberg, M. D. (1998). *Advanced Engineering Mathematics*, 2nd edition, Prentice-Hall, Inc., Upper Saddle River, New Jersey, USA.

Jeffrey, A. (2002). *Advanced Engineering Mathematics*, Harcourt Academic Press, Burlington, Massachusetts, USA.

Kijima, M. (2003). *Stochastic Processes with Applications to Finance*, Chapman & Hall/CRC, Boca Raton, FL, USA.

Lawler, G. F. (2006). *Introduction to Stochastic Processes*, 2nd edition, Chapman & Hall/CRC, Boca Raton, FL, USA.

Luenberger, D. G. (1998). *Investment Science*, Oxford University Press, New York, USA.

Marsden, J. E. and Hoffman, M. J. (1987). *Basic Complex Analysis*, W. H. Freeman and Company, New York, NY, USA.

McDonald, R. L. (2006). *Derivatives Markets*, Addison-Wesley, Boston, MA, USA.

Mikosch, T. (1998). *Elementary Stochastic Calculus With Finance in View*, Vol. 6 of *Advanced Series on Statistical Science & Applied Probability*, World Scientific Publishing, River Edge, New Jersey, USA.

Neftci, S. N. (2000). *An Introduction to the Mathematics of Financial Derivatives*, 2nd edition, Academic Press, San Diego, CA, USA.

Noble, D. and Daniel, J. W. (1988). *Applied Linear Algebra*, 3rd edition, Prentice Hall, Englewood Cliffs, NJ, USA.

Redner, S. (2001). *A Guide to First Passage Processes*, Cambridge University Press, Cambridge, UK.

Ross, S. M. (1999). *An Introduction to Mathematical Finance: Options and Other Topics*, Cambridge University Press, Cambridge, UK.

Ross, S. M. (2003). *Introduction to Probability Models*, 8th edition, Academic Press, San Diego, CA, USA.

Ross, S. M. (2006). *A First Course in Probability*, 7th edition, Prentice Hall, Inc., Upper Saddle River, NJ, USA.

Selvin, S. (1975). A problem in probability, *American Statistician*, **29**, 1, p. 67.

Seydel, R. (2002). *Tools for Computational Finance*, Springer-Verlag, New York, NY, USA.

Shodor   Education   Foundation, Inc., "Graphing and Interpreting Bivariate Data", 20 May 2008 ⟨http://www.shodor.org/interactivate/discussions/GraphingData/⟩.

Shreve, S. E. (2004). *Stochastic Calculus for Finance I*, Springer-Verlag, New York, NY, USA.

Smith, R. T. and Minton, R. B. (2002). *Calculus*, 2nd edition, McGraw-Hill, Boston, MA, USA.

Steele, J. M. (2001). *Stochastic Calculus and Financial Applications*, Volume 45 in Applications of Mathematics, Springer-Verlag, New York, NY, USA.

Stewart, J. (1999). *Calculus*, 4th edition, Brooks/Cole Publishing Company, Pacific Grove, CA, USA.

Strang, G. (1986). *Introduction to Applied Mathematics*, Wellesley-Cambridge Press, Wellesley, MA, USA.

Taylor, A. E. and Mann, W. R. (1983). *Advanced Calculus*, 3rd edition, John Wiley & Sons, Inc., New York, NY, USA.

vos Savant, M. (February 1990). "Ask Marilyn" column, *Parade Magazine*, p. 12.

Wilmott, P. (2006). *Paul Wilmott on Quantitative Finance*, 2nd edition, John Wiley & Sons, Inc., Hoboken, NJ, USA.

Wilmott, P., Howison, S. and Dewynne, J. (1995). *The Mathematics of Financial Derivatives: A Student Introduction*, Cambridge University Press, Cambridge, UK.

Index

absorbing boundary condition, 96
Addition rule, 20
additivity, 44
American option, 139
American options, 251
amount due, 2
annuity, 9
arbitrage, 71, 146
Arbitrage Theorem, 71, 86, 140
area as probability, 41
arithmetic mean, 248
ask price, 124
average, *see* expected value

backwards parabolic equation, 147
backwards substitution, 102
bell curve, 46
Bernoulli random variable, 25, 93
beta, 239
betting strategy, 86
bid price, 124
bid/ask spread, 124
bid/offer spread, 124
binary model, 143
binomial model, 153, 165, 251
binomial probability, 172
binomial random variable, 25, 46
Black-Scholes Option Pricing
   Formula, 177
Black-Scholes equation, 140, 146
boundary condition, 148
   absorbing, 96

European call option, 150
Brownian motion, 91, 106, 241

call option, 139
canonical form, 78
capital, 2
Capital Asset Pricing Model, 236
capital market line, 235, 237
CAPM, *see* Capital Asset Pricing
   Model
centered difference formula, 50
Central Limit Theorem, 54, 229
certainty equivalent, 225
chain rule, 115, 182
   multivariable, 116
closing price, 118
compound amount, 2
compound interest, 3
compounding period, 3
concavity, 185, 220
conditional exit time, 103
conditional probability, 21
constrained optimization, 232
constraints, 75
   equality, 74, 77
   inequality, 74
continuous distribution, 40
continuous probability, 40
continuous random variable, 39
continuous random walk, 105
continuously compounded interest, 5
convex set, 76